AF181946

The IMA Volumes in Mathematics and its Applications

Volume 78

Series Editors
Avner Friedman   Willard Miller, Jr.

Springer
New York
Berlin
Heidelberg
Barcelona
Budapest
Hong Kong
London
Milan
Paris
Santa Clara
Singapore
Tokyo

Institute for Mathematics and its Applications IMA

The **Institute for Mathematics and its Applications** was established by a grant from the National Science Foundation to the University of Minnesota in 1982. The IMA seeks to encourage the development and study of fresh mathematical concepts and questions of concern to the other sciences by bringing together mathematicians and scientists from diverse fields in an atmosphere that will stimulate discussion and collaboration.

The IMA Volumes are intended to involve the broader scientific community in this process.

Avner Friedman, Director

Willard Miller, Jr., Associate Director

* * * * * * * * * *

IMA ANNUAL PROGRAMS

1982–1983	Statistical and Continuum Approaches to Phase Transition
1983–1984	Mathematical Models for the Economics of Decentralized Resource Allocation
1984–1985	Continuum Physics and Partial Differential Equations
1985–1986	Stochastic Differential Equations and Their Applications
1986–1987	Scientific Computation
1987–1988	Applied Combinatorics
1988–1989	Nonlinear Waves
1989–1990	Dynamical Systems and Their Applications
1990–1991	Phase Transitions and Free Boundaries
1991–1992	Applied Linear Algebra
1992–1993	Control Theory and its Applications
1993–1994	Emerging Applications of Probability
1994–1995	Waves and Scattering
1995–1996	Mathematical Methods in Material Science
1996–1997	High Performance Computing
1997–1998	Emerging Applications of Dynamical Systems

Continued at the back

Boris S. Mordukhovich Hector J. Sussmann
Editors

Nonsmooth Analysis and Geometric Methods in Deterministic Optimal Control

With 18 Illustrations

Springer

Boris S. Mordukhovich
Department of Mathematics
Wayne State University
Detroit, MI 48202 USA

Hector J. Sussmann
Department of Mathematics
Rutgers University
New Brunswick, NJ 08903 USA

Series Editors:
Avner Friedman
Willard Miller, Jr.
Institute for Mathematics and its Applications
University of Minnesota
Minneapolis, MN 55455 USA

Mathematics Subject Classifications (1991): 49-03, 49J30, 49J52, 49K15, 49K20, 49K24, 49K40, 49L25, 49M25, 49N25, 49N35, 93-02, 93B03, 93B25, 93B27, 93B40, 93B50, 93B52, 93D15

Library of Congress Cataloging-in-Publication Data
Nonsmooth analysis and geometric methods in deterministic optimal control/Boris S. Mordukhovich, Hector J. Sussmann, editors.
p. cm. – (The IMA volumes in mathematics and its applications ; v. 78)
Proceedings of a workshop held in Feb. 1993.
Includes bibliographical references.
ISBN-13:978-1-4613-8491-5 e-ISBN-13:978-1-4613-8489-2
DOI: 10.1007/978-1-4613-8489-2

1. Nonsmooth optimization—Congresses. 2. Geometry, Differential—Congresses. 3. Control theory—Congresses. I. Mordukhovich, B. Sh. (Boris Sholimovich) II. Sussmann, Hector J., 1946–
III. Series.
QA402.5.N68 1996
515′.64—dc20 96-13083

Printed on acid-free paper.

© 1996 Springer-Verlag New York, Inc.
Softcover reprint of the hardcover 1st edition 1996

All rights reserved. This work may not be translated or copied in whole or in part without the written permission of the publisher (Springer-Verlag New York, Inc., 175 Fifth Avenue, New York, NY 10010, USA), except for brief excerpts in connection with reviews or scholarly analysis. Use in connection with any form of information storage and retrieval, electronic adaptation, computer software, or by similar or dissimilar methodology now known or hereafter developed is forbidden.
The use of general descriptive names, trade names, trademarks, etc., in this publication, even if the former are not especially identified, is not to be taken as a sign that such names, as understood by the Trade Marks and Merchandise Marks Act, may accordingly be used freely by anyone.
Permission to photocopy for internal or personal use, or the internal or personal use of specific clients, is granted by Springer-Verlag, Inc., for libraries registered with the Copyright Clearance Center (CCC), provided that the base fee of $5.00 per copy, plus $0.20 per page, is paid directly to CCC, 222 Rosewood Drive, Danvers, MA 01923, USA. Special requests should be addressed directly to Springer-Verlag New York, 175 Fifth Avenue, New York, NY 10010, USA.
ISBN-13:978-1-4613-8491-5/1996 $5.00 + 0.20

Production managed by Hal Henglein; manufacturing supervised by Joe Quatela.
Camera-ready copy prepared by the IMA.

9 8 7 6 5 4 3 2 1

ISBN-13:978-1-4613-8491-5 SPIN 10524690

FOREWORD

This IMA Volume in Mathematics and its Applications

NONSMOOTH ANALYSIS AND GEOMETRIC METHODS IN DETERMINISTIC OPTIMAL CONTROL

is based on the proceedings of a workshop that was an integral part of the 1992–93 IMA program on "Control Theory." The purpose of this workshop was to concentrate on powerful mathematical techniques that have been developed in deterministic optimal control theory after the basic foundations of the theory (existence theorems, maximum principle, dynamic programming, sufficiency theorems for sufficiently smooth fields of extremals) were laid out in the 1960s. These advanced techniques make it possible to derive much more detailed information about the structure of solutions than could be obtained in the past, and they support new algorithmic approaches to the calculation of such solutions.

We thank Boris S. Mordukhovich and Hector J. Sussmann for organizing the workshop and editing the proceedings. We also take this opportunity to thank the National Science Foundation and the Army Research Office, whose financial support made the workshop possible.

Avner Friedman

Willard Miller, Jr.

PREFACE

This volume contains the proceedings of the workshop on Nonsmooth Analysis and Geometric Methods in Deterministic Optimal Control held at the Institute for Mathematics and its Applications on February 8-17, 1993 during a special year devoted to Control Theory and its Applications. The workshop—whose organizing committee consisted of V. Jurdjevic, B.S. Mordukhovich, R.T. Rockafellar, and H.J. Sussmann—brought together internationally recognized authorities in both geometric and nonsmooth analysis methods in optimal control and its applications. Some of the papers presented at the workshop are published in this volume.

The topics in this volume include nonsmooth analysis and related techniques in necessary optimality conditions for abstract semilinear optimal control problems with applications to distributed time delay systems and semilinear elliptic equations involving domain and boundary controls; optimization of nonconvex differential inclusions with free time and endpoint constraints; discrete approximations of constrained optimal control systems with convergence analysis and error estimates; approximation schemes for infinite horizon optimal control problems with state constraints and viscosity solutions of the corresponding Hamilton–Jacobi–Bellman equation; perturbation analysis and well-posedness in optimal control; Lie algebraic methods applied to the study of the structure of reachable sets, optimal feedback control, dynamic feedback stabilization, high-order optimality conditions, and controlled invariance of distributions; and a survey of recent developments in the theory of impulsive control systems.

Boris S. Mordukhovich
Hector J. Sussmann

CONTENTS

IMPULSIVE CONTROL SYSTEMS

ALBERTO BRESSAN*

1. Introduction. This paper has a tutorial character. Its purpose is to provide an introduction to the theory of impulsive control systems, described by equations such as

$$\dot{x} = \Phi(t, x, u, \dot{u}), \tag{1.1}$$

$$x(0) = \bar{x}. \tag{1.2}$$

Here $x \in \mathbb{R}^n$ is the state variable, the control u ranges in a set $U \subset \mathbb{R}^m$ and, for simplicity, we assume that Φ is continuously differentiable w.r.t. all variables. When the control $u = u(t)$ is absolutely continuous, its derivative $\dot{u}$ is an integrable function, defined almost everywhere. A solution of (1.1) can thus be defined in the usual Carathéodory sense, i.e. as an absolutely continuous function which satisfies (1.1) at almost every t. On the other hand, when the control u is discontinuous, its derivative must be interpreted as a distribution. This gives to the system (1.1) an impulsive character, because the corresponding trajectory may then be discontinuous as well. In this case, the previous concept of Carathéodory solution is no longer applicable, and alternative definitions must be sought.

The construction of generalized solutions will be carried out here for systems of the form

$$\dot{x} = f(t, x, u) + \sum_{i=1}^{m} g_i(t, x, u)\dot{u}_i, \tag{1.3}$$

linear with respect to $\dot{u}$. This is the framework most frequently encountered in applications.

Observe that, in the special case where u is right continuous with bounded variation, the derivatives $\dot{u}_i$ can be regarded as measures; (1.3) is then equivalent to the control system driven by measures

$$\dot{x} = f(t, x, u) + \sum_{i=1}^{m} g_i(t, x, u)\, d\mu_i,$$

with $\mu_i\big((a, b]\big) = u_i(b) - u_i(a)$. Most of our discussion, however, will not be restricted to this case.

As a motivation for subsequent material, in §2 we introduce a class of controlled Lagrangean systems whose equations have impulsive character.

* S.I.S.S.A., Via Beirut 4, Trieste 34014, Italy.

Such systems were first studied in [7], determining, in particular, under which conditions the equations take the special form (1.3).

In order to define a generalized concept of solution for (1.3) which is consistent with the classical one when u is absolutely continuous, a natural approach is to approximate the measurable control u by a sequence of more regular controls $u^{(\nu)}$ (with respect to the $\mathcal{L}^1$ distance) and study the limits of the corresponding trajectories $x^{(\nu)} = x(\cdot, u^{(\nu)})$. In connection with the Cauchy problem (1.2), (1.3), two possibilities may then arise:

CASE 1: As $\nu \to \infty$, the sequence of Carathéodory solutions $x^{(\nu)}$ converges to a unique limit $\hat{x}$, which does not depend on the choice of the approximating sequence $u^{(\nu)}$. It thus makes good sense to define $\hat{x} = x(\cdot, u)$ as the *generalized solution* of the Cauchy problem (1.2), (1.3) corresponding to the control u.

CASE 2: As $\nu \to \infty$, the sequence $x^{(\nu)}$ may converge to different limits, or diverge, depending on the choice of the approximating sequence $u^{(\nu)}$.

In §3 we will show that the first case actually occurs, provided that all Lie brackets of the vector fields g_i vanish identically. Indeed, when this crucial commutativity assumption holds, one can prove a representation formula for solutions of (1.2), (1.3) in terms of a standard (nonimpulsive) auxiliary control system. This formula, given at (3.16), is perfectly meaningful even for discontinuous controls u. It can therefore be used as a the basis for a definition of generalized trajectory, in the case where the control u in (1.3) is only measurable. This approach has been pursued in several papers [2, 4, 12, 19, 21], in connection with control systems driven by measures and with stochastic differential equations.

On the other hand, if the vector fields g_i do not commute, then the previous construction breaks down. Various approximating sequences $u^{(\nu)} \to u$ may now yield sequences of trajectories $x^{(\nu)}$ with different limits. In this second case, knowing the values of the control u itself is not sufficient, in order to determine a unique solution to the Cauchy problem. In §5 we indicate how an alternative concept of solution can still be provided, at least for controls u with bounded variation. As shown in [3], however, a unique determination of the trajectory is now possible only if a "graph completion" of u is given. More precisely, at each time τ where u is discontinuous, one should specify the path, joining $u(\tau-)$ with $u(\tau+)$, along which the instantaneous jump of u takes place. Contrary to the previous (commutative) case, different paths may now lead to different trajectories $x(\cdot, u)$.

The last section of this paper is concerned with optimization problems for impulsive systems. When the commutativity assumptions hold, we show how a Mayer problem for (1.2) (1.3) can be easily reduced to a standard optimization problem for a suitable nonimpulsive control system. This second variational problem can then be analyzed by well known techniques, such as dynamic programming or the Pontryagin maximum principle. For additional results and applications we refer to [4, 8, 13, 14, 17, 18, 20].

In closing, we remark that a theory of "generalized solutions" for the general Cauchy problem (1.1), (1.2) can be developed along two somewhat different lines.

Let $\mathcal{U}$ be a family of sufficiently regular, admissible control functions, such that the corresponding trajectory $x(\cdot, u)$ of (1.1), (1.2) can be uniquely defined, in the usual Carathéodory sense, for every $u \in \mathcal{U}$. In the first approach, the key step consists in finding some topology on $\mathcal{U}$ which renders continuous the input-output map $\phi : u \mapsto x(\cdot, u)$. If δ is a distance on $\mathcal{U}$ which generates this new topology, call $\widetilde{\mathcal{U}}$ the completion of the metric space $(\mathcal{U}, \delta)$ and let $\tilde{\phi}$ be the unique continuous extension of ϕ to the space $\widetilde{\mathcal{U}}$. Elements $\tilde{u} \in \widetilde{\mathcal{U}}$ can now be regarded as "generalized inputs". It is then natural to call $x(\cdot, \tilde{u}) \doteq \tilde{\phi}(\tilde{u})$ the generalized solution of (1.1), (1.2) corresponding to $\tilde{u}$. For results related to this point of view, see [3, 22].

Following an alternative approach, one considers the family $\mathcal{S}$ of all Carathéodory solutions of (1.1), (1.2) with controls $u \in \mathcal{U}$, together with its closure $\overline{\mathcal{S}}$ with respect to the $\mathcal{L}^1$ distance. In analogy with the well known construction of chattering controls, one now seeks a representation of $\overline{\mathcal{S}}$ as the set of all admissible trajectories for some auxiliary control system. In the present setting, the auxiliary system may still be impulsive, but will typically be linear in $\dot{u}$ and commutative, so that its trajectories can be defined unambiguously for arbitrary measurable controls. Results in this direction can be found in [6] for noncommutative systems of the form (1.3), and in [5, 9] for equations in which the derivative $\dot{u}$ enters quadratically.

2. Lagrangean systems with impulsive behavior. To start with a simple example, consider a man standing on a swing, who wishes to increase the width of his oscillations by raising or lowering his barycenter at suitable times. Neglecting friction and the mass of the swing itself, the motion can be described in terms of two Lagrangean coordinates: the angle θ formed by the swing and the downward vertical direction, and the radius of oscillation r, measured by the distance between the barycenter P of the swinger and the center of rotation O. Denoting by g the gravity acceleration and normalizing the mass to a unit, the Lagrangean associated with this system is

$$L = L(r, \theta, \dot{r}, \dot{\theta}) \doteq \frac{1}{2}(\dot{r}^2 + r^2\dot{\theta}^2) + gr\cos\theta. \tag{2.1}$$

Assume that, by bending his knees, the swinger can vary his radius of oscillation. This amounts to the addition of a constraint $r = u(t)$, implemented by forces acting on P, parallel to the vector OP. The function $u(\cdot)$ can be regarded here as a control, whose values are chosen at will by the man riding on the swing, within certain physical bounds, say $u(t) \in [r_-, \ r_+]$

with $0 < r_- < r_+$. The motion of the remaining free coordinate $\theta = \theta(t)$ is determined by the equation

$$\frac{d}{dt}\frac{\partial L}{\partial \dot\theta} = \frac{\partial L}{\partial \theta},$$

which in this case yields

$$\ddot\theta = -\frac{g\sin\theta}{r} - \frac{2\dot r\dot\theta}{r}. \tag{2.2}$$

Writing (2.2) as a first order system for the variables $x_1 = \theta$, $x_2 = \dot\theta$ and substituting $r = u(t)$, we thus recover an impulsive control system where the time derivative $\dot u$ of the control enters linearly:

$$\begin{cases} \dot x_1 &= x_2, \\ \dot x_2 &= -\dfrac{g\sin x_1}{u} - \dfrac{2x_2}{u}\dot u. \end{cases} \tag{2.3}$$

More generally, consider a system described by $n+m$ Lagrangean coordinates, say $q_1, \ldots, q_n, q_{n+1}, \ldots, q_{n+m}$. Let

$$T(q,\dot q) = \frac{1}{2}\sum_{i,j=1}^{n+m} A_{i,j}(q)\dot q_i \dot q_j, \tag{2.4}$$

be its kinetic energy, and assume that the system is affected by external forces having components $Q_i = Q_i(t,q,\dot q)$. The motion of the (uncontrolled) system with $n+m$ degrees of freedom is thus determined by the equations

$$\frac{d}{dt}\frac{\partial T}{\partial \dot q_i} = \frac{\partial T}{\partial q_i} + Q_i(q,\dot q) \qquad i = 1,\ldots,n+m. \tag{2.5}$$

Assume now that the last m coordinates $q_{n+1},\ldots,q_{n+m}$ are *controllized*, i.e. their values are prescribed at will by a controller, who has the capability of implementing m additional frictionless constraints. Here, "frictionless" means that the forces produced by the constraints make zero work in connection with any virtual displacement of the remaining free coordinates $q_1,\ldots,q_n$. The evolution of $q_1,\ldots,q_n$ can then be determined by the first n equations in (2.5), inserting the prescribed values

$$q_{n+1} = u_1(t), \quad \ldots \quad , q_{n+m} = u_m(t) \tag{2.6}$$

for the last m coordinates. Using (2.4) in (2.5) and multiplying by the components of the inverse matrix $A^{-1}_{\ell,i}$, one obtains

$$\sum_{j=1}^{n+m} A_{i,j}(q)\ddot q_j + \frac{1}{2}\sum_{j,h=1}^{n+m}\frac{\partial A_{i,j}(q)}{\partial q_h}\dot q_j\dot q_h = \frac{1}{2}\sum_{j,h=1}^{n+m}\frac{\partial A_{j,h}(q)}{\partial q_i}\dot q_j\dot q_h + Q_i(q,\dot q),$$

$$(2.7)\qquad \ddot{q}_\ell = \frac{1}{2}\sum_{i,j,h=1}^{n+m} A_{\ell,i}^{-1}\left(\frac{\partial A_{j,h}}{\partial q_i} - \frac{\partial A_{i,j}}{\partial q_h}\right)\dot{q}_j\dot{q}_h + \sum_{i=1}^{n+m} A_{\ell,i}^{-1} Q_i.$$

The substitution (2.6) in (2.7) thus yields an impulsive control system of the form (1.1), involving the control functions $u_i(\cdot)$ as well as their first order derivatives with respect to time.

In general, the right hand side of (2.7) will be a quadratic polynomial with respect to the derivatives $\dot{u}_1, \ldots, \dot{u}_m$. It is interesting to isolate those cases in which the derivatives $\dot{u}_i$ enter linearly in the equations. In [7], impulsive systems with this property were called "fit for jumps", since their evolution can be well defined even when the control u is discontinuous. By examining the terms which appear on the right hand side of (2.7), one easily obtains

THEOREM 2.1. *Let the matrix $A = A(t,q)$ in (2.4) be strictly positive definite and assume that*

$$(2.8)\qquad \begin{gathered}\sum_{i=1}^{n+m} A_{\ell,i}^{-1}\left(\frac{\partial A_{i,j}}{\partial q_h} - \frac{\partial A_{j,h}}{\partial q_i}\right) \equiv 0 \\ \ell \in \{1,\ldots,n\}, \quad i,j \in \{n+1,\ldots,n+m\}.\end{gathered}$$

Moreover, assume that the functions Q_i in (2.5) depend linearly on the derivatives $\dot{q}_{n+1}, \ldots, \dot{q}_{n+m}$. Then, inserting the values (2.6) in the first n equations in (2.5), one obtains a second order system for the variables $q_1, \ldots, q_n$ which is linear w.r.t. the derivatives $\dot{u}_1, \ldots, \dot{u}_m$.

More general results in this direction, for systems written in Hamiltonian form, can be found in [7, 15, 16].

Example 2.2. If the radial coordinate in (2.2) is controllized, the assignment $r = u(t)$ yields the system (2.3) which is linear w.r.t. $\dot{u}$. This would follow from the above theorem, observing that the component of the gravity force $Q = -gr^{-1}\sin\theta$ does not depend on $\dot{r}$ and that the matrices A, A^{-1} here take the form

$$A = \begin{pmatrix} 1 & 0 \\ 0 & r^2 \end{pmatrix} \qquad A^{-1} = \begin{pmatrix} 1 & 0 \\ 0 & r^{-2} \end{pmatrix}.$$

On the other hand, if we controllize the angular coordinate θ by choosing some control function v and implementing the constraint $\theta = v(t)$, from

$$\frac{d}{dt}\frac{\partial L}{\partial \dot{r}} = \frac{\partial L}{\partial r}$$

we obtain $\ddot{r} = g\cos\theta + r\dot{\theta}^2$. For the variables $x_1 = r$, $x_2 = \dot{r}$, this yields the system

$$\begin{cases} \dot{x}_1 &= x_2, \\ \dot{x}_2 &= gv + x_1\dot{v}^2, \end{cases}$$

which is quadratic w.r.t. $\dot{v}$. Of course, the key assumption (2.8) in Theorem 2.1 now fails.

3. Generalized trajectories: the commutative case. The aim of this section is to provide a definition of generalized solution to (1.3), in the case where the control $u(\cdot)$ is a bounded, measurable function, possibly discontinuous and with unbounded variation. As a preliminary, we observe that, by introducing the additional variables $x_0, x_{n+1}, \ldots, x_{n+m}$ with equations

$$\dot{x}_0 = 1, \quad \dot{x}_{n+1} = \dot{u}_1, \quad \ldots \quad , \quad \dot{x}_{n+m} = \dot{u}_m, \tag{3.1}$$

the system (1.3) can be transformed into

$$\dot{x} = F(x) + \sum_{i=1}^{m} G_i(x)\dot{u}_i, \tag{3.2}$$

where the new vector fields F, G_i on $\mathbb{R}^N$ $(N = 1 + n + m)$ no longer depend on t, u. For simplicity, we shall thus consider the Cauchy problem determined by (3.2), together with the initial condition

$$x(0) = \bar{x} \in \mathbb{R}^N. \tag{3.3}$$

To carry out our program, a crucial commutativity hypothesis on the vector fields G_i is needed. Precisely, we shall assume that all of their Lie brackets vanish identically:

$$[G_i,\ G_j](x) \equiv 0 \qquad \forall i, j = 1, \ldots m, \tag{3.4}$$

We recall that the *Lie bracket* of two vector fields f, g is defined as

$$[f,\ g] \doteq (D_x g) \cdot f \ - \ (D_x f) \cdot g,$$

where $D_x f$ denotes the Jacobian matrix of first order partial derivatives of f. Moreover, we say that a vector field f is *complete* if, for every $\bar{x}$, the Cauchy problem

$$\dot{x}(\tau) = f(x(\tau)), \qquad x(0) = \bar{x}$$

has a solution defined for all $\tau \in \mathbb{R}$. The starting point for the construction of trajectories is

THEOREM 3.1. *Let the vector fields $G_1, \ldots, G_m$ on $\mathbb{R}^N$ be continuously differentiable, complete, and satisfy (3.4). Then, for any given $\bar{x}$, there exists a unique $\mathcal{C}^1$ map $\varphi : \mathbb{R}^m \mapsto \mathbb{R}^N$ such that*

$$(3.5)\quad \varphi(0) = \bar{x}, \qquad \frac{\partial \varphi(u)}{\partial u_i} = G_i(\varphi(u)) \qquad i = 1, \ldots, m, \quad u \in \mathbb{R}^m.$$

This is indeeed a corollary of Frobenius' theorem [11, pp.303–307], where the completeness of the vector fields G_i guarantees that the solution φ is globally defined. For the value $\varphi(u)$ of the solution of (3.5), it is convenient to use the notation

$$(3.6)\qquad \left(\exp \sum_{i=1}^{m} u_i G_i\right)(\bar{x}) \doteq \varphi(u).$$

To compute $\varphi(u)$, consider any smooth (or piecewise smooth) path $\gamma : [0,1] \mapsto \mathbb{R}^m$ joining the origin with u, so that $\gamma(0) = 0,\ \gamma(1) = u$. Next, solve the Cauchy problem

$$(3.7)\qquad x(0) = \bar{x}, \qquad \dot{x} \doteq \frac{dx}{d\sigma} = \sum_{i=1}^{m} G_i(x(\sigma))\dot{\gamma}_i(\sigma) \qquad \sigma \in [0,1].$$

When $\sigma = 1$, the value $x(1)$ of this solution is precisely $\varphi(u)$. Indeed, (3.5) and (3.7) imply

$$\begin{aligned} \frac{d}{d\sigma}[\varphi(\gamma(\sigma)) - x(\sigma)] &= \sum_{i=1}^{m} G_i(\varphi(\gamma(\sigma)))\,\dot{\gamma}_i(\sigma) - \dot{x}(\sigma) \\ &= \sum_{i=1}^{m} \left[G_i(\varphi(\gamma(\sigma))) - G_i(x(\sigma))\right]\dot{\gamma}_i(\sigma). \end{aligned}$$

Observing that $\varphi(\gamma(0)) = x(0)$, we can use Gronwall's lemma and conclude

$$\varphi(\gamma(\sigma)) - x(\sigma) = 0 \qquad \forall \sigma \in [0,1].$$

Hence, in particular, $\varphi(u) = \varphi(\gamma(1)) = x(1)$.

By Theorem 3.1, the value $\varphi(u)$ does not depend on the choice of the path γ used in the above construction, as long as $\gamma(0) = 0$, $\gamma(1) = u$. Letting γ be the polygonal line in $\mathbb{R}^m$ with vertices $V_i = (u_1, u_2, \ldots, u_i, 0, \ldots, 0)$, $i = 0, \ldots, m$, one obtains the identity

$$(3.8)\qquad \left(\exp \sum_{i=1}^{m} u_i G_i\right)(\bar{x}) = (\exp u_m G_m) \cdots (\exp u_1 G_1)(\bar{x}),$$

where $(\exp u_i G_i)(\eta)$ denotes the value at time $\tau = u_i$ of the solution to the Cauchy problem

$$(3.9)\qquad \dot{x}(\tau) = G_i(x(\tau)), \qquad x(0) = \eta.$$

In other words, (3.8) can be computed as follows. Start from $\bar{x}$. Move along the flow of the vector field G_1 during a time interval of length u_1, then move along the flow of G_2 for a time u_2, etc... At the m-th step, after following the flow of G_m for a time u_m, the point $\varphi(u)$ is reached.

We remark that, because of the commutativity assumptions (3.4), any permutation of the order of the exponentials $(\exp u_i G_i)$ in (3.8) would yield exactly the same result. More generally, for any $u, u' \in \mathbb{R}^m$ one has the identity

$$\begin{aligned}\Big(\exp\sum_{i=1}^{m}(u_i+u_i')G_i\Big)(\bar{x}) &= \Big(\exp\sum_{i=1}^{m}u_iG_i\Big)\Big(\exp\sum_{i=1}^{m}u_i'G_i\Big)(\bar{x})\\ &= \Big(\exp\sum_{i=1}^{m}u_i'G_i\Big)\Big(\exp\sum_{i=1}^{m}u_iG_i\Big)(\bar{x}).\end{aligned} \tag{3.10}$$

Next, consider an absolutely continuous control function $u(\cdot)$ and let $x(\cdot)$ be the corresponding Carathéodory solution of the Cauchy problem (3.2), (3.3). Fix any constant value $u^* \in \mathbb{R}^m$ and consider the auxiliary trajectory

$$\xi(t) \doteq \Big(\exp\sum_{i=1}^{m}\big(u_i^*-u_i(t)\big)G_i\Big)\big(x(t)\big). \tag{3.11}$$

We shall exhibit a differential equation satisfied by ξ. Denoting by $D_x\big(\exp\sum u_iG_i\big)$ the $N\times N$ Jacobian matrix of the diffeomorphism $x \to \big(\exp\sum u_iG_i\big)(x)$, define the function

(3.12)

$$F^*(\xi,u) \doteq \Big[D_x\Big(\exp\sum_{i=1}^{m}(u_i^*-u_i)G_i\Big)\Big]\cdot F\Big(\Big(\exp\sum_{i=1}^{m}(u_i-u_i^*)G_i\Big)(\xi)\Big).$$

Otherwise stated: the value of $F^*(\xi,u)$ is obtained by

1. computing the vector F at the point $x = \big(\exp\sum(u_i-u_i^*)G_i\big)(\xi)$,
2. pulling back this vector from x to ξ, using the diffeomorphism

$$x \mapsto \Big(\exp\sum_{i=1}^{m}(u_i^*-u_i)G_i\Big)(x). \tag{3.13}$$

Observe that (3.13) defines a differentiable map, which sends the point x into ξ. Therefore, its Jacobian matrix maps tangent vectors at x into tangent vectors at ξ. The function F^* is thus well defined. We now have

THEOREM 3.2. *Let the vector fields F, G_i be continuously differentiable, complete, and satisfy (3.4). Let u be an absolutely continuous control function on $[0,T]$ and let $x(\cdot,u)$ be the corresponding solution of (3.2), (3.3). Then the function ξ defined at (3.11) is a Carathéodory solution of the Cauchy problem*

$$\dot{\xi}(t) = F^*\big(\xi, u(t)\big), \tag{3.14}$$

$$(3.15)\qquad \xi(0)=\bar{\xi}\doteq\left(\exp\sum_{i=1}^{m}(u_i^*-u_i(0))G_i\right)(\bar{x}).$$

The proof of Theorem 3.2 will postponed to the end of this section.

The remarkable property of the differential equation (3.14) is that its right hand side does not involve any of the derivatives $\dot{u}_i$. Indeed, as a result of the transformation (3.11), the contribution of the terms $G_i\dot{u}_i$ in (3.2) has been cancelled out. A Carathéodory solution of (3.14) is thus well defined even when the control u is only measurable. Moreover, as soon as $\xi(t)$ is known, the value of $x(t)$ can be recovered by inverting (3.11):

$$(3.16)\qquad x(t)=\left(\exp\sum_{i=1}^{m}(u_i(t)-u_i^*)G_i\right)(\xi(t)).$$

Motivated by the previous analysis, the following concept of generalized trajectory can now be introduced.

DEFINITION 3.3. Let the assumptions of Theorem 3.2 on F, G_i hold. Let $u:[0,T]\mapsto \mathbb{R}^m$ be a measurable control function. Then we say that $x:[0,T]\mapsto\mathbb{R}^N$ is a *generalized solution* of the Cauchy problem (3.2), (3.3) if, for some constant $u^*\in\mathbb{R}^m$, the function ξ in (3.11) is a Carathéodory solution of (3.14), (3.15).

Remark 3.4. An easy computation shows that, if x satisfies the above conditions for some constant u^*, then the corresponding equations (3.14), (3.15) still hold if u^* is replaced by any other constant $u^\dagger\in\mathbb{R}^m$ in (3.11)-(3.15). The above definition of generalized trajectory is thus independent of the choice of u^*.

Example 3.5. Adding the variable $x_3=u$, the impulsive system (2.3) takes the standard form

(3.17)

$$(\dot{x}_1,\dot{x}_2,\dot{x}_3)=\left(x_2,\ -\frac{g\sin x_1}{x_3},\ 0\right)+\left(0,\ -\frac{2x_2}{x_3},\ 1\right)\dot{u}\doteq F(x)+G(x)\dot{u}.$$

In this case, solving the differential equation $\dot{x}=G(x)$, we find

$$(3.18)\qquad (\exp uG)(x_1,x_2,x_3)=\left(x_1,\ \frac{x_2x_3^2}{(x_3+u)^2},\ x_3+u\right).$$

Choosing $u^*=1$ as reference value for the control, (3.11) yields

$$(3.19)\qquad (\xi_1,\xi_2,\xi_3)=\left(x_1,\ \frac{x_2x_3^2}{(x_3+1-u)^2},\ x_3+1-u\right).$$

At this stage, it is useful to recall the physical meaning of our variables: $x_1=\xi_1=\theta$ is the angle formed by the swing with the downward vertical direction, $x_2 = \dot\theta$ is the angular velocity, while $x_3 = u = r$ is the radius of oscillation. In (3.19) we therefore have $\xi_3 \equiv 1$, while $\xi_2 = \dot\theta r^2$ is the angular momentum. The differential equation satisfied by ξ can of course be recovered from the general formula (3.12). However, it is more convenient to derive it directly from (2.2). Observing that

$$\text{(3.20)}\quad \frac{d}{dt}(\dot\theta r^2) = \ddot\theta r^2 + 2\dot\theta r\dot r = \left(-\frac{g\sin\theta}{r} - \frac{2\dot r\dot\theta}{r}\right) r^2 + 2\dot\theta r\dot r = -gr\sin\theta,$$

we obtain for ξ the nonimpulsive system

$$\text{(3.21)}\qquad (\dot\xi_1, \dot\xi_2, \dot\xi_3) = \left(\frac{\xi_2}{u^2},\ -gu\sin\xi_1,\ 0\right) \doteq F^*(\xi, u).$$

Observe how the terms involving $\dot r = \dot u$ cancel each other out in (3.20). As soon as the solution ξ of (3.21) is found, the evolution of the original variables x_i can be traced back, using the identity

$$\text{(3.22)}\qquad (x_1, x_2, x_3)(t) = \left(\xi_1(t),\ \frac{\xi_2(t)}{u^2(t)},\ u(t)\right).$$

Remark 3.6. Assume that the control u is piecewise continuously differentiable, with jumps occurring at finitely many times $\tau_1 < \tau_2 < \cdots < \tau_k$. In this case, a piecewise continuous map $t \mapsto x(t)$ is a generalized solution of (3.2) provided that

(i) $x(\cdot)$ is a classical solution of (3.2) inside each subinterval $[\tau_{j-1},\ \tau_j]$ where u is smooth.

(ii) At each time τ where u has a jump, the left and right limits of $x(t)$ as $t \to \tau$ satisfy

$$\text{(3.23)}\qquad x(\tau+) = \left(\exp\sum_{i=1}^m (u_i(\tau+) - u_i(\tau-))G_i\right)(x(\tau-)).$$

Indeed, observing that ξ remains continuous even at those times τ where u jumps, the representation formula (3.16) yields

$$\begin{aligned} x(\tau+) &= \left(\exp\sum_{i=1}^m (u_i(\tau+) - u_i^*)G_i\right)(\xi(\tau)) \\ &= \left(\exp\sum_{i=1}^m (u_i(\tau+) - u_i(\tau-))G_i\right) \\ &\qquad \left(\exp\sum_{i=1}^m (u_i(\tau-) - u_i^*)G_i\right)(\xi(\tau)) \\ &= \left(\exp\sum_{i=1}^m (u_i(\tau+) - u_i(\tau-))G_i\right)(x(\tau-)). \end{aligned}$$

We remark that, instead of (3.23), naive intuition might suggest the wrong formula

$$x(\tau+) = x(\tau-) + \sum_{i=1}^{m} \big(u_i(\tau+) - u_i(\tau-)\big) G_i\big(x(\tau-)\big). \tag{3.24}$$

To appreciate the difference, consider the system (3.17). Assume that the radius of oscillation $r = u(t)$ makes a jump at time τ, say from $r(\tau-)$ to $r(\tau+)$. Using (3.23) and (3.18), in terms of the physical variables $\theta, \dot\theta$ we obtain

$$\theta(\tau+) = \theta(\tau-), \qquad \dot\theta(\tau+) = \frac{\dot\theta(\tau-) r^2(\tau-)}{r^2(\tau+)}. \tag{3.25}$$

Observe that the second equation in (3.25) is equivalent to the conservation of the angular momentum $\dot\theta r^2$. Of course, this is true because the constraint $r = u(t)$ is implemented by impulsive forces acting along OP, which cannot produce any change in the angular momentum. On the other hand, recalling the definition (3.17) of G, using (3.24) one would obtain

$$\dot\theta(\tau+) = \dot\theta(\tau-) - \big(r(\tau+) - r(\tau-)\big) \cdot \frac{2\dot\theta(\tau-)}{r(\tau-)},$$

which is physically unacceptable, being inconsistent with the conservation of momentum.

The reader is thus cautioned that (3.24) can be used only when all vector fields G_i are constant, in which case the two expressions (3.23), (3.24) coincide.

Remark 3.7. The assumption (3.4) is trivially satisfied if $m = 1$, i.e. if the control u is a scalar.

In the case where the system (3.2) is derived from (1.2) by adding the new variables (3.1), a straightforward computation shows that the commutativity assumptions (3.4) hold if and only if the identity

$$\frac{\partial}{\partial u_j} g_i + \big(D_x g_i\big) \cdot g_j = \frac{\partial}{\partial u_i} g_j + \big(D_x g_j\big) \cdot g_i \tag{3.26}$$

holds at all points (t, x, u). Here $D_x g_i$ denotes the $n \times n$ matrix of derivatives of g_i with respect to $x_1, \ldots, x_n$.

Remark 3.8. The previous analysis has taken place in the Euclidean space $\mathbb{R}^N$. However, everything remains valid if the state variable x ranges on a manifold. Indeed, the commutativity assumptions (3.4) as well as the definitions (3.11), (3.12) are invariant under changes of coordinates. Notice that this invariance property is also true of (3.23), but fails for (3.24).

Remark 3.9. If we drop the assumption that all vector fields are complete, it may happen that, for a given trajectory $x : [0, T] \mapsto \mathbb{R}^N$, the function ξ in (3.11) is well defined for some choices of the constant u^* but not for others. To cope with this more general situation, we can modify Definition 3, allowing that (3.11)-(3.16) be satisfied with different constants u^* on different subintervals of $[0,T]$. More precisely, one can call x a *generalized solution* of the Cauchy problem (3.2), (3.3) if there exists a partition $t_0 = 0 < t_1 < \cdots < t_k = T$ of $[0,T]$ and constant vectors $u^{j*} \in \mathbb{R}^m$, $j = 1,\ldots,k$, such that the following holds. For each j, replacing u^* by u^{j*} in (3.11)-(3.15), the corresponding function ξ defined at (3.11) is a Carathéodory solution of (3.14), (3.15) on the subinterval $[t_{j-1},\, t_j]$.

Proof of Theorem 3.2. By the commutativity property (3.10), for any $v \in \mathbb{R}^m$ and $i \in \{1,\ldots,m\}$ one has

$$
\begin{aligned}
\Big(\exp \textstyle\sum_{j=1}^m v_j G_j\Big)(x) &= \Big(\exp \textstyle\sum_{j\neq i} v_j G_j\Big)\big(\exp v_i G_i\big)(x) \\
&= \big(\exp v_i G_i\big)\Big(\exp \textstyle\sum_{j\neq i} v_j G_j\Big)(x).
\end{aligned}
\tag{3.27}
$$

Differentiating (3.27) with respect to v_i, one establishes the identity

$$
\begin{aligned}
\tfrac{\partial}{\partial v_i}\Big(\textstyle\sum_{j=1}^m v_j G_j\Big)(x) &= \Big[D_x\Big(\exp \textstyle\sum_{j=1}^m v_j G_j\Big)\Big]\cdot G_i(x) \\
&= G_i\Big(\Big(\exp \textstyle\sum_{j=1}^m v_j G_j\Big)(x)\Big).
\end{aligned}
\tag{3.28}
$$

After this preliminary, we observe that the absolute continuity of x, u, together with the regularity of the vector fields G_i, implies that the function ξ in (3.11) is absolutely continuous as well. Next, we differentiate (3.11), recalling that the exponential map is a solution of (3.5). This yields

$$
\begin{aligned}
\dot\xi(t) &= -\sum_{i=1}^m G_i\left(\Big(\exp\sum_{j=1}^m \big(u_j^* - u_j(t)\big)G_j\Big)\big(x(t,u)\big)\right)\dot u_i(t) \\
&\quad + \Big[D_x\Big(\exp \textstyle\sum_{j=1}^m \big(u_j^* - u_j(t)\big)G_j\Big)\Big]\,\dot x(t,u) \\
&= -\sum_{i=1}^m G_i\left(\Big(\exp\sum_{j=1}^m \big(u_j^* - u_j(t)\big)G_j\Big)\big(x(t,u)\big)\right)\dot u_i(t) \\
&\quad + \Big[D_x\Big(\exp \textstyle\sum_{j=1}^m \big(u_j^* - u_j(t)\big)G_j\Big)\Big]\cdot F\big(x(t,u)\big)
\end{aligned}
$$

$$+\left[D_x\left(\exp\sum_{j=1}^m (u_j^* - u_j(t))G_j\right)\right]\cdot\sum_{i=1}^m G_i(x(t,u))\dot u_i(t)$$

$$= \left[D_x\left(\exp\sum_{i=1}^m (u_i^* - u_i(t))G_i\right)\right]$$

$$\cdot F\left(\left(\exp\sum_{i=1}^m (u_i(t) - u_i^*)G_i\right)(\xi(t))\right).$$

Indeed, the two summations involving the derivatives $\dot u_i$ cancel out each other, because of (3.28). □

4. Existence and continuous dependence. The existence and uniqueness of generalized solutions of generalized solutions of (3.2), (3.3), as well as their dependence on $\bar x, u$, can be determined using the representation formula (3.16). Observe that, if the vector fields G_i are $\mathcal{C}^2$ (i.e., twice continuously differentiable), then the same is true of the exponential map (3.13). The Jacobian of (3.13) is thus $\mathcal{C}^1$ and the function F^* in (3.12) is continuously differentiable with respect to both ξ and u. Applying the standard theory of O.D.E. to the Cauchy problem (3.14), (3.15), one obtains

THEOREM 4.1. *Let F be continuously differentiable. Moreover, assume that the vector fields G_i are $\mathcal{C}^2$, complete, and satisfy (3.4). Let u be any bounded, measurable control function, defined for $t \geq 0$. Then the Cauchy problem (3.2), (3.3) has a unique generalized solution, defined on some maximal interval* $[0,\tau)$.

The next result shows that the concept of generalized solution given in Definition 3 is robust: a generalized solution $x(\cdot,u)$ is obtained as the unique limit of Carathéodory solutions $x(\cdot,u^{(\nu)})$, as the control function u is approximated by a sequence of more regular functions $u^{(\nu)}$.

THEOREM 4.2. *Under the same assumptions on F, G_i, u as in Theorem 4.1, let $u^{(\nu)} : [0,T] \mapsto \mathbb{R}^m$ be a sequence of absolutely continuous, uniformly bounded control functions, such that*

(4.1)
$$\lim_{\nu\to\infty} u^{(\nu)}(0) = u(0),\ \lim_{\nu\to\infty} u^{(\nu)}(T) = u(T),\ \lim_{\nu\to\infty}\int_0^T \left|u^{(\nu)}(t) - u(t)\right| dt = 0.$$

Moreover, assume that the corresponding Carathéodory solutions $x^{(\nu)} = x(t,u^{(\nu)})$ of (3.2), (3.3) exist and remain inside a fixed compact set, as $t\in[0,T]$, $\nu \geq 1$. Then the generalized solution $x = x(t,u)$ is well defined on $[0,T]$, and one has

(4.2)
$$\lim_{\nu\to\infty}\int_0^T \left|x^{(\nu)}(t) - x(t,u)\right| dt = 0, \qquad \lim_{\nu\to\infty}\left|x^{(\nu)}(T) - x(T,u)\right| = 0.$$

Indeed, for a fixed u^*, the sequence $\xi^{(\nu)}$ of solutions of

(4.3)
$$\dot\xi^{(\nu)}(t) = F^*\big(\xi^{(\nu)}(t), u^{(\nu)}(t)\big), \xi^{(\nu)}(0) = \Big(\exp\sum_{i=1}^{m}(u_i^* - u_i^{(\nu)}(0))G_i\Big)(\bar x),$$

converges to the unique solution of (3.14), (3.15) corresponding to the control u, uniformly on $[0,T]$. The limits in (4.2) are thus an immediate consequence of (4.1) and of the representation formula (3.16).

5. Construction of trajectories: the noncommutative case. In this section, we consider again the impulsive control system (3.2), but we drop the commutativity assumption (3.4). In this case, Theorem 3.1 no longer holds and the exponential map (3.6) is not well defined. As a consequence, if u is a discontinuous control, the strategy of approximating u with more regular controls and computing the limit of the corresponding trajectories no longer works.

Example 5.1. Consider the impulsive system on $\mathbb{R}^2$:

$$(\dot x_1, \dot x_2) \;=\; (1,0)\dot u_1 + (0, x_1)\dot u_2 \;\doteq\; G_1(x)\dot u_1 + G_2(x)\dot u_2, \tag{5.1}$$

with initial condition

$$(x_1, x_2)(0) = (0,0). \tag{5.2}$$

Observe that in this case

$$[G_1, G_2] \equiv (0,1),$$

hence (3.4) fails. Consider the discontinuous control function $u : [0,2] \mapsto \mathbb{R}^2$:

$$(u_1, u_2)(t) = \begin{cases} (0,0) & \text{if} \quad t < 1, \\ (1,1) & \text{if} \quad t > 1. \end{cases} \tag{5.3}$$

The control u can be approximated by the sequence of continuous, piecewise linear functions $v^{(n)}$:

$$(v_1^{(n)}, v_2^{(n)})(t) = \begin{cases} (0,0) & \text{if} \quad t \in [0,\ 1 - 1/n], \\ (0,\ 1 + n(t-1)) & \text{if} \quad t \in [1 - 1/n,\ 1], \\ (n(t-1),\ 1) & \text{if} \quad t \in [1,\ 1 + 1/n], \\ (1,1) & \text{if} \quad t \in [1 + 1/n,\ 2]. \end{cases} \tag{5.4}$$

The corresponding Carathéodory solutions of (5.1), (5.2) are found to be

$$(x_1, x_2)(t, v^{(n)}) = \begin{cases} (0,0) & \text{if} \quad t \in [0,1], \\ (n(t-1),\ 0) & \text{if} \quad t \in [1,\ 1 + 1/n], \\ (1,0) & \text{if} \quad t \in [1 + 1/n,\ 2]. \end{cases} \tag{5.5}$$

As $n \to \infty$, the sequence in (5.5) converges (pointwise and in $\mathcal{L}^1$) to the limit trajectory

$$(5.6) \qquad (x_1, x_2)(t) = \begin{cases} (0,0) & \text{if} \quad t < 1, \\ (1,0) & \text{if} \quad t > 1. \end{cases}$$

Next, consider a second approximating sequence $w^{(n)}$:

$$(5.7) \quad (w_1^{(n)}, w_2^{(n)})(t) = \begin{cases} (0,0) & \text{if} \quad t \in [0,\ 1-1/n], \\ (1+n(t-1),\ 0) & \text{if} \quad t \in [1-1/n,\ 1], \\ (1,\ n(t-1)) & \text{if} \quad t \in [1,\ 1+1/n], \\ (1,1) & \text{if} \quad t \in [1+1/n,\ 2]. \end{cases}$$

This time, the corresponding solutions of (5.1), (5.2) are

$$(5.8) \quad (x_1, x_2)(t, w^{(n)}) = \begin{cases} (0,0) & \text{if} \quad t \in [0,\ 1-1/n], \\ (1+n(t-1), 0) & \text{if} \quad t \in [1-1/n,\ 1], \\ (1,\ n(t-1)) & \text{if} \quad t \in [1,\ 1+1/n], \\ (1,1) & \text{if} \quad t \in [1+1/n,\ 2]. \end{cases}$$

Observe that, in this second case, the limit trajectory

$$(5.9) \qquad (x_1, x_2)(t) = \begin{cases} (0,0) & \text{if} \quad t < 1, \\ (1,1) & \text{if} \quad t > 1, \end{cases}$$

still exists, but is different from (5.6).

The above example shows that, in this noncommutative case, the limit of the approximating trajectories depends not only on the control u itself, but also on the way in which we approximate u by more regular controls. Observe that, in the first case, the values of $v^{(n)}$ shift from $(0,0)$ to $(0,1)$ and then to $(1,1)$. In the second case, the values of $w^{(n)}$ vary from $(0,0)$ to $(1,0)$ and then to $(1,1)$. This suggests that a generalized trajectory can now be determined only if, at every time τ where u has a jump, we specify along which path the instantaneous motion of u from $u(\tau-)$ to $u(\tau+)$ takes place. The next definition makes this more precise.

DEFINITION 5.2. A *graph-completion* of a function $u : [0,T] \mapsto \mathbb{R}^m$ is an absolutely continuous path $\gamma = (\gamma_0, \gamma_1, \ldots, \gamma_m) : [0,S] \mapsto [0,T] \times \mathbb{R}^m$, such that

(i) $\gamma(0) = \big(0, u(0)\big)$, $\gamma(S) = \big(T, u(T)\big)$,
(ii) $\gamma_0(s_1) \le \gamma_0(s_2)$, for all $0 \le s_1 \le s_2 \le S$,
(iii) for each $t \in [0,T]$, there exists some s such that $\gamma(s) = \big(t, u(t)\big)$.

The path γ thus provides a continuous parametrization of the graph of u in the (t,u)-space, which "bridges" the values $u(\tau-)$, $u(\tau+)$ with a continuous curve, at each time τ where u has a jump. We remark that a graph-completion of u exists if and only if the total variation of u is bounded. However, it is by no means unique.

In connection with γ, consider the Cauchy problem

$$\frac{dy}{ds} = F(y(s))\dot\gamma_0(s) + \sum_{i=1}^{m} G_i(y(s))\dot\gamma_i(s), \qquad y(0) = \bar x. \tag{5.10}$$

DEFINITION 5.3. Let $y(\cdot,\gamma)$ be the Carathéodory solution of (5.10). Then the (possibly multivalued) function

$$x(t,\gamma) = \{y(s);\quad \gamma_0(s) = t\} \tag{5.11}$$

will be called the *generalized trajectory of* (3.2), (3.3) *determined by the graph-completion* γ *of* u.

Observe that, by definition, the path γ is absolutely continuous, hence a solution of (5.10) can be understood in the classical sense.

It can be shown that the trajectory $x(\cdot,\gamma)$ depends on the curve γ itself, but not on the way in which it is parametrized. Otherwise stated, if $\gamma' : [0,S'] \mapsto [0,T] \times \mathbb{R}^m$ is another graph-completion of u, such that

$$\gamma'(\phi(s)) = \gamma(s) \qquad s \in [0,S]$$

for some absolutely continuous, strictly increasing $\phi : [0,S] \mapsto [0,S']$, then the generalized trajectory $x(\cdot,\gamma')$ is the same as $x(\cdot,\gamma)$.

Example 5.4. If $u : [0,T] \mapsto \mathbb{R}^m$ is absolutely continuous, a canonical graph-completion of u is the path

$$\gamma(s) = \big(s,\ u(s)\big) \qquad s \in [0,T].$$

In this case, one easily checks that the generalized trajectory defined at (5.11) coincides with the usual Carathéodory solution.

Example 5.5. Let $u : [0,T] \mapsto \mathbb{R}^m$ be any right- or left-continuous function with bounded variation. A simple way for constructing a graph-completion of u consists in bridging every jump of u with a straight segment. More precisely, in the (t,u)-space, consider the set

$$\Gamma \doteq \{(t,u(t));\quad t \in [0,T]\} \cup \{\big(\tau,\ \lambda u(\tau-) + (1-\lambda)u(\tau+)\big);\\ \lambda \in [0,1],\quad u(\tau-) \neq u(\tau+)\},$$

obtained by adding to the graph of u all segments with endpoints $\big(\tau,\ u(\tau-)\big)$, $\big(\tau,\ u(\tau+)\big)$, as τ ranges over the countable set of times where u is discontinuous. One can show that Γ is then a continuous curve, with finite length. Parametrizing Γ by its arclength, we obtain a graph-completion of u. A more detailed study of this "canonical" graph-completion can be found in [10].

Remark 5.6. For any fixed control function u with bounded variation, one can construct infinitely many graph-completions γ. In general, they will yield different generalized trajectories (5.11). The construction outlined in Example 5 singles out a particular graph-completion, uniquely determined by the control u. This construction, however, is not invariant under a change of coordinates in the u-space.

For practical applications, it appears that no general rule can be given: the use of any particular graph-completion must be justified by physical considerations, case by case.

Example 5.1 (continued). For the discontinuous control function u in (5.3), consider the graph-completion $\gamma : [0,4] \mapsto [0,2] \times \mathbb{R}^2$, defined by

$$\gamma(s) = \begin{cases} (s,0,0) & \text{if} \quad t \in [0,1], \\ (1,0,s-1) & \text{if} \quad t \in [1,2], \\ (1,s-2,1) & \text{if} \quad t \in [2,3], \\ (s-3,1,1) & \text{if} \quad t \in [3,4]. \end{cases} \tag{5.12}$$

The generalized trajectory $t \mapsto x(t,\gamma)$ then coincides with (5.6) for all $t \in [0,2]$, $t \neq 1$, while $x(1,\gamma)$ is multivalued. Observe that the curve described by γ is precisely the limit of the graphs of the approximating functions $v^{(n)}$ in (5.4).

On the other hand, the "canonical" graph-completion of u, obtained as in Example 5 by bridging the jump from $(0,0)$ to (1.1) with a straight segment, yields here the path $\eta : [0,3] \mapsto [0,2] \times \mathbb{R}^2$, defined by

$$\eta(s) = \begin{cases} (s,0,0) & \text{if} \quad s \in [0,1], \\ (1,s-1,s-1) & \text{if} \quad s \in [1,2], \\ (s-1,1,1) & \text{if} \quad s \in [2,3]. \end{cases}$$

The corresponding generalized trajectory of (5.1), (5.2) satisfies

$$x(t,\eta) = \begin{cases} (0,0) & \text{if} \quad t < 1, \\ (1,1/2) & \text{if} \quad t > 1, \end{cases}$$

while $x(1,\eta)$ is multivalued.

For additional results concerning the continuous dependence of these trajectories (in a suitable topology), as well as their approximation by classical ones, we refer to [3].

6. Optimization problems for impulsive systems. Consider again the impulsive Cauchy problem

$$\dot{x} = F(x) + \sum_{i=1}^{m} G_i(x)\dot{u}_i, \qquad x(0) = \bar{x} \in \mathbb{R}^N. \tag{6.1}$$

Assume that the control values $u(t)$ are constrained within some compact set $U \subset \mathbb{R}^m$, and let the initial value $u(0) = \bar{u} \in U$ be assigned. The family of admissible control functions will be denoted by

$$\mathcal{U} \doteq \{u : [0,T] \mapsto U, \quad u \text{ measurable}, \quad u(0) = \bar{u}, \quad T > 0\}.$$

Given a continuous function $J = J(t,x)$ and a closed set $S \subset I\!R \times I\!R^N$, consider the following optimization problem of Mayer, with terminal constraints and variable terminal time:

$$\max_{u\in\mathcal{U}} J\big(T, x(T,u)\big), \qquad \text{subject to } \big(T, x(T,u)\big) \in S. \tag{6.2}$$

Assuming that the hypotheses of Theorem 3.2 hold, the terminal point $x(T,u)$ is computed by the formula (3.16). This allows us to transform the problem (6.1), (6.2) into a standard Mayer problem for the auxiliary variable ξ, whose evolution is governed by the nonimpulsive equations (3.14).

First, define the functions

$$J_S(t,x) = \begin{cases} J(t,x) & \text{if} \quad (t,x) \in S, \\ -\infty & \text{if} \quad (t,x) \notin S, \end{cases}$$

$$J^*(t,\xi) \doteq \max_{\omega\in U} J_S\left(T, \left(\exp \sum_{i=1}^{m} \omega_i G_i\right)(\xi)\right). \tag{6.3}$$

Observe that the maximum in (6.3) is always attained (unless it equals $-\infty$), because J_S is upper semicontinuous and U is compact.

Next, consider the optimization problem for the control system (3.14), (3.15):

$$\max_{u\in\mathcal{U}} J^*\big(T, \xi(T,u)\big). \tag{6.4}$$

As a consequence of the representation formula (3.16), one obtains

THEOREM 6.1. *Let the hypotheses of Theorem 3.2 hold, and assume that at least one admissible trajectory of (5.1) satisfies the constraint $(T, x(T,u)) \in S$. Then, in the above setting, a control function $u \in \mathcal{U}$ is optimal for the Mayer problem (6.2), (6.1) if and only if u is optimal for the optimization problem (6.4), (3.14), (3.15) and the assignment $\omega = u(T)$ yields the maximum value of J_S in (6.3).*

If the function J^* is sufficiently regular, the new optimization problem for the variable ξ can be studied by standard techniques. We give below two simple examples. Variational problems of this type, for Lagrangean mechanical systems, were first studied in [8].

Example 6.2. Consider a man riding on a swing, who can vary at will his radius of oscillation. Using the same notations as in Example 2, let his initial position and angular velocity be described by $r(0) = \rho$, $\theta(0) = \alpha$, $\dot\theta(0) = \beta$. Suppose that, at a fixed time T, he wishes to maximize the (forward) angular velocity of the swing.

In this case, the set of admissible controls is

$$\mathcal{U} \doteq \Big\{u : [0,T] \mapsto [r_-, \ r_+]; \quad u \ \text{ measurable}, \quad u(0) = \rho\Big\},$$

and the optimization problem can be written as

$$\max_{u \in \mathcal{U}} x_2(T, u), \tag{6.5}$$

relative to the system (3.17), with initial conditions

$$(x_1, x_2, x_3)(0) = (\alpha, \beta, \rho). \tag{6.6}$$

In terms of the auxiliary variables $(\xi_1, \xi_2, \xi_3) \doteq (x_1,\ x_2 x_3^2,\ 1)$, the above problem takes the form

$$\max_{u \in \mathcal{U}} J^*(\xi(T)), \tag{6.7}$$

where

$$J^*(\xi) = \begin{cases} \xi_2 / r_-^2 & \text{if} \quad \xi_2 \geq 0, \\ \xi_2 / r_+^2 & \text{if} \quad \xi_2 < 0. \end{cases}$$

According to Theorem 6.1, a control $u \in \mathcal{U}$ is optimal for the impulsive maximization problem (6.5), (6.6), (3.17) if and only if u is optimal for the standard Mayer problem (6.7), relative to the system (3.21) with initial conditions

$$(\xi_1, \xi_2, \xi_3)(0) = (\alpha,\ \beta\rho^2,\ 1), \tag{6.8}$$

and the final value of u satisfies

$$u(T) = \begin{cases} r_- & \text{if} \quad \xi_2(T) > 0, \\ r_+ & \text{if} \quad \xi_2(T) < 0. \end{cases}$$

The optimization problem (6.7), (3.21), (6.8) can now be solved by applying the well known Pontryagin Maximum Principle.

Example 6.3. For the same system (3.17), (6.6) as in the previous example, assume that the initial angular velocity β is positive, and consider the problem of maximizing the amplitude of the first half-oscillation:

$$\max_{u \in \mathcal{U}} x_1(T), \qquad \text{subject to} \quad x_2(T) = 0, \qquad x_2(t) \geq 0 \ \ \forall t \in [0, T]. \tag{6.9}$$

In terms of the auxiliary variables ξ_i, this is equivalent to

$$\max_{u \in \mathcal{U}} \xi_1(T), \qquad \text{subject to} \quad \xi_2(T) = 0, \qquad \xi_2(t) \geq 0 \ \ \forall t \in [0, T]. \tag{6.10}$$

In this case, the terminal value $u(T)$ of the control is irrelevant, because a jump in the radius of oscillation does not instantaneously change the angle $\theta = x_1$. We claim that the optimal feedback control for the above problem is

$$u^\dagger(\xi_1, \xi_2) = \begin{cases} r_+ & \text{if} \quad \xi_1 < 0, \\ r_- & \text{if} \quad \xi_1 > 0. \end{cases} \tag{6.11}$$

In other words, the radius of oscillation should be maximum when the swing moves downward, and minimum when it moves upward. To prove our claim, observe that, assuming $0 \leq \xi_1(T) \leq \pi/2$, the problem (6.10) is equivalent to

$$\min_{u \in \mathcal{U}} \{gr_- \cos\xi_1(T)\}, \qquad \text{subject to } \xi_2(T) = 0. \tag{6.12}$$

If the feedback strategy (6.11) is implemented, given any initial condition $(\xi_1, \xi_2)(0) = (\bar{\xi}_1, \bar{\xi}_2)$ with $\bar{\xi}_2 \geq 0$, the value attained by $gr_- \cos\xi_1(T)$ at the first time when $\xi_2(T) = 0$ is

$$V(\bar{\xi}_1, \bar{\xi}_2) = \begin{cases} gr_- \cos\bar{\xi}_1 - \frac{1}{2}\frac{\bar{\xi}_2^2}{r_-^2} & \text{if } \bar{\xi}_1 \geq 0, \\ gr_- - \frac{1}{2}\frac{\bar{\xi}_2^2}{r_-^2} - \frac{gr_+^3}{r_-^2}(1 - \cos\bar{\xi}_1) & \text{if } \bar{\xi}_1 < 0, \end{cases} \tag{6.13}$$

provided that

$$V(\bar{\xi}_1, \bar{\xi}_2) \geq -gr_-. \tag{6.14}$$

Actually, if (6.14) fails, then the control $u^\dagger$ enables the swinger to make a 360° revolution (assuming that the swing is suspended with rigid bars) with a uniformly positive angular velocity. In this case, the problem (6.9) is physically not well posed.

It is clear that the value function V is continuous across the line $\bar{\xi}_1 = 0$. Moreover, V satisfies the dynamic programming equation

$$\min_{u \in [r_-,\, r_+]} \left\{ \frac{\partial V}{\partial \bar{\xi}_1} \cdot \frac{\bar{\xi}_2}{u^2} + \frac{\partial V}{\partial \bar{\xi}_2} \cdot (-gu \sin\bar{\xi}_1) \right\} = 0, \tag{6.15}$$

together with the boundary condition

$$V(\bar{\xi}_1, \bar{\xi}_2) = gr_- \cos\bar{\xi}_1 \qquad \text{whenever} \qquad \bar{\xi}_2 = 0,\ \bar{\xi}_1 \geq 0.$$

Assuming (6.14), if $u \in \mathcal{U}$ is any admissible control, as long as $\xi_2 > 0$ we thus have

$$\frac{d}{dt} V\big(\xi_1(t,u), \xi_2(t,u)\big) = \frac{\partial V}{\partial \bar{\xi}_1}\dot{\xi}_1 + \frac{\partial V}{\partial \bar{\xi}_2}\dot{\xi}_2 \geq 0, \tag{6.16}$$

because of (6.15). Therefore, if T is the first time when $\xi_2(T,u) = 0$, from (6.16) it follows

$$V\big(\xi_1(T,u), \xi_2(T,u)\big) = gr_- \cos\xi_1(T,u) \geq V(\bar{\xi}_1, \bar{\xi}_2),$$

showing that the performance of the control u cannot be better then the performance of the feedback control $u^\dagger$.

The above analysis also shows that the control $u^\dagger$ is the unique optimal one. If $\xi_1(0) = \alpha < 0$, then $u^\dagger$ is discontinuous, switching from r_+ to r_-

when $\xi_1 = 0$. In this case, the variational problem (6.9) admits no optimal solution within the class of absolutely continuous control functions. This illustrates the relevance of generalized controls, for the optimization of impulsive systems.

REFERENCES

[1] G. Barles, *Deterministic impulse control problems*, SIAM J. Control **23** (1985), pp. 419–432.

[2] Alberto Bressan, *On differential systems with impulsive controls*, Rend. Sem. Mat. Univ. Padova **77** (1987), pp. 1–9.

[3] Alberto Bressan and F. Rampazzo, *On differential systems with vector-valued impulsive controls*, Boll. Un. Mat. Ital. (7) **2-B** (1988), pp. 641–656.

[4] Alberto Bressan and F. Rampazzo, *Impulsive systems with commutative vector fields*, J. Optim. Theory Appl. **71** (1991), pp. 67–83.

[5] Alberto Bressan and F. Rampazzo, *On systems with quadratic impulses and their application to Lagrangean mechanics*, SIAM J. Control (1993) (to appear).

[6] Alberto Bressan and F. Rampazzo, *Impulsive control systems without commutativity assumptions*, J. Optim. Theory Appl., (submitted).

[7] Aldo Bressan, *Hyper-impulsive motions and controllizable coordinates for Lagrangian systems*, Atti Accad. Naz. Lincei, Memoirs, Ser. VIII, **XIX** (1990), pp. 197–246.

[8] Aldo Bressan, *On some control problems concerning the ski or swing*, Atti Accad. Naz. Lincei, Memoirs **IX-1** (1991), pp. 149–196.

[9] Aldo Bressan and M. Favretti, *On motions with bursting characters for Lagrangean mechanical systems with a scalar control. Parts I and II*, Atti Accad. Naz. Lincei, Rend. Mat. **IX-2** (1991), pp. 339–343 and **IX-3**, pp. 35–42.

[10] G. Dal Maso and F. Rampazzo, *On systems of ordinary differential equations with measures as controls*, Differential and Integral Equat. **4** (1991), pp. 739–765.

[11] J. Diéudonne, *Foundations of Modern Analysis*, Academic Press, New York, 1960.

[12] M.A. Krasnoselskii and A.V. Pokrovskii, *Vibrostable differential equations with a continuous right-hand side*, Proc. Moscow Math. Soc. **27** (1972), pp. 739–765.

[13] J.M. Murray, *Existence theorems for optimal control and calculus of variations problems where the states can jump*, SIAM J. Control **24** (1986), pp. 412–438.

[14] R.W. Rishel, *An extended Pontryagin Maximum Principle for control systems containing measures*, SIAM J. Control **3** (1965), pp. 191–205.

[15] F. Rampazzo, *On Lagrangean systems with some constraints as controls*, Atti Accad. Naz. Lincei, Rend. Mat. **82** (1988), pp. 685–695.

[16] F. Rampazzo, *On the Riemannian structure of a Lagrangean system and the problem of adding time-dependent constraints as controls*, European J. Mechanics A/Solids **10-4** (1991), pp. 405–431.

[17] F. Rampazzo, *Optimal impulsive controls with a constraint on the total variation*, in *New Trends in Systems Theory* (G. Conte, A.M. Perdon and B.F. Wynman, eds.), Birkhäuser, Boston, 1991, pp. 606–613.

[18] R.W. Rishel, *An extended Pontryagin Maximum Principle for control systems containing measures*, SIAM J. Control **3** (1965), pp. 191–205.

[19] A.V. Sarychev, *Nonlinear systems with impulsive and generalized function controls*, Proc. Conf. on Nonlinear Synthesis, Sopron, Hungary, 1989.

[20] W.W. Schmaedeke, *Optimal control theory for nonlinear vector differential equations containing measures*, SIAM J. Control **3** (1965), pp. 231–280.

[21] H.J. Sussmann, *On the gap between deterministic and stochastic ordinary differ-*

ential equations, Ann. of Probability 6 (1978), pp. 17–41.

[22] H.J. SUSSMANN AND W. LIU, *A characterization of continuous dependence of trajectories with respect to the input for control-affine systems*, SIAM J. Control (to appear).

APPROXIMATION OF OPTIMAL CONTROL PROBLEMS WITH STATE CONSTRAINTS: ESTIMATES AND APPLICATIONS*

FABIO CAMILLI† AND MAURIZIO FALCONE†‡

Abstract. We present some a priori estimates for the rate of convergence of two approximation schemes related to the deterministic infinite horizon problem with state constraints. A first order and a second order scheme are studied in detail and some hints on the construction of higher order methods are given. We prove that the schemes converge to the constrained viscosity solution of the related Hamilton–Jacobi–Bellman equation and we show that they can also be used to produce approximate optimal trajectories. The above results are applied to the numerical solution of a Vidale-Wolfe advertising model with state constraint.

Key words. state constraints, numerical methods, a priori estimates, viscosity solutions

AMS(MOS) subject classifications. Primary 65N12; Secondary 65N55, 49L20.

1. Introduction. We deal with the numerical approximation of the infinite horizon control problem with state constraints. Let Ω be an open bounded and convex subset of $\mathbb{R}^n$ and let the controlled dynamics be given by

$$(1.1)\qquad \begin{cases} \dot{y}(t) = b(y(t), \alpha(t)) & t \geq 0 \\ y(0) = x \end{cases}$$

Here $x \in \overline{\Omega}$ and the *control* $\alpha(t)$ belongs to the set of admissible control functions $\mathcal{A}$, defined as

$$(1.2)\qquad \mathcal{A} \equiv \{\alpha : [0, +\infty) \to A, \text{measurable}\}$$

where A is a compact subset of $\mathbb{R}^m$. The state constraint for (1.1) corresponds to require that the state remains in $\overline{\Omega}$ for all $t \geq 0$. This reduces the set of admissible controls to the following subset of $\mathcal{A}$

$$(1.3)\qquad \mathcal{A}_x \equiv \{\alpha(\cdot) \in \mathcal{A} : y(t, \alpha(t)) \in \overline{\Omega},\ \forall t \geq 0\}, \qquad \text{for any } x \in \overline{\Omega}.$$

where $y(\cdot, \alpha(\cdot))$ is the solution trajectory of (1.1) corresponding to the control $\alpha(\cdot)$.

* This work has been completed while the second author was visiting the Institute for Mathematics and its Applications (IMA) of the University of Minnesota.

† Dipartimento di Matematica, Università di Roma "La Sapienza", P. Aldo Moro 2, 00185, Roma, Italy.

‡ This author wishes to thank the IMA and the Istituto per le Applicazioni del Calcolo I.A.C.-C.N.R. of Rome for their support to this research.

Given the cost functional

$$J_x(\alpha) \equiv \int_0^{+\infty} f(y(t), \alpha(t)) e^{-\lambda t} dt, \tag{1.4}$$

(where λ is strictly positive) the problem is to determine the *value function*

$$v(x) = \inf_{\alpha \in \mathcal{A}_x} J_x(\alpha), \tag{1.5}$$

and possibly an *optimal control* (or at least an approximate optimal control). The above model problem has been studied by Soner [S]; he proved that the corresponding value function is the "constrained" viscosity solution of the dynamic programming equation

$$\lambda v(x) = \inf_{a \in A} \{b(x,a) \cdot \nabla v(x) + f(x,a)\} \text{ in } \overline{\Omega}, \tag{HJB}$$

(see Section 2 for a precise definition). Further developments in this direction can be found in Capuzzo Dolcetta-Lions [CDL] and in the review paper by Crandall-Ishii-Lions [CIL]. As far as the dynamics is concerned, we should mention that the existence of so called *viable trajectories*, i.e. trajectories which satisfy (1.1) and the constraint $\overline{\Omega}$, is a necessary requirement to set the optimal control problem. We limit ourselves to a rather simple situation assuming that Ω is convex and has a smooth boundary, however a collection of results for more general classes of constraints (including the non convex case) can be found in Aubin [A]. We should also mention that some ideas coming from viability theory are at the origin of the scheme studied in Falcone-Digrisolo [FD] where some tools of nonsmooth analysis have also been used to establish its convergence.

In this paper we mainly focus our attention on the order of convergence in L^∞ of the approximation schemes for v coming from discrete dynamic programming. We will consider at first the scheme studied in [FD] and then pass to more accurate schemes. They are all built making first a discretization in time of the original control problem and proving that a discrete dynamic programming principle holds true. A further discretization in space is necessary to get a finite dimensional problem which can be actually solved by a fixed point algorithm. In this respect our results in Sections 2 and 3 extend to the constrained problem those obtained by Capuzzo Dolcetta [CD], Capuzzo Dolcetta-Ishii [CDI] and Falcone [F] (see also the survey paper Capuzzo Dolcetta-Falcone [CDF] for other numerical methods related to optimal control problems). It is interesting to notice that the procedure used to build this discretization is quite general and can also be applied to other optimal control problems (e.g. finite horizon and optimal stopping problems) although we will not consider here these extensions.

Since one of the typical drawbacks of the dynamic programming approach to the numerical solution of optimal control problems is the "rise of dimension", the development of numerical schemes which converge fast to the solution seems to be crucial for its application to real problems. This is our main motivation for studying a second order scheme. Some hints on the construction of higher order methods for constrained problems can be obtained coupling the technique used in Section 3 with the results in Falcone-Ferretti [FF1], where a detailed analysis of a class of high–order methods for unconstrained problems is presented.

Another possibility to reduce the dimension of the problem is to split the global problem given in Ω into a number of small size problems adopting a domain decomposition strategy. We will also discuss this strategy which has been developed by Falcone-Lanucara-Seghini [FLS] and Sun [Su].

The outline of the paper is the following.

In Section 2 we introduce our basic assumptions, characterize the value function in terms of the Hamilton–Jacobi–Bellman equation and introduce the first order time-discrete approximation scheme. Section 3 is devoted to the proof of a priori estimates in L^∞ establishing the order of convergence for the first and second order scheme. The space discretization (step k) is studied in Section 4 where we prove the convergence of the fully discrete scheme to v and we present some features of the algorithm. Some results related to the domain decomposition strategy are also contained in this section. Finally, in Section 5 we discuss the application of the above techniques to the numerical solution of a Vidale-Wolfe advertising model with state constraints.

2. Convergence of the approximation schemes. We start recalling some preliminary results. We will denote by $y_x(t, \overline{\alpha}(t))$ the position at time t of the solution trajectory of (1.1) corresponding to the control $\overline{\alpha} \in \mathcal{A}$. Whenever this will be possible without ambiguity we will adopt the simplified notations $y_x(t)$ or $y(t)$ instead of $y_x(t, \overline{\alpha}(t))$. We want to minimize J_x with respect to the controls in $\mathcal{A}_x$ so that, to have a meaningful problem, we need to assume at least that

$$\mathcal{A}_x \neq \emptyset \qquad \text{for any } x \in \overline{\Omega}. \tag{2.1}$$

It is important to notice that (2.1) is not sufficient in general to guarantee the continuity of $v(x)$ in $\overline{\Omega}$. This is due to the structure of the multivalued map $x \longmapsto \mathcal{A}_x$. Soner has shown that the value function is continuous on $\overline{\Omega}$ if $\partial\Omega$ is sufficiently smooth and the following boundary condition on the vectorfield is satisfied

$$\exists \mu > 0 \ : \forall x \in \partial\Omega, \ \exists a \in A \text{ such that } b(x,a) \cdot \nu(x) \leq -\mu < 0 \tag{2.2}$$

where $\nu(x)$ is the outward normal to Ω at the point x.

We will make the following assumptions:

$$\Omega \text{ is a bounded, open convex subset of } \mathbb{R}^n, \tag{A0}$$

$$(A1)\qquad \begin{cases} \text{there exists a function } \varphi \in C_b^{1,1}(\mathbb{R}^n) : \\ \Omega = \{x \in \mathbb{R}^n : \varphi(x) > 0\} \\ \partial\Omega = \{x \in \mathbb{R}^n : \varphi(x) = 0\} \\ \text{and } \nabla\varphi(x) \neq 0, \quad \forall x \in \partial\Omega \end{cases}$$

$$(A2)\qquad \text{A is a compact subset of } \mathbb{R}^m,$$

$$(A3)\qquad \begin{cases} b : \mathbb{R}^n \times A \longrightarrow \mathbb{R}^n \text{ continuous,} \\ f : \mathbb{R}^n \times A \longrightarrow \mathbb{R} \text{ continuous,} \\ \sup_{a\in A} |b(x,a) - b(y,a)| \leq L_b|x-y|, \\ \sup_{a\in A} |f(x,a) - f(y,a)| \leq L_f|x-y|. \end{cases}$$

Clearly there exist two positive constants M_b, M_f such that

$$\sup_{a\in A} |b(x,a)| \leq M_b, \tag{2.3}$$

$$\sup_{a\in A} |f(x,a)| \leq M_f, \tag{2.4}$$

for any $x \in \overline{\Omega}$. Notice that, under the above assumptions, the value function is bounded in $\overline{\Omega}$ by M_f/λ as one can easily check.

We define for any $x, y \in \overline{\Omega}$ and $g \in W^{1,\infty}(\Omega)$

$$L_g \equiv \sup_{x\neq y} \frac{|g(x) - g(y)|}{|x-y|}. \tag{2.5}$$

Moreover, by (A1) we can write (2.2) as

$$\exists \mu > 0 : \forall x \in \partial\Omega, \exists a \in A \text{ such that } b(x,a) \cdot \frac{\nabla\varphi(x)}{|\nabla\varphi(x)|} \geq \mu > 0. \tag{2.6}$$

Remark 2.1 The following results can also be easily extended to the case of an unbounded constraint set provided its boundary is sufficiently smooth and b, f are bounded in Ω. However, for numerical purposes we will always focus our attention on a bounded Ω.

By the Dynamic Programming Principle, Soner has shown that v is the unique *constrained viscosity solution* of (HJB). This means that v satisfies

$$H(x, u(x), \nabla u(x)) \leq 0, \qquad \text{for } x \in \Omega \tag{2.7}$$

$$H(x, u(x), \nabla u(x)) \geq 0, \qquad \text{for } x \in \overline{\Omega} \tag{2.8}$$

where

$$(2.9)\quad H(x,u(x),\nabla u(x)) \equiv \lambda u(x) + \max_{a\in A}\{-b(x,a)\cdot\nabla u(x) - f(x,a)\}$$

and the above inequalities should be understood in the viscosity sense (cfr. [S] and [CDL]). A function satisfying (2.7) (respectively (2.8)) is called a *viscosity subsolution* (respectively *supersolution*) of $H(x,u(x),\nabla u(x)) = 0$. More precisely, the result proved in [S] is

Theorem 2.1. *Let (2.2), (A0)–(A3) be satisfied. Then $v \in C(\overline{\Omega})$ and it is the unique constrained viscosity solution of (HJB) in $\overline{\Omega}$.*

Following [FD] we turn now to the construction of our first discrete-time scheme for (HJB). We fix a positive parameter h, the time step, and consider the approximation scheme for (1.1) and (1.4)

$$(2.10)\quad \begin{cases} y_{n+1} = y_n + hb(y_n, a_n), & n = 0,1,2,\dots \\ y_0 = x \end{cases}$$

$$(2.11)\quad J_x^h(\{a_n\}) \equiv h\sum_{n=0}^{+\infty} f(y_n, a_n)\,\beta^n,$$

where $x \in \overline{\Omega}$, $a_n \in A$ and $\beta \equiv e^{-\lambda h}$.
The corresponding value function is

$$(2.12)\quad v_h(x) = \inf_{\{a_n\}\in\mathcal{A}_x^h} J_x^h(\{a_n\}),\quad x \in \overline{\Omega},$$

where

$$(2.13)\quad \mathcal{A}_x^h \equiv \{\{a_n\} : a_n \in A \text{ and } y_n \in \overline{\Omega},\ \forall n = 1,2,\dots\}.$$

The above definition is meaningful only provided there exists a step h such that $\mathcal{A}_x^h \neq \emptyset$. We look for conditions which guarantee the existence of viable discrete trajectories. We will denote by $y_x(n,a)$ the n-th point of the discrete trajectory corresponding to a control sequence $\{a_n\}$ and when possible we will adopt the simplified notations $y(n,a)$, y_n. Let us introduce the following multivalued map

$$(2.14)\quad A_h(x) \equiv \{a \in A : x + hb(x,a) \in \overline{\Omega}\}.$$

Clearly $\{a_n\} \in \mathcal{A}_x^h$ if and only if $a_n \in A_h(y_n)$ for any $n = 0,1,2,\dots$.

We start proving that the uniform boundary condition (2.6) on the continuous trajectories implies a uniform boundary condition for the discrete trajectories.

Lemma 2.2. *Let $\Gamma_\varepsilon \equiv \{x \in \overline{\Omega} : d(x, \partial\Omega) < \frac{1}{\varepsilon}\}$ and let (A1) and (2.6) be satisfied. Then there exists three positive constants c_0, ε_0 and h_0 such that*

$$\text{(2.15)} \qquad \begin{cases} \text{for any } x \in \Gamma_{\varepsilon_0} \text{ there exists } a \in A \text{ such that} \\ \varphi(x + hb(x,a)) \geq c_0 h \ , \quad \forall h < h_0 . \end{cases}$$

Proof. Since b and φ are Lipschitz continuous there exist two positive constants ε_0 and c such that

$$\text{(2.16)} \qquad \forall x \in \Gamma_{\varepsilon_0} \quad \exists a \in A : \quad b(x,a) \cdot \nabla\varphi(x) \geq c > 0 .$$

Then, for $x \in \Gamma_{\varepsilon_0}$ and for some $\xi \in [0,1]$, we have

$$\begin{aligned} \varphi(x + hb(x,a)) &= \varphi(x) + hb(x,a) \cdot \nabla\varphi(x + \xi hb(x,a)) \geq \\ &\geq hb(x,a) \cdot \nabla\varphi(x) - L_{\nabla\varphi} h^2 |b(x,a)|^2 \geq hc - L_{\nabla\varphi} h^2 M_b^2 \end{aligned}$$

and we get (2.15) setting $c_0 = \frac{c}{2}$ and $h_0 = \frac{c}{2L_{\nabla\varphi} M_b^2}$. □

Notice that the more restrictive choice

$$\text{(2.17)} \qquad h_0 \equiv \min\left\{ \frac{c}{2L_{\nabla\varphi} M_b^2}, \frac{\varepsilon_0}{M_b} \right\}$$

guarantees that

$$\text{(2.18)} \qquad \forall x \in \overline{\Omega} \text{ and } h < h_0 \ , \ \exists a \in A : x + hb(x,a) \in \overline{\Omega}.$$

Let $N \in \mathbb{N}$ and $\{a_n\}$ satisfy (2.20). We define

$$\text{(2.19)} \qquad J^h(N, \{a_n\}) \equiv h \sum_{n=0}^{N} f(y_n, a_n) \beta^n .$$

The following lemma shows how to project a general control sequence

$$\text{(2.20)} \qquad \{a_n\} \text{ such that } a_n \in A, \quad \forall n \in \mathbb{N},$$

onto $\mathcal{A}_x^h$, i.e. how to obtain an admissible sequence for the constrained discrete problem from a sequence which is *not* admissible, still having a bound on the difference between the corresponding costs.

Lemma 2.3. *Let (A0)–(A3) and (2.6) hold true. Then, there exist an $N^* \in \mathbb{N}$ and a positive real constant L such that for every $x \in \overline{\Omega}$ and every control sequence $\{a_n\}$ satisfying (2.20) there exists $\{\overline{a}_n\} \in \mathcal{A}_x^h$ satisfying*

$$\text{(2.21)} \qquad |J^h(N^*, \{a_n\}) - J^h(N^*, \{\overline{a}_n\})| \leq L\varepsilon$$

where $\varepsilon \equiv \sup_{n \le N^*} d(y_x(n,a), \overline{\Omega})$.

Proof. Given $\{a_n\}$ and a couple of integers $N, K_0 \in \mathbb{N}$ we define

$$N_0 \equiv \begin{cases} \min\{n \le N : y_{n+1} \notin \overline{\Omega}\} \\ N \text{ if } y_n \in \overline{\Omega} \text{ for any } n \le N \end{cases} \tag{2.22}$$

$$\overline{a}_n \equiv \begin{cases} a_n \text{ if } n \le N_0 - 1 \\ \hat{a}_{N_0} \text{ if } N_0 \le n \le N_0 + K_0 - 1 \\ a_{n-K_0} \text{ if } n \ge N_0 + K_0 \end{cases} \tag{2.23}$$

where $\hat{a}_{N_0}$ is the control satisfying (2.15) in $y_x(N_0, a_{N_0})$.

Moreover, we will denote by y_m and $\overline{y}_m$ the discrete trajectories corresponding respectively to the control sequences $\{a_n\}$, $\{\overline{a}_n\}$ and

$$z_m \equiv \varphi(y_m), \quad \overline{z}_m \equiv \varphi(\overline{y}_m). \tag{2.24}$$

We want to prove that there is a choice of $N = N^*$ and K_0 guaranteeing that $\overline{y}_m \in \overline{\Omega}$, $\forall m \le N^*$. By (A1) and the definition (2.24) this is equivalent to prove that $\overline{z}_m \ge 0$, for any $m \le N^*$.

We can write

$$z_{m+1} = z_m + hp(z_m, a_m) \tag{2.25}$$

where $p : \mathbb{R} \times A \to \mathbb{R}$ is defined as $p(z,a) = (\varphi(x + hb(x,a)) - \varphi(x))/h$ for $z = \varphi(x)$. The following properties of p are straightforward:

$$\begin{cases} \text{i)} & |p(z,a)| \le L_\varphi M_b \quad \forall z, \forall a; \\ \text{ii)} & p(\cdot, a) \text{ is Lipschitz continuous with constant} \\ & L_p = L_0^2(M_b + L_b + hM_bL_b) \\ & \text{where } \; L_0 = \max\{L_\varphi, L_{\nabla\varphi}, L_{\varphi^{-1}}\}; \\ \text{iii)} & p(z, a^*) \ge L_0 \cdot c_0 \text{ for } z = \varphi(x) \text{ with } x \in \Gamma_{\varepsilon_0} \\ & \quad \text{and } a^* \text{ satisfying (2.15).} \end{cases} \tag{2.26}$$

By the definitions (2.22), (2.23), (2.25) and by (2.26) we have for any $m \ge N_0$,

$$\begin{aligned} \overline{z}_{m+K_0} - z_m &= \overline{z}_{N_0+K_0} - z_{N_0} + \sum_{i=N_0}^{m-1} h\left[p(\overline{z}_{i+K_0}, \overline{a}_{i+K_0}) + \right. \\ &\left. - p(z_i, a_i)\right] = \overline{z}_{N_0+K_0} - z_{N_0} + \sum_{i=N_0}^{m-1} h\left[p(\overline{z}_{i+K_0}, a_i) - p(z_i, a_i)\right] \ge \\ &\ge \overline{z}_{N_0+K_0} - z_{N_0} - \sum_{i=N_0}^{m-1} hL_p|\overline{z}_{i+K_0} - z_i| \ge \\ &\ge \overline{z}_{N_0+K_0} - z_{N_0} - hL_p|\overline{z}_{N_0+K_0} - z_{N_0}| \sum_{i=N_0}^{m-1} (1 + hL_p)^{i-N_0} \ge \\ &\ge \overline{z}_{N_0+K_0} - z_{N_0} - hM_pK_0\left[(1 + hL_p)^{m-N_0} - 1\right]. \end{aligned} \tag{2.27}$$

Now we want to prove a lower bound for $\overline{z}_{N_0+K_0} - z_{N_0}$. We set $w_m \equiv h(m-N_0)p(z_{N_0}, \overline{a}_{N_0})$. Since $z_{N_0} = \varphi(y_{N_0})$ and $y_{N_0} \in \Gamma_{\varepsilon_0}$, by (2.26) we have

$$w_m \geq h(m-N_0)L_0c_0. \tag{2.28}$$

Since $\overline{z}_{N_0} = z_{N_0}$, we have for any $m \leq N_0 + K_0$

$$\begin{aligned}
|\overline{z}_m - z_{N_0} - w_m| &= |\sum_{i=N_0}^{m-1} h[p(\overline{z}_i, \overline{a}_i) - p(\overline{z}_{N_0}, \overline{a}_{N_0}]| = \\
&= |\sum_{i=N_0}^{m-1} h[p(\overline{z}_i, \overline{a}_{N_0}) - p(z_{N_0}, \overline{a}_{N_0})]| \leq \\
&\leq L_p \sum_{i=N_0}^{m-1} h^2 M_p (i - N_0) = \\
&= \frac{1}{2} L_p M_p h^2 (m-N_0)(m-N_0-1).
\end{aligned}$$

By applying the last inequality and (2.28), for $m = N_0 + K_0$, we get

$$\overline{z}_{N_0+K_0} - z_{N_0} \geq hK_0L_0c_0 - \frac{1}{2}h^2 L_p M_p K_0^2. \tag{2.29}$$

Substituting (2.29) in (2.27) we have

$$\begin{aligned}
\overline{z}_{m+K_0} - z_m &\geq hK_0L_0c_0 - \frac{1}{2}h^2 L_p M_p K_0^2 + \\
&- hM_pK_0[(1+hL_p)^{m-N_0} - 1].
\end{aligned}$$

Let N^* be such that

$$hN^* \leq \frac{1}{L_p} \log\left(1 + \frac{c_0 L_0}{4M_p}\right)$$

then

$$\overline{z}_{m+K_0} - z_m \geq hK_0 \left(\frac{3}{4}c_0L_0 - \frac{1}{2}hL_pM_pK_0\right). \tag{2.30}$$

Let us fix K_0 such that

$$hK_0 = \frac{2\varepsilon}{c_0},$$

then we have

$$\frac{1}{2}hK_0L_0c_0 = L_0\varepsilon$$

Requiring that

$$\frac{1}{4}c_0 L_0 - \frac{1}{2}hM_p L_p K_0 \geq 0$$

by the above definition of K_0 and by the inequality $\varepsilon \leq M_b N^* h$, we deduce the following bound on N^*

$$N^* h \leq \frac{1}{4}\frac{c_0 L_0}{L_b M_b^2}$$

Choosing N^* such that

$$N^* h \equiv \min\left[\frac{1}{L_p}\log\left(1 + \frac{c_0 L_0}{4M_p}\right), \frac{1}{4}\frac{c_0 L_0}{L_p M_p^2}\right]. \tag{2.31}$$

by (2.30) we conclude that

$$z_{m+K_0} \geq z_m + L_0\varepsilon \geq 0, \qquad \text{for any } N_0 \leq m \leq N^*$$

since

$$L_0\varepsilon \geq \sup\left\{-z_m : 0 \leq m \leq N^*\right\}.$$

We turn now to the proof of (2.21) which will follow by the estimate of the discrete trajectories related to $\{a_n\}$, $\{\overline{a}_n\}$ and from (2.23)

$$\begin{aligned}
|J^h(N^*, a) - J^h(N^*, \overline{a})| \leq & \sum_{n=N_0}^{N_0+K_0-1} h|f(\overline{y}_n, \overline{a}_n)|\beta^n + \\
&+ \sum_{n=N_0}^{N^*-K_0} h|f(y_n, a_n) - f(\overline{y}_{n+K_0}, a_n)|\beta^{n+K_0} + \\
&+ \sum_{n=N_0}^{N^*-K_0} h|f(y_n, a_n)|(\beta^n - \beta^{n+K_0}) + \sum_{n=N^*-K_0+1}^{N^*} h|f(y_n, a_n)|\beta^n \leq \\
\leq hM_f \cdot K_0 + & \sum_{n=N_0}^{N^*-K_0} hL_f|y_n - \overline{y}_{n+K_0}|\beta^{n+K_0} + \\
&+ hM_f\lambda hK_0 \sum_{n=N_0}^{N^*-K_0} \beta^n + hM_f K_0.
\end{aligned}$$

We observe that

$$\begin{aligned}
|y_n - \overline{y}_{n+K_0}| &\leq (1 + hL_b)^{n-N_0}\,|y_{N_0} - \overline{y}_{N_0+K_0}| \leq \\
&\leq (1 + hL_b)^n\,|\overline{y}_{N_0} - \overline{y}_{N_0+K_0}| \leq (1 + hL_b)^n hM_b K_0
\end{aligned}$$

Therefore, by the definition of K_0, we can conclude

$$|J^h(N^*,a) - J^h(N^*,\overline{a})| \leq 2M_f\frac{2}{c_0}\varepsilon + \lambda M_f\frac{2}{c_0}\varepsilon + \\ +L_f M_b\frac{2}{c_0}\varepsilon\sum_{n=N_0}^{N^*-K_0} h(1+hL_b)^n\beta^{n+K_0} \leq L\varepsilon.$$

□

The proof of the following result can be obtained adapting to the constrained case the arguments in [CDF].

Proposition 2.4. *Let v_h be defined as in (2.12). Then,*

$$v_h(x) = \inf_{\{a_n\}\in\mathcal{A}_x^h}\left(h\sum_{n=0}^{p-1} f(y_n,a_n)\,\beta^n + \beta^p v_h(y_p)\right),$$

for any $x \in \overline{\Omega}$ and $p \geq 1$.

We will refer to the above formula as the Discrete Dynamic Programming Principle (DDPP). For $p = 1$, it gives the following discrete version of (HJB)

$$(HJB_h) \qquad v(x) = \inf_{a\in A_h(x)}\{\beta v(x+hb(x,a)) + hf(x,a)\}, \qquad x \in \overline{\Omega}.$$

The proof of the following result can be easily obtained by a fixed point argument (see e.g. [CDF]).

Theorem 2.5. *For any $h \in (0,\frac{1}{\lambda}]$ there exists a unique solution $v_h \in L^\infty(\Omega)$ of (HJB_h).*

The continuity of v_h for small time steps is established in the following theorem.

Theorem 2.6. *Let the assumptions (A0)–(A3) and (2.6) be satisfied. Then there exists $h_1 > 0$ such that for any $h < h_1$, v_h is bounded and uniformly continuous in $\overline{\Omega}$.*

Proof. Let $x, z \in \overline{\Omega}$ be such that $|x - z| < r$. By Proposition 2.4, for every $\delta > 0$ there exists a control sequence $\{a_n\} \in \mathcal{A}_z^h$ such that

$$(2.32) \qquad J_z^h(p,\{a_n\}) + \beta^{p+1}v_h(y_z(p+1,a)) \leq v_h(z) + \delta$$

for any $p \geq 0$.

Let $N^*, \{\overline{a}_n\}$ and K_0 be defined as in Lemma 2.3 for the trajectory $y_x(n, a_n)$ and let

$$\varepsilon \equiv \max_{n \leq N^*} \text{ dist } (y_x(n,a), \overline{\Omega}).$$

Since

$$\begin{aligned} \varepsilon &\leq \max_{n \leq N^*} |y_x(n,a) - y_z(n,a)| \leq \\ &\leq \max_{n \leq N^*} |x - z|(1 + hL_b)^n \leq C_1 r \end{aligned} \tag{2.33}$$

by (2.21) we get

$$|J_x^h(N^*, \{a_n\}) - J_x^h(N^*, \{\overline{a}_n\})| \leq LC_1 r \tag{2.34}$$

By the definition of $\{\overline{a}_n\}$, for $n \geq N_0 + K_0 + 1$ we have

$$\begin{aligned} |y_x(n,a) - y_x(n,\overline{a})| &\leq 2M_b h(K_0 + n - (N_0 + K_0)) + \\ &+ h \sum_{m=N_0+K_0}^{n-1} L_b |y_x(m,a) - y_x(m,\overline{a})|. \end{aligned}$$

By applying the same arguments of Lemma 2.3 we get for $N_0 + K_0 + 1 \leq n \leq N^*$

$$|y_x(n,a) - y_x(n,\overline{a})| \leq C_2 r$$

where

$$C_2 \equiv C_2(N^* h, K_0, L_b, M_b) > 0.$$

Since for any $n \leq N^*$ and any control sequence $\{a_n\}$

$$|y_x(n,a) - y_z(n,a) \leq |x - z|(1 + hL_b)^{N^*} \tag{2.35}$$

there exists a positive constant C_3 such that

$$|y_x(n,\overline{a}) - y_z(n,a)| \leq C_3 r, \quad \forall n \leq N^* \tag{2.36}$$

so that we can conclude

$$|J_x^h(N^* - 1, \{\overline{a}_n\}) - J_z^h(N^* - 1, \{a_n\})| \leq Cr \tag{2.37}$$

for some positive constant C. Coupling (2.32), (2.34) and (2.37), we get

$$\begin{aligned} v_h(x) - v_h(z) &\leq J_x^h(N^* - 1, \{\overline{a}_n\}) - J_z^h(N^* - 1, \{a_n\}) + \\ &+ \beta^{N^*}[v_h(y_x(N^*, \overline{a})) - v_h(\overline{y}_z(N^*, a)] + \delta \leq \\ &\leq Cr + \beta^{N^*}[v_h(y_x(N^*, \overline{a}) - v_h(y_z(N^*, a))] + \delta. \end{aligned}$$

We denote by $\omega_h(\cdot)$ the modulus of continuity of v_h. Since δ is arbitrary we have

$$\begin{aligned}\omega_h(r) &\leq Cr + \beta^{N^*}\omega_h(C_3 r) \leq \\ &\leq Cr + \beta^{N^*}[C_3 Cr + \beta^{N^*}\omega_h(C_3^2 r)] \leq \\ &\leq \beta^{N^* n}\omega_h(C_3^n r) + Cr\sum_{m=0}^{n-1}(C_3\beta^{N^*})^m.\end{aligned}$$

Setting $r = C_3^{-n}$ and assuming $C_3 \neq 1$ it is immediate to prove that

$$\lim_{r\to 0}\omega_h(r) = 0.$$

□

Theorem 2.7. *Let (A0), (A2), (A3) and (2.6) be satisfied and let v be the unique constrained viscosity solution of (HJB). Then $v_h \to v$ uniformly in $\overline{\Omega}$ for $h \to 0$.*

Proof. In order to prove the result, we shall use the convergence theorem in [BS]. We write the equation (HJB_h) in compact form as

$$S(h, x, v_h(x), v_h) = 0 \text{ in } \overline{\Omega} \tag{2.38}$$

where $S(h,x,t,v) : \mathbb{R}^+ \times \overline{\Omega} \times \mathbb{R} \times L^\infty(\Omega) \to L^\infty(\Omega)$ is defined by

$$S(h,x,t,v) = (1-\beta)t + \sup_{a\in A_h(x)}\left\{-\beta(v(x+hb(x,a)) - v(x)) - hf(x,a)\right\}.$$

Since the continuous problem verifies a comparison principle (see [BP],[IK]), the sequence of the solutions of (2.38) converges towards the solution of (HJB) if the scheme is monotone, stable and consistent.

It is straightforward to verify that the scheme defined by S is monotone, i.e.

$$S(h,x,t,v) \leq S(h,x,t,w), \qquad \text{if } w \leq v.$$

By Proposition 2.5, the scheme is also stable (i.e. for all h, there exists a unique solution v_h of (2.38)). Moreover, v_h has a bound independent of h, namely $\|v_h\|_\infty \leq M_f/\lambda$.

To verify that the scheme is consistent with the equation (HJB), we have to prove that for any $\phi \in C^\infty(\overline{\Omega})$

$$\underline{\lim}_{\substack{h\to 0\\ y\to x}} \frac{S(h,y,\phi(y),\phi)}{h} \geq H(x,\phi,D\phi) \quad \text{for any } x \in \Omega \tag{2.39}$$

$$\overline{\lim}_{\substack{h\to 0\\ y\to x}} \frac{S(h,y,\phi(y),\phi)}{h} \leq H(x,\phi,D\phi) \quad \text{for any } x \in \overline{\Omega}. \tag{2.40}$$

We start proving (2.39). Fix $x \in \Omega$, for any $a \in A$ there exists $\overline{h}$ such that $x + hb(x,a) \in \overline{\Omega}$ for $h < \overline{h}$. From the continuity of b, we have that there exists a neighborhood $I(x)$ of x in Ω such that $y + hb(y,a) \in \overline{\Omega}$ for all $y \in I(x)$ and $h < \overline{h}$, which implies $a \in A_h(y)$. We have

$$\begin{aligned}
&\underline{\lim}_{\substack{h\to 0\\ y\to x}} \frac{S(h,y,\phi(y),\phi)}{h} \geq \\
&\geq \underline{\lim}_{\substack{h\to 0\\ y\to x}} \frac{1-\beta}{h}\phi(y) - \beta\frac{\phi(x+hb(x,a)) - \phi(x)}{h} - f(y,a) = \\
&= \quad \lambda\phi(x) - b(x,a)\nabla\phi(x) - f(x,a).
\end{aligned}$$

Repeating the same argument for any $a \in A$, we have (2.39).

Let us prove now (2.40). For any $\varepsilon > 0$, there exists a control $a^* \in A_h(y)$, depending on h and y, such that

$$\text{(2.41)}\quad \frac{S(h,y,\phi(y),\phi)}{h} \leq \frac{(1-\beta)}{h}\phi(y) - \beta\frac{\phi(y+hb(y,a^*) - \phi(y))}{h} + - f(y,a^*) + \varepsilon.$$

Since $\phi \in C^\infty(\overline{\Omega})$, for all $y \in \overline{\Omega}$ we have

$$\text{(2.42)}\qquad \left|b(y,a)\nabla\phi(y) - \frac{\phi(y+hb(y,a)) - \phi(y)}{h}\right| \leq Ch.$$

Then (2.41) and (2.42) imply that

$$\begin{aligned}
&\overline{\lim}_{\substack{h\to 0\\ y\to x}} \frac{S(h,y,\phi(y),\phi)}{h} \leq \\
&\leq \overline{\lim}_{\substack{h\to 0\\ y\to x}} \frac{1-\beta}{h}\phi(y) - \beta\frac{\phi(y+hb(y,a^*)) - \phi(y)}{h} - f(y,a^*) + \varepsilon \leq \\
&\leq \overline{\lim}_{\substack{h\to 0\\ y\to x}} \frac{1-\beta}{h}\phi(y) - \beta b(y,a^*)\cdot\nabla\phi(y) - f(y,a^*) + \varepsilon + Ch \leq \\
&\leq \overline{\lim}_{\substack{h\to 0\\ y\to x}} \frac{1-\beta}{h}\phi(y) + \sup_{a\in A}\{-b(y,a)\cdot\nabla\phi(y) - f(y,a)\} + \varepsilon + Ch \leq \\
&\leq H(x,\phi,\nabla\phi) + \varepsilon.
\end{aligned}$$

□

3. A priori estimates and algorithms. As we said in the introduction, we focus our attention on the rate of convergence of v_h to v. This is a crucial point when applying the algorithms to the solution of real problems.

It is known that under our assumptions $v \in C^{0,\gamma}(\overline{\Omega})$ where $\gamma = \gamma(\lambda, L_f)$ (see [CDL]). Moreover in Loreti-Tessitore [LT] and Ishii-Koike

[IK] it has been shown that $v \in C^{0,1}(\overline{\Omega})$ for sufficiently large discount rates, $\lambda \geq C(L_f)$. We will use the above regularity results to obtain our estimates.

Lemma 3.1. *Let $v \in C^{0,\gamma}(\overline{\Omega})$. Under the same assumptions of Lemma 2.3 there exists two positive constant C and T^* such that*

$$v(x) - v_h(x) \leq Ch^\gamma + e^{-\lambda T^* h}\|v - v_h\|_\infty \tag{3.1}$$

for every $x \in \overline{\Omega}$.

Proof. By Lemma 3.2 in [S] we know that there exist a time $t^* > 0$ and a constant $C_1 > 0$ such that for any $x \in \overline{\Omega}$ and $\alpha \in \mathcal{A}$ there is a control $\overline{\alpha} \in \mathcal{A}_x$ satisfying

$$|J_x(t^*, \overline{\alpha}) - J_x(t^*, \alpha)| \leq C_1 \sup_{t \leq t^*} d(y_x(t, \alpha), \overline{\Omega}) \tag{3.2}$$

where

$$J_x(t^*, \alpha) \equiv \int_0^{t^*} f(y_x(s, \alpha(t)), \alpha(t))e^{-\lambda t} dt. \tag{3.3}$$

We define $T^* \equiv \left[\frac{t^*}{h}\right] + 1$. For any $\delta > 0$ there exists $\{a_n\} \in \mathcal{A}_x^h$ such that

$$J_x^h(T^* - 1, a) + \beta^{T^*} v_h(y_x(T^*, a)) \leq v_h(x) + \delta. \tag{3.4}$$

For any sequence $\{a_n\}$ we can define a piecewise constant trajectory

$$\tilde{y}(t) \equiv y_x(n, a) \quad \text{for } t \in (nh, (n+1)h) \tag{3.5}$$

and a measurable control

$$\alpha_h(t) \equiv a_n \quad \text{for } t \in [nh, (n+1)h)\ . \tag{3.6}$$

By (3.2) there exists a control $\overline{\alpha}_h \in \mathcal{A}_x$ such that

$$|J_x(t^*, \alpha_h) - J_x(t^*, \overline{\alpha}_h)| \leq C_1 \varepsilon \tag{3.7}$$

where

$$\varepsilon \leq \sup_{t \leq t^*} |y_x(t, \alpha_h) - \tilde{y}_x(t, \overline{\alpha}_h)| \leq M_b h e^{L_b t^*} \equiv C_2 h. \tag{3.8}$$

By (3.4) and the definition of v, for any $\delta > 0$ and for a suitable sequence $\{a_n\} \in \mathcal{A}_x^h$ we have

$$\begin{aligned} v(x) - v_h(x) \ &\leq \ J_x(t^*, \overline{\alpha}_h) - J_x^h(T^* - 1, a) + \\ &+ \ e^{-\lambda t^*} v(y_x(t^*, \overline{\alpha}_h)) - \beta^{T^*} v_h(y_x(T^*, a)) + \delta \end{aligned} \tag{3.9}$$

where $\overline{\alpha}_h$ is the control satisfying (3.7).

We can easily prove that there exist two positive constants C_3 and C_4 such that

$$\begin{aligned} (3.10) \qquad |J_x(t^*, \alpha_h) - J_x^h(T^* - 1, a)| &\le C_3 h \\ (3.11) \qquad |y_x(T^*, a) - \tilde{y}_x(t^*, \overline{\alpha}_h)| &\le C_4 h. \end{aligned}$$

Let us denote by z the point of the discrete trajectory $y_x(T^*, a)$, (notice that this point is in $\overline{\Omega}$). By (3.9), (3.10), (3.11) and by our regularity assumption on v we conclude that

$$\begin{aligned} (3.12) \quad v(x) - v_h(x) \quad &\le C_3 h + e^{-\lambda T^* h}[v(z) - v_h(y_x(T^*, a))] + \\ &+ e^{-\lambda T^* h}[v(y_x(t^*, \overline{\alpha}_h)) - v(z)] + \\ &+ (e^{-\lambda t^*} - e^{-\lambda T^* h})\, v(y_x(t^*, \overline{\alpha}_h)) + \delta \le \\ &\le C_3 h + e^{-\lambda T^*} \|v - v_h\|_\infty + \\ &+ e^{-\lambda T_h^*} |y_x(t^*, \overline{\alpha}_h) - z|^\gamma + \frac{M_f}{\lambda} h + \delta \le \\ &\le C h^\gamma + e^{-\lambda T^* h} \|v - v_n\|_\infty + \delta \end{aligned}$$

for any positive δ, which ends the proof. □

To prove the estimate for $v_h - v$ we need some further assumptions. Let us define

$$\begin{aligned} (3.13) \qquad \overline{\mathcal{A}}_x \equiv \quad &\{\alpha \in \mathcal{A}_x : \exists \{s_i\} \text{ such that } 0 \le s_i \le s_{i+1} \\ &\text{and } \alpha(s) = \text{ constant for } s \in [s_i, s_{i+1}]\} \end{aligned}$$

If a control $\alpha \in \overline{\mathcal{A}}_x$ has a bounded variation in $[0, R]$ we denote its total variation by

$$(3.14) \qquad V(\alpha, R) \equiv \sum_{0 < s_{i+1} \le R} |\alpha(s_{i+1}) - \alpha(s_i)|$$

where s_i are the points of discontinuity of α.

We make the following assumptions:

$$\begin{aligned} (3.15) \qquad &|g(x, a_1) - g(y, a_2)| \le L_g(|x - y| + |a_1 - a_2|) \\ &\text{for } g = f, b \text{ and for any } x, y \in \overline{\Omega},\ a_1, a_2 \in A \end{aligned}$$

$$(BV) \qquad \begin{cases} \text{there exist a control sequence } \{\alpha_n\} \in \overline{\mathcal{A}}_x \text{ and a positive} \\ \text{constant } M \text{ such that:} \\ i)\, V(\alpha_n, (N^* + 1)h) \le M \\ ii)\, v(x) = \lim\limits_{n \to +\infty} J_x(\alpha_n) \end{cases}$$

Lemma 3.2. *Let $v \in C^{0,\gamma}(\overline{\Omega})$. Assume that (3.15) and (BV) are satisfied. Then there exist two positive constants C and $N^* \in \mathbb{N}$ such that*

$$v_h(x) - v(x) \leq Ch^\gamma + e^{-\lambda(N^*+1)h}\|v_h - v\|_\infty. \tag{3.16}$$

Proof. By (BV, ii) for any $\delta > 0$ there exists a control $\alpha \in \overline{\mathcal{A}}_x$ such that

$$J_x((N^*+1)h, \alpha) + e^{-\lambda(N^*+1)h} v(y((N^*+1)h)) \leq v(x) + \delta. \tag{3.17}$$

By Lemma 2.2 in [BF] there exists a control $\alpha_h \in \mathcal{A}$ such that

$$\alpha_h(t) = \text{constant} = a_n \text{ for } t \in [nh, (n+1)h[\tag{3.18}$$

and

$$\int_0^{(N^*+1)h} |\alpha(s) - \alpha_h(s)| ds \leq hV(\alpha, (N^*+1)h). \tag{3.19}$$

Then (3.15), (3.19) and (BV, i) imply

$$|y_x(t, \alpha) - y_x(t, \alpha_h)| \leq C_1 h \quad \text{for } t \leq (N^*+1)h. \tag{3.20}$$

By Lemma 2.3 there exists a control sequence $\{\overline{a}_n\} \in \mathcal{A}_x^h$ satisfying (2.21) with $\{a_n\} = \{\alpha_h(nh)\}$. Moreover,

$$\begin{aligned} \varepsilon &\equiv \max_{n \leq N^*} d(y_x(n, a), \overline{\Omega}) \leq \\ &\leq \max_{n \leq N^*} |y_x(n, a_n) - y_x(nh, \alpha(nh))| \leq C_1 h \end{aligned} \tag{3.21}$$

and, by the Lipschitz continuity of f,

$$\begin{aligned} |J^h(x, a) &- J_x((N^*+1)h, \alpha)| \leq \\ &\leq \sum_{k=0}^{N^*} \Bigg[L_f \int_{kh}^{(k+1)h} |y_x(t, u) - \tilde{y}_x(t, u_h)| + \\ &\quad + M_f [e^{-\lambda t} - e^{-\lambda hk}] dt \Bigg] \leq C_2 h \end{aligned} \tag{3.22}$$

where the constant C_2 depends on $N^* h$.

Working on the integral representation of the trajectories as in the preceding results and recalling that $\varepsilon \leq C_1 h$ we can prove

$$|y_x(N^*+1, \overline{a}) - y_x((N^*+1)h, \alpha)| \leq C_3 h \tag{3.23}$$

Then, we can conclude

$$(3.24)\quad \begin{aligned} v_h(x) - v(x) \le\ & J_x^h(N^*, \overline{a}) - J_x((N^*+1)h, \alpha) + \\ & + e^{-\lambda(N^*+1)h} v_h(y_x(N^*+1, \overline{a})) + \\ & - e^{-\lambda(N^*+1)h} v(y_x((N^*+1)h, \alpha)) + \delta \le \\ & \le C_4 h + e^{-\lambda(N^*+1)h} \|v - v_h\|_\infty + \\ & + e^{-\lambda(N^*+1)h} |v(y_x(N^*+1, \overline{a})) - v(y_x((N^*+1)h, \alpha))| + \delta \le \\ & \le C h^\gamma + e^{-\lambda(N^*+1)h} \|v - v_h\|_\infty + \delta \end{aligned}$$

for any positive δ, which ends the proof. □

We can now prove our main result on the order of convergence.

Theorem 3.3. *Let $v \in C^{0,\gamma}(\overline{\Omega})$. Assume that (3.15) and (BV) are satisfied. Then there exists a positive constant C such that*

$$(3.25)\qquad \|v - v_h\|_\infty \le C h^\gamma .$$

Proof. We know a uniform lower bound for $T^* h$, namely

$$T^* h \ge t^*$$

where t^* has been defined in the proof of Lemma 3.1. Moreover, $N^* h$ is a constant as it is shown in (2.31).

Then

$$L \equiv \max[e^{-\lambda t^*}, e^{-\lambda N^* h}] < 1$$

and by Lemma 3.1 and Lemma 3.2 we get

$$|v(x) - v_h(x)| \le C h^\gamma + L \|v - v_h\|_\infty, \quad \forall x \in \overline{\Omega}$$

which ends the proof. □

The above scheme has been obtained making a discretization of the continuous control problem, namely approximating the dynamics in (1.1) by an Euler scheme and the cost functional by the rectangle rule. We will show how it is possible to set up an approximation scheme of higher order for v coupling more efficient one step scheme for the dynamics and quadrature formulae for the cost functional. To this end, we shall extend the results in [FF1] to the constrained problem.

For simplicity, let us restrict our attention to a specific scheme obtained coupling the Heun's method for the dynamics with the trapezoid rule for the cost functional. However, the following results can be extended to the general class of approximation schemes introduced and studied in [FF1].

We shall define a new discrete control problem setting

$$\begin{cases} y_{n+1} = y_n + h\Phi(y_n, a_n, h) \\ y_0 = x \end{cases} \tag{3.26}$$

where

$$\Phi(x,a,h) = \frac{1}{2}b(x,a^0) + \frac{1}{2}b(x + hb(x,a^0), a^1) \tag{3.27}$$

and $a = (a^0, a^1) \in A \times A$. A control law $\{a_n\}$ is admissible for the constrained control problem if $a_n \in \tilde{A}_h(y_n)$, for all n, where

$$\tilde{A}_h(x) = \{a = (a^0, a^1) \in A \times A : (x + hb(x,a^0), x + h\Phi(x,a,h)) \in \overline{\Omega} \times \overline{\Omega}\}.$$

We introduce the discrete cost functional

$$\tilde{J}^h_x(\{a_n\}) = \sum_{n=0}^{+\infty} \beta^n \left[\frac{h}{2}f(y_n, a_n^0) + \frac{h}{2}f(y_n + h\Phi(y_n, a_n, h))\right].$$

It is easy to prove that the value function $\tilde{v}_h$ of the above discrete control problem satisfies the following discrete dynamic programming principle

$$(\widetilde{HJB}_h) \qquad \begin{aligned} \tilde{v}_h(x) &= \inf_{a \in \tilde{A}_h(x)} \Big\{ \beta \tilde{v}_h(x + h\Phi(x,a,h)) + \\ &+ \frac{h}{2}f(x,a_0) + \frac{h}{2}f(x + h\Phi(x,a,h), a_1) \Big\}, \quad \text{for } x \in \overline{\Omega}. \end{aligned}$$

We will show how we can adapt the technique of Section 2 to get a better estimate for this scheme. The proof of following lemma is a simple adaptation of that of Lemma 2.1.

Lemma 3.4. *Let (A.1) and (2.6) be satisfied. Then there exist* ε_0, c_0 *such that*

$$\begin{cases} \forall x \in \Gamma_{\varepsilon_0}, \exists (a_0, a_1) \in A \times A \text{ such that} \\ \varphi(x + h\Phi(x, (a_0, a_1), h)) \geq c_0 h \quad \forall h < h_0 \end{cases} \tag{3.28}$$

where Γ_{ε_0} *is defined as in Lemma 2.2.*

Notice that choosing h_0 according to (2.17) we will have $\tilde{A}_h(x) \neq \emptyset$ for all $x \in \overline{\Omega}$ and $h < h_0$.

Lemma 3.5. *Let (A0)–(A3) and (2.6) hold. Then, there exist* $N^* \in \mathbb{N}$ *and* $L > 0$ *such that for any* $x \in \overline{\Omega}$ *and any sequence* $\{(a_n^0, a_n^1)\}$ *with values in* $A \times A$, *we can determine* $\{\overline{a}_n\} \in \tilde{A}_h(x)$ *such that*

$$|\tilde{J}_x^h(N^*, \{\overline{a}_n\}) - \tilde{J}_x^h(N^*, \{a_n^0, a_n^1\})| \le L\varepsilon \tag{3.29}$$

where $\varepsilon = \sup_{n \le N^*} d(y_n(x, \{a_n^0, a_n^1\}), \overline{\Omega})$.

Proof. We will prove the result by adapting the argument of Lemma 2.3. We define

$$N_0 = \begin{cases} \inf\{n \le N^* : (y_n + hb(y_n, a_n^0), y_n + h\Phi(y_n, (a_n^0, a_n^1), h) \notin \overline{\Omega} \times \overline{\Omega}\} \\ N \text{ if } y_n \in \overline{\Omega} \quad \forall n \le N \end{cases}$$

$$\overline{a}_n = \begin{cases} (a_n^0, a_n^1) \text{ if } n \le N_0 - 1 \\ (\overline{a}_{N_0}^0, \overline{a}_{N_0}^1) \text{ if } N_0 \le n \le N_0 + K\varepsilon - 1 \\ (a_{n-K_0}^0, a_{n-K_0}^1) \text{ if } n \ge N_0 + K_0 \end{cases}$$

where $(\overline{a}^0, \overline{a}^1)$ satisfy (3.28) in $x = y_{N_0}$. We want to determine N^* such that $\overline{y}_m \in \overline{\Omega}$ for all $m \le N^*$.

We can write

$$z_{m+1} = z_m + h\tilde{p}(z_m, (a_m^0, a_m^1))$$

where $\tilde{p} : \mathbb{R}^n \times A \times A \to \mathbb{R}$ is defined by $\tilde{p}(z, (a^0, a^1)) \equiv \frac{1}{h}[\varphi(x + h\Phi(x, (a^0, a^1), h) - \varphi(x)]$ for $z = \varphi(x)$. Notice that $\tilde{p}$ has the following properties, which are analogous to (2.26):

$$\begin{cases} \text{i)} & |\tilde{p}(z, a_0, a_1)| \le L_\varphi M_b; \\ \text{ii)} & \tilde{p}(\cdot, a_0, a_1) \text{ is Lipschitz continuous with constant} \\ & L_{\tilde{p}} = L_0^2\left(M_b + L_b + h\left(L_b + \frac{L_b^2}{2}\right) + \frac{1}{2}h^2 L_b\right) \\ & \text{where } L_0 \text{ is defined in (2.6) ;} \\ \text{iii)} & \tilde{p}(z, (\overline{a}_0, \overline{a}_1)) \ge L_0 c_0 \text{ for } z = \varphi(x) \text{ with } x \in \Gamma_{\varepsilon_0} \\ & \text{and } (\overline{a}_0, \overline{a}_1) \text{ satisfying (3.28).} \end{cases} \tag{3.30}$$

We sketch the proof of ii). Let $z = \varphi(x)$ and $w = \varphi(y)$ then

$$\begin{aligned} |\tilde{p}(z, a_0, a_1) - \tilde{p}(w, a_0, a_1)| = \Big| \int_0^1 & [\Phi(x, a, h)\nabla\varphi(x + sh\Phi(x, a, h)) + \\ & -\Phi(y, a, h)\nabla\varphi(y + sh\Phi(y, a, h))]ds\Big| \le \\ & \le \int_0^1 |\nabla\varphi(x + sh\Phi(x, a, h))|\, |\Phi(x, a, h) - \Phi(y, a, h)| ds + \\ & + \int_0^1 |\Phi(y, a, h)|\, |\nabla\varphi(x + h\Phi(x, a, h)) - \nabla\varphi(y + h\Phi(y, a, h))| ds \end{aligned}$$

and we can conclude by the Lipschitz continuity of $b, D\varphi, \varphi^{-1}$.

We can repeat then the same proof of Lemma 2.2. In fact, that proof relies only on the properties of $\tilde{p}$ and on simple estimates on y_n. □

We omit the proofs of the continuity of $\tilde{v}_h$ and of the convergence of $\{\tilde{v}_h\}$ to the viscosity solution of (HJB), since they are straightforward adaptations of Theorem 2.6 and Theorem 2.7. To obtain the a priori estimate for the rate of convergence we need some additional hypotheses. We will assume that the following conditions hold true:

(i) For all $z \in \overline{\Omega}$ and $\alpha : [0, h) \to A$ measurable, there exist $a = (a^0, a^1) \in A \times A$ and $K_1, K_2 \in \mathbb{R}_+$ such that

(3.31) $|y_z(h, \alpha) - z - h\Phi(z, (a^0, a^1), h)| \leq K_1 h^2,$

(3.32) $$\left| \int_0^h f(y_z(s), \alpha(s))e^{-\lambda s} ds - \frac{h}{2}[f(z, a^0) + f(z + h\Phi(z, a, h), a^1)] \right| \leq$$
$$\leq K_2 h^2.$$

(ii) For all $z \in \overline{\Omega}$, $a = (a^0, a^1) \in A \times A$, there exists a measurable control $\alpha : [0, h) \to A$, such that (3.32) and (3.31) hold true.

We refer to [FF1] for the motivations and some sufficient conditions which guarantee (3.31) and (3.32). For example, they hold if there exists a regular optimal control and/or if b and f are linear with respect to the control.

Theorem 3.6. *Let the assumptions (3.31), (3.32), (A0)-(A3) and (2.6) hold true and let $v \in C^{0,\gamma}(\overline{\Omega})$. Then*

$$\|v - \tilde{v}_h\|_\infty \leq Ch^{2\gamma}$$

for some constant $C > 0$.

Proof. We will only prove that

$$\tilde{v}_h(x) - v(x) \leq ch^{2\gamma} + e^{-\lambda(N^*+1)h}\|v - v_h\|_\infty$$

leaving to the reader the proof of the estimate on $v - \tilde{v}_h$ which follows the same arguments. By the dynamic programming principle for any $\delta > 0$ there exists a control law $\alpha \in \mathcal{A}_x$ such that

(3.33) $J_x((N^* + 1)h, \alpha) + e^{-\lambda(N^*+1)h} v(y((N^* + 1)h, \alpha)) \leq v(x) + \delta.$

Let $a = \{(a_n^0, a_n^1)\}$ be such that (3.31) and (3.32) hold for $z = y_x(nh, \alpha)$ if $n \leq N^*$. By Lemma 2.2 there exists a control law $\overline{a} \in \mathcal{A}_x^h$ such that

(3.34) $$|\tilde{J}_x^h(N^*, a) - \tilde{J}_x^h(N^*, \overline{a})| \leq L\varepsilon$$

where

$$\varepsilon = \sup_{n \leq N^*} \{\text{dist } (y_x(n,a), \overline{\Omega})\}.$$

By (3.31), we have

$$\varepsilon \leq \sup_{n \leq N^*} |y_x(n,a) - y_x(nh, \alpha(nh))| \leq C_1 h^2 \tag{3.35}$$

Moreover, by (3.31) and (3.32) it follows

$$|\tilde{J}_x^h(N^*, a) - J_x((N^*+1)h, \alpha)| \leq C_2 h^2 \tag{3.36}$$

Coupling (3.33)-(3.36), for every $\delta > 0$ we have

$$\begin{aligned}
&\tilde{v}_h(x) - v(x) \leq \tilde{J}_x^h(N^*, a) - J_x((N^*+1)h, \alpha) + \\
&e^{-\lambda(N^*+1)h}[\tilde{v}_h(y_x(N^*+1, \overline{a})) - v(y_x((N^*+1)h, \alpha)] + \delta \leq \\
&\leq Ch^2 + e^{-\lambda(N^*+1)h}\|v - \tilde{v}_h\|_\infty + \\
&+e^{-\lambda(N^*+1)h}|v(y_x((N^*+1)h, \overline{a})) - v(y_x(N*+1, \alpha))| + \delta \leq \\
&\leq Ch^2 + e^{-\lambda(N^*+1)h}\|v - \tilde{v}_h\|_\infty + L_v h^{2\gamma}.
\end{aligned}$$

□

4. Fully discrete schemes and algorithms. The numerical solution of (HJB) requires also a discretization in space of (HJB_h). In this section, we will concentrate on the first order scheme but similar results can also be obtained for the second order scheme. We refer to [FF2] for the construction of fully discrete high-order schemes for unconstrained problems.

It is interesting to notice that when adapting the algorithm for the unconstrained problem presented in [F] to the constrained case we will take advantage of the constraint reducing the set of admissible controls, i.e. using the map $A_h(x)$ at the nodes of the grid. This procedure is very effective since it reduces the number of computations and comparisons to be made (in particular in a neighbourhood of the boundary of $\overline{\Omega}$).

Let us set a regular triangulation of $\overline{\Omega}$ made up by a finite family of simplices $\{S_j\}$ in $\mathbb{R}^n$. Assume, for simplicity, that $\overline{\Omega}$ is a convex polyhedral set so that $\overline{\Omega} = \cup_j S_j$. We will denote by x_i, $i = 1, \ldots, N$, the nodes of the triangulation and by k the maximum diameter of the simplices of the triangulation. Let us define

$$\mathcal{W}^k \equiv \{w : C(\overline{\Omega}) \to \mathbb{R}, \text{ such that } \nabla w = \textit{constant} \text{ in } S_j\}.$$

We look for a solution in $\mathcal{W}^k$ of the following system of N nonlinear equations

$$(HJB_h^k) \ v(x_i) = \inf_{a \in A_h(x_i)} \{\beta v(x_i + hb(x_i, a)) + hf(x_i, a)\}, \qquad i = 1, \ldots, N$$

Remember that, by (2.17) and (2.18) the boundary condition (2.2) implies

$$A_h(x_i) \neq \emptyset, \qquad i = 1, \ldots, N.$$

for h sufficiently small.

Theorem 4.1. *Assume (A0)–(A3) and let $h \in (0, \frac{1}{\lambda}]$. Then there exists a unique solution v_h^k of (HJB_h^k) in $\mathcal{W}^k$.*

Proof. For any $w \in \mathcal{W}^k$, we can write (HJB_h^k) as

$$(4.1)\quad w(x_i) = \inf_{a \in A_h(x_i)} \left\{ \beta \sum_{j=1}^{N} \lambda_j(x_i, a) w(x_j) + h f(x_i, a) \right\}, \quad i = 1, \ldots, N$$

where

$$0 \leq \lambda_j(x_i, a) \leq 1, \quad \sum_{j=1}^{N} \lambda_j(x_i, a) = 1$$

$$x_i + h b(x_i, a) = \sum_{j=1}^{N} \lambda_j(x_i, a) x_j.$$

The vector V^* of the values $v_h^k(x_j)$ is the fixed point of the operator $T : \mathbb{R}^N \to \mathbb{R}^N$, defined componentwise by

$$(4.2)\quad (T(V))_i = \inf_{a \in A_h(x_i)} \{\beta \sum_{j=1}^{N} \lambda_j(x_i, a) V + h f(x_i, a)\}_i, \qquad i = 1, \ldots, N$$

The proof in [F] that T is a contraction operator with coefficient β when $A_h(x_i) \equiv A$ can be easily adapted to this case. □

Theorem 4.2. *For any fixed $h \in (0, \frac{1}{\lambda}]$, the following estimate holds true*

$$\|v_h^k - v_h\|_\infty \leq \frac{\omega(k)}{1 - e^{-\lambda h}}$$

where ω is the modulus of continuity of v_h.

Proof. Notice that by construction $v_h^k(x_j) = V_j^*$ for any node x_j of the grid. Using the fact that v_h is the solution of (HJB_h) and V^* is the fixed point of T we obtain

$$|v_h(x_j) - V_j^*| \leq \beta \|v_h - v_h^k\|_\infty . \tag{4.3}$$

For any $x \in \overline{\Omega}$ we can always find a node x_j such that $|x - x_j| \leq k$. Then by (4.3) and the continuity of v_h we obtain

$$\begin{aligned} |v_h(x) - v_h^k(x)| &\leq \sum_{j=1}^{N} \lambda_j |v_h(x) - v_h(x_j)| + \sum_{j=1}^{N} \lambda_j |v_h(x_j) - V_j^*| \leq \\ &\leq \omega(k) + \beta \|v_h - v_h^k\|_\infty . \end{aligned}$$

Since $\beta \equiv e^{-\lambda h}$ we can conclude that the statement holds true. □

The following result is straightforward by Theorem 3.3 and Theorem 4.2.

Corollary 4.3. *v_h^k converges uniformly to v in $\overline{\Omega}$ for $h \to 0^+$, $k \to 0^+$ and*

$$\frac{\omega(k)}{1 - e^{-\lambda h}} \to 0^+$$

where ω is the modulus of continuity of v_h.

The *algorithm* is based on the fixed point operator T. We compute the sequence $V^{n+1} = T(V^n)$ accelerating the convergence with a technique described in [CDF]. At each step we compute the i-th component of V^n by (4.2) and we look for the minimum over the set of admissible controls at the node x_i. The essential feature of the constrained problem algorithm is the construction of a selection map which describes the sets $A_h(x_i)$, $i = 1, \ldots, N$. Assume to have M controls, the selection is represented by a boolean $N \times M$ matrix Q such that the element q_{ij}, $i = 1, \ldots, N$, $j = 1, \ldots, M$ is defined as

$$q_{ij} = \begin{cases} 1 & \text{if } x_i + hb(x_i, a_j) \in \overline{\Omega} \\ 0 & \text{elsewhere.} \end{cases}$$

Since $A_h(x_i) \neq \emptyset$ there is at least one element equal to 1 in each row.

The above estimates suggest that an accurate approximation of the problem would require a huge number of grid points. As we mentioned this difficulty can be solved using higher order methods and/or a domain decomposition strategy.

We describe the essential features of the 2-domain decomposition considered in [FLS] but the results can be extended to an m-domain decomposition although the corresponding algorithm would be quite complicated. We split up Ω into two subdomains with overlapping in order to reduce the problem to a couple of problems of manageable size. To this end we will define two discrete operators which corresponds to two constrained problems in the subdomains and we impose a linking condition in the overlapping region.

Let $\overline{\Omega}$ be partitioned into two open subdomains $\overline{\Omega}_1$ and $\overline{\Omega}_2$, such that $\overline{\Omega} = \overline{\Omega}_1 \cup \overline{\Omega}_2$ and the *overlapping region* is non empty, $\overline{\Omega}_0 \equiv \overline{\Omega}_1 \cap \overline{\Omega}_2 \neq \emptyset$.

We introduce the following notation,

$$A_h^r(x) \equiv \{a \in A : x + hb(x,a) \in \Omega_r\}, \quad x \in \overline{\Omega}_r\,, \quad r = 1,2\,. \tag{4.4}$$

and we will need the following assumptions

$$0 < h \leq \frac{1}{M_b} \inf_{\substack{x \in \Omega_1 \backslash \Omega_0 \\ y \in \Omega_2 \backslash \Omega_0}} \operatorname{dist}(x,y) \tag{4.5}$$

$$A_h^r(x) \neq \emptyset, \qquad \forall x \in \overline{\Omega}_r, r = 1,2 \tag{4.6}$$

$$\begin{cases} \text{the regular triangulation of } \Omega \text{ is such that each simplex} \\ \text{is not crossing the interface between } \Omega_1 \backslash \Omega_0 \text{ and } \Omega_0 \\ \text{and the interface between } \Omega_2 \backslash \Omega_0 \text{ and } \Omega_0. \end{cases} \tag{4.7}$$

Assumption (4.6) is a restriction to the behaviour of the discrete trajectories in Ω_0 and should be understood as a discrete analogue of (2.2) for the constrained problems in the subdomains.

We will divide the nodes x_i, $i = 1, \ldots N$, into three classes depending on the region to which they belong defining

$$\begin{aligned} I_r &= \{i : x_i \in \overline{\Omega}_r \backslash \overline{\Omega}_0\}\,, \quad r = 1,2 \\ I_0 &= \{i : x_i \in \overline{\Omega}_0\} \end{aligned} \tag{4.8}$$

Let N_r, $r = 1,2$, be the number of nodes in $\overline{\Omega}_r$. We define the "discrete" restriction operators

$$R_r : \mathbb{R}^N \to \mathbb{R}^{N_r}, \qquad R_r(U) = \{U_i\}_{i \in I_r \cup I_0}\,, \quad r = 1,2 \tag{4.9}$$

and two discrete operators D_1 and D_2, $D_r : \mathbb{R}^{N_r} \to \mathbb{R}^{N_r}$, $r = 1,2$, related to the subdomains Ω_1 and Ω_2

$$[D_r(U)]_i \equiv \min_{a \in A_h^r(x_i)} \{\beta \sum_{j \in I_r \cup I_0} \lambda_{ij}^{(r)}(a) U_j + hF_i(a)\}\,, \quad i \in I_r \cup I_0\,. \tag{4.10}$$

Finally, by D_1 and D_2 we define the operator $D : \mathbb{R}^N \to \mathbb{R}^N$ related to the splitting algorithm in Ω,

$$[D(U)]_i = \begin{cases} [D_1(U^1)]_i & i \in I_1 \\ [D_2(U^2)]_i & i \in I_2 \\ \min\{[D_1(U^1)]_i\,,\ [D_2(U^2)]_i\} & i \in I_0\,. \end{cases} \tag{4.11}$$

where $U^1 = R_1(U)$ and $U^2 = R_2(U)$. Notice that the definition of $[D(U)]_i$ for $i \in I_0$ plays an important role making the link between the subdomains.

Given two vectors V^0 and W^0 in $\mathbb{R}^N$, we define by recursion

$$V^n = T(V^{n-1}), \qquad W^n = D(W^{n-1}) . \tag{4.12}$$

Proposition 4.4. *Assume (4.5)–(4.7) and let $V^0 = W^0$, then $V^n = W^n$ for any $n \in \mathbb{N}$.*

Proof. We prove that $T(U) = D(U)$ for all $U \in \mathbb{R}^N$. Assumption (4.5) and the definition of $A_h^r(x)$, $r = 1, 2$, imply that

$$\begin{aligned} A_h(x) &= A_h^r(x) && \forall x \in \overline{\Omega}_r \setminus \overline{\Omega}_0 \quad r = 1,2 \\ A_h(x) &= A_h^1(x) \cup A_h^2(x) && \forall x \in \overline{\Omega}_0 . \end{aligned} \tag{4.13}$$

Moreover, by (4.5) and (4.7) we have that

$$\forall i \in I_r,\ r = 1,2,\ x_i + hb(x_i, a) = \sum_{j \in I_r \cup I_0} \lambda_{ij}^{(r)}(a) x_j,\ a \in A_h^r(x_i) \tag{4.14}$$

$$\forall i \in I_0,\ x_i + hb(x_i, a) = \begin{cases} \sum_{j \in I_1 \cup I_0} \lambda_{ij}^{(1)}(a) x_j , & a \in A_h^1(x_i) \\ \sum_{j \in I_2 \cup I_0} \lambda_{ij}^{(2)}(a) x_j , & a \in A_h^2(x_i) . \end{cases} \tag{4.15}$$

Then the statement follows by the definition of D (and in particular by its definition for $i \in I_0$), (4.13), (4.14) and (4.15). □

A direct consequence of the above proposition is that the sequence W^n converges to V^* for any $W^0 \in \mathbb{R}^N$, where V^* is the fixed point for T. By Theorem 4.3 this implies the convergence of the approximate solution obtained by the domain decomposition algorithm to v.

Remark 4.5. Assumption (4.6) plays a role only in the definition of D_r. It can be eliminated adopting the following definition

$$[D_r(U)]_i \equiv \begin{cases} \min\limits_{a \in A_h^r(x_i)} \{\beta \sum_{j \in I_r \cup I_0} \lambda_{ij}^{(r)}(a) U_j + hF_i(a)\} & \text{if } A_h^r(x_i) \neq \emptyset \\ \frac{M_f}{\lambda} & \text{if } A_h^r(x_i) = \emptyset \end{cases}$$

where $i \in I_r \cup I_0$, $r = 1, 2$.

Here is the numerical *splitting algorithm* corresponding to the definition of W^n in (4.12).

Step 0. Given $W^0 \in \mathbb{R}^N$, define

$$W^{1,0} = R_1(W^0) \in \mathbb{R}^{N_1}$$
$$W^{2,0} = R_2(W^0) \in \mathbb{R}^{N_2}$$

and set $n = 0$.

Step 1. Compute for $r = 1, 2$,

$$W_i^{r,n+1/2} = \min_{a \in A_h^r(x_i)} \left\{ \beta \sum_{j \in I_r \cup I_0} \lambda_{ij}^{(r)}(a) W_j^{r,n} + hF_i(a) \right\}, \quad i \in I_r \cup I_0.$$

Step 2. Compute for $r = 1, 2$

$$W_i^{r,n+1} = \begin{cases} W_i^{r,n+1/2}, & i \in I_r \\ \min\{W_i^{1,n+1/2}, \ W_i^{2,n+1/2}\} & i \in I_0. \end{cases}$$

Step 3. Check a stopping criterion.

IF it is satisfied THEN STOP
ELSE

Increase n by 1 and GO TO *Step 1.*

Notice that the definition of D guarantees $W_i^{1,n+1} = W_i^{2,n+1}$ for each $i \in I_0$. The above algorithm allows to split the computations in the subdomains making a link at the end of each iteration (Step 2). However, its speed of convergence to the fixed point is quite slow since the contraction mapping coefficient is always β. This difficulty can be solved by adapting the acceleration technique described in [CDF].

5. Numerical solution of an advertising model. This section is devoted to the numerical resolution of an economic model by means of our approximation scheme. We consider a two products advertising model of Vidale–Wolfe type. The problem for the manager is to decide the best advertising policy for his products with the goal to maximize his revenue in the long run. We refer to Sethi [Se] for a general survey on Vidale–Wolfe models and to Dorroh–Ferreyra [DF] for a similar model using impulsive controls.

Model 1.

Following [Se], let $x(t)$ and $y(t)$ be respectively the captured fractions of the potential market for the first and the second product at time t. They evolve according to the system

$$\text{(5.1)} \qquad \begin{cases} \dot{x}(t) = (1 - x)u(t)d(x, y) - \delta x, & x(0) = x_0 \\ \dot{y}(t) = (1 - \frac{y}{x})v(t)d(x, y) - \eta y, & y(0) = y_0 \end{cases}$$

where $(u(t), v(t)) \in L^\infty([0, +\infty[; A)$ are the control laws and

$$A \equiv \{(u, v) \in [0, 1]^2 : u + v \leq 1\}$$

The terms $1 - x$ and $1 - y/x$ are the portions of potential sales rates on which advertising has effect, d is a positive real function representing the maximum investment for advertising and $u(t)d(t)$,$v(t)d(t)$ are the shares destined to advertise each product, $-\delta x(t)$,$-\eta y(t)$ represent the forgetting effect of the market.
Notice that, if

$$\delta \leq \eta \tag{5.2}$$

$T = \{(x, y) : 0 \leq x \leq 1, y \in [0, x]\}$ is invariant with respect to the dynamics given by (5.1). Moreover, starting from the interior of T the set $\{(x, y) \in T : x = 1\}$ cannot be reached for any choice of the control since the right-hand side of the first equation is negative near that part of the boundary. We observe that on the line $y = 0$ the dynamics has an equilibrium, which depend on u, in $x = 1 - \delta(uk_1)^{-1}$. On the line $y = x$, $\dot{y} = -\eta y$ and the dynamics cannot stop there. Note also that, without advertising, the system is rapidly driven in $(0, 0)$.

Let $k(t)$ be the capital of the manager. We impose a balance constraint requiring

$$k(t) \geq c(t) + d(t), \qquad \forall t > 0 \tag{5.3}$$

where $c(t)$ is the total production cost related to x and y.
The objective is to maximize the net profit, which depends on sales, costs and advertising

$$J(u, v) \equiv \int_0^{+\infty} [k(t) - c(t) - d(t)]\, e^{-\lambda t}\, dt.$$

We will treat the following case

$$\begin{cases} k(t) &= k_1 x(t) + k_2 y(t) \\ c(t) &= c_0 + c_1 x(t) + c_2 y(t) \\ d(t) &= \alpha k(t) \end{cases}$$

where k_1, k_2, c_0, c_1, c_2 are positive constants and $\alpha \in [0, 1]$. Condition (5.3) corresponds to the following constraint on the state

$$[(1 - \alpha)k_1 - c_1]x(t) + [(1 - \alpha)k_2 - c_2]y(t) \geq c_0 \tag{5.4}$$

Therefore the constrained Hamilton–Jacobi–Bellman equation is defined in the set

$$\overline{\Omega} = \{(x, y) \in T : [(1 - \alpha)k_1 - c_1]x + [(1 - \alpha)k_2 - c_2]y \geq c_0\}.$$

We assume a natural compatibility condition

$$[(1-\alpha)k_1 - c_1] + [(1-\alpha)k_2 - c_2] \geq c_0 \tag{5.5}$$

meaning that at least the point $(1,1)$ belongs to Ω. We set

$$\xi_1 \equiv (1-\alpha)k_1 - c_1, \qquad \xi_2 \equiv (1-\alpha)k_2 - c_2$$

According to the signs of ξ_1, ξ_2 we have three possible types of constraint sets Ω.

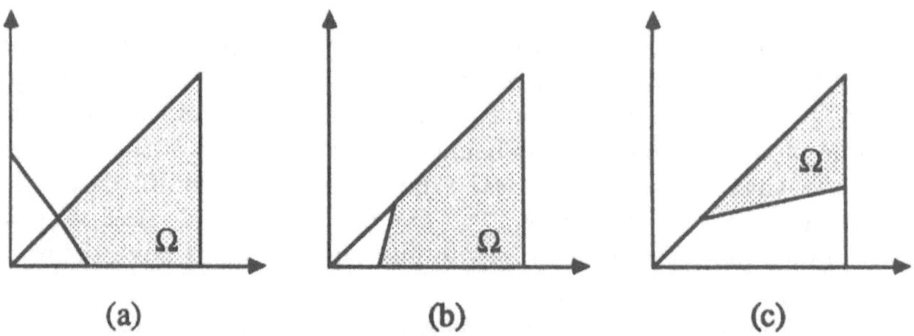

Figure 1.

If $\xi_i \leq 0$, the corresponding product have a negative balance; in this case, the sales of the other product has to stay large enough to get a total positive balance.

Model 2.

We modify the dynamics (5.1) considering

$$\begin{cases} \dot{x}(t) & = (1-x)u(t)d(x,y) - \delta(1-x)^2 \\ \dot{y}(t) & = \left(1-\frac{y}{x}\right) v(t)d(x,y) - \eta\left(1-\frac{y}{x}\right)^2. \end{cases} \tag{5.6}$$

The difference with respect to the previous model is in the forgetting effect terms. In (5.1) the negative terms grow with x and y/x, but in (5.6) they go exactly in the opposite way. Notice that in this model the state $(1,1)$ can be reached and it is an equilibrium point. On the line $y = x$ we have $\dot{y} = 0$ (in this case the forgetting effect for the second product is absent) and we have $\dot{x} = 0$ if $x = \delta(\alpha(k_1+k_2)u+\delta)^{-1}$. On the line $y = 0$ the dynamics has an equilibrium if there exist u and v such that $x = \eta(\alpha k_1 v)^{-1} = \delta(\alpha u k_1 + \delta)^{-1}$. Finally, for $x = 1$ there is an equilibrium, depending on v, in $y = \eta(\alpha(k_1+k_2)v+\eta)^{-1}$.

In this model, the negative terms in (5.6) could also be interpreted as a friction against the manager policy: it tends to vanish as far as he is reaching a monopolistic control on the market.

Naturally many other aspects can be included in the above models, e.g. an additional state constraint representing an "anti-monopolistic law" requiring that $x \leq \overline{x}$ and $y \leq \overline{y}$ and/or non linear capital functions which give rise to non linear balance constraints.

Test 1.
In the first numerical experiment, we have considered Model 1 with the following set of parameters

$$\begin{cases} k_1 = 1, & k_2 = 1 \\ c_0 = 0.1, & c_1 = 0.25, \quad c_2 = 0.25 \\ \delta = 0.05, & \eta = 0.05, \quad \alpha = 0.25, \quad \lambda = 1. \end{cases}$$

The constraint set Ω is the polygon $\{(0.2,0),(0.1,0.1),(1,0),(1,1)\}$. We have also discretized the continuos control set A with 50 points.

By the approximate value function (Figure 2) we have computed the approximate optimal trajectories and feedbacks (Figure 3a,b). Looking at Figure 3a one can see that Ω can be divided into two regions. In the upper region the approximate optimal control is (0,1) (in black) and in the lower region is (1,0), excepted for a small neighbourhood of the constraint line

$$[(1-\alpha)k_1 - c_1] + [(1-\alpha)k_2 - c_2] = c_0 \tag{5.7}$$

where the approximate optimal control takes values different from (0,1) and (1,0) because of the state constraint. Some oscillations in the control are also due to the fact that the set A is discretized by a finite number of points.

Looking at the results of the numerical experiment, we conjecture that the optimal control is bang-bang and the fact that it switches between the two values (1,0) and (0,1) means that the manager invests only on one product depending on his initial position.

All the trajectories in the lower region are going to the point $(1-\delta/k_1, 0)$ which becomes an equilibrium point setting the control equal to (1,0). The trajectories in the upper region are driven by the optimal control $(0,1)$ to the line $y = x$ and then the investor rapidly loses fractions of the market up to the point $(0.1, 0.1)$ where he changes his policy.

Test 2.
In the second experiment, we have studied Model 2 with the following set of parameters

$$\begin{cases} k_1 = 1., & k_2 = 0.8 \\ c_0 = 0.1, & c_1 = 0.5, \quad c_2 = 0.4 \\ \delta = 0.05, & \eta = 0.05, \quad \alpha = 0.3, \quad \lambda = 1 \end{cases}$$

In this case, Ω is the polygon $\{(0.5,0),(\frac{5}{18},\frac{5}{18}),(1,0),(1,1)\}$. The control set A is discretized with 50 points.

The behaviour of the optimal control is similar to that observed in Test 1. There are again two regions, separated by a switching curve, where the manager invests only on one product. Similar considerations can be also made near the constraint line.

However, some remarkable differences can be observed on the behaviour of the trajectories (see figure 4a,b). The optimal trajectories in the lower region stop in the point where they reach the x-axis and all the points of the x-axis are equilibrium points for the optimal dynamics. The trajectories in the upper region are driven to the line $y = x$, moving with the optimal control $(0, 1)$: when they reach this line, the optimal control switches to a new value for which the system gets an equilibrium.

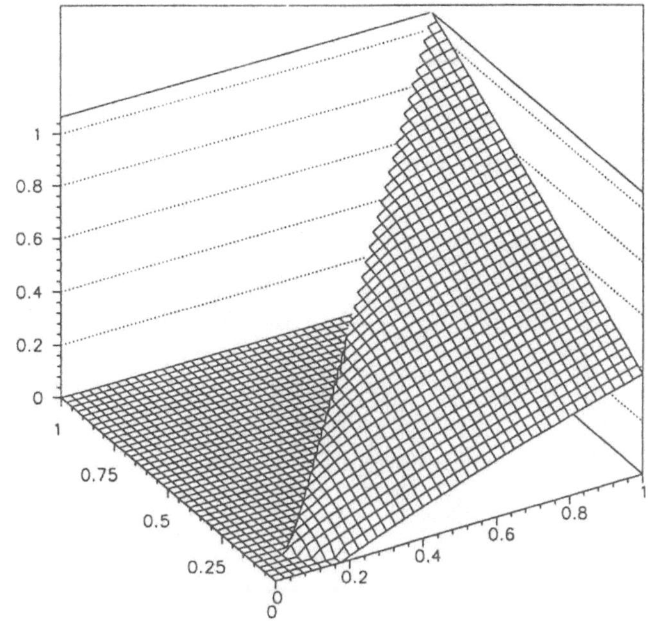

(a)

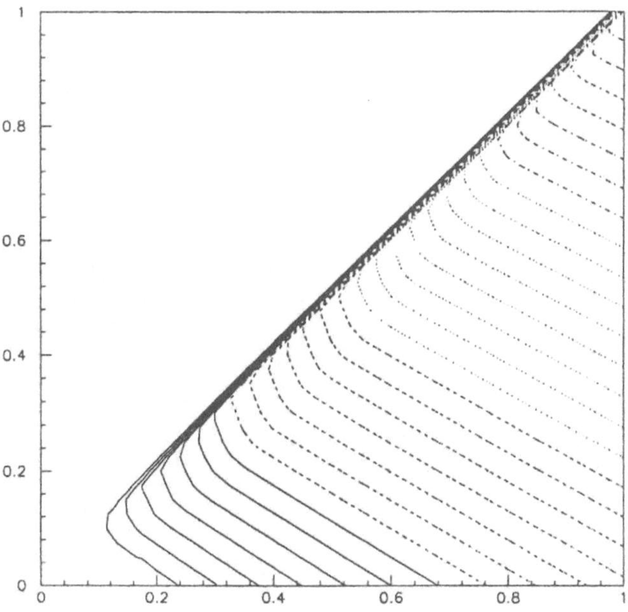

(b)

Figure 2. Test 1 ($h = 0.05$, $k = 0.03$)
(a) the approximate value function, (b) its level curves.

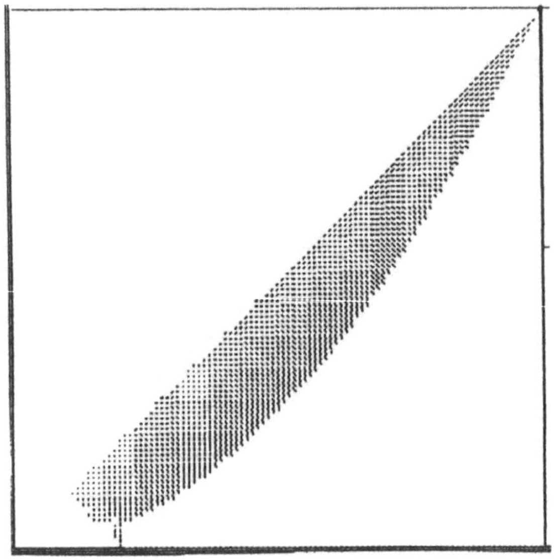

(a)

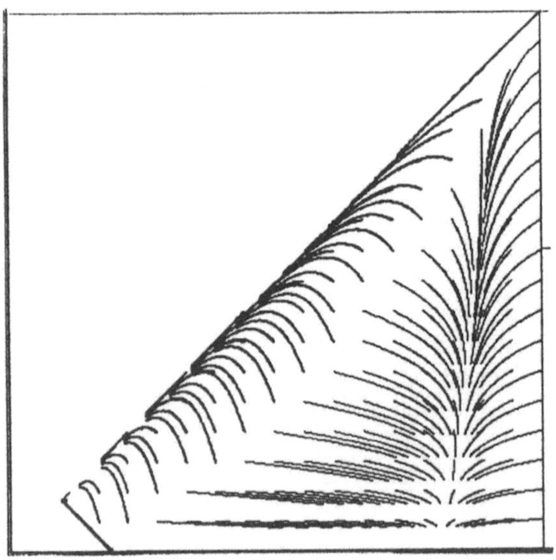

(b)

Figure 3. Test 1
(a) approximate optimal feedbacks
(b) approximate optimal trajectories

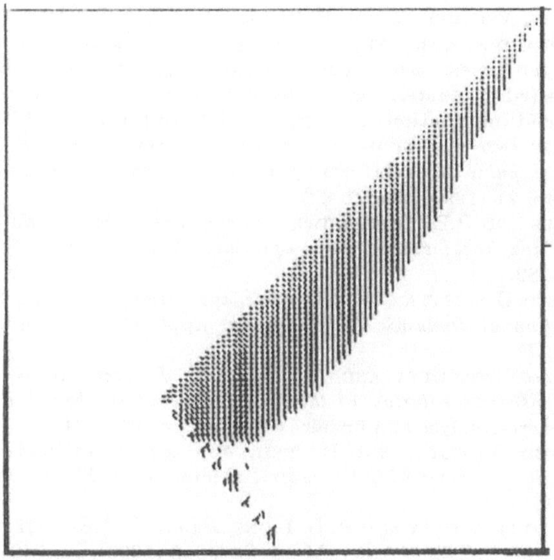

(a)

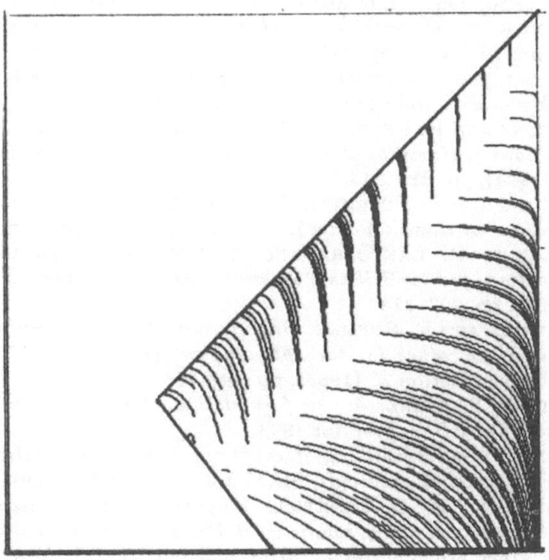

(b)

Figure 4. Test 2
(a) approximate optimal feedbacks
(b) approximate optimal trajectories

REFERENCES

[A] J.P. Aubin, *Viability Theory*, Birkhäuser Verlag, Basel, 1991.

[BF] M. Bardi and M. Falcone, *Discrete approximation of the minimal time function for systems with regular optimal trajectories*, in A. Bensoussan, J.L. Lions (eds.), *Analysis and Optimization of Systems*, Lecture Notes in Control and Information Sciences, n. 144, Springer-Verlag, 1990, pp. 103–112.

[BP] G. Barles and B. Perthame, *Comparison principle for Dirichlet-type H.-J. and singular perturbations of degenerate elliptic equations*, Appl. Math. Optim. 21 (1990), pp. 21–44.

[BS] G. Barles and P.E. Souganidis, *Convergence of approximation schemes for fully non linear second order equations*, Asymptotic Analysis 4 (1991), pp. 271–282.

[CD] I. Capuzzo Dolcetta, *On a discrete approximation of the Hamilton-Jacobi equation of Dynamic Programming*, Appl. Math. Optim. 10 (1983), pp. 367–377.

[CDF] I. Capuzzo Dolcetta and M. Falcone, *Discrete dynamic programming and viscosity solution of the Bellman equation*, Annales de l'Institut H. Poincaré-Analyse non lineaire 6 (1989), pp. 161–184.

[CDI] I. Capuzzo Dolcetta and H. Ishii, *Approximate solutions of the Bellman equation of deterministic control theory*, Appl. Math. Optim. 11 (1984), pp. 161–181.

[CDL] I. Capuzzo Dolcetta and P. L. Lions, *Hamilton-Jacobi equations with state-constraints*, Trans. Amer. Math. Soc. 318 (1990), pp. 643–683.

[CIL] M. G. Crandall H. Ishii and P. L. Lions, *User's guide to viscosity solutions of Hamilton-Jacobi equations*, Bull. Amer. Math. Soc. 27 (1992), pp. 1–68.

[DF] J. R. Dorroh and G. Ferreyra, *A multi-state multi-control problem with unbounded controls*, to appear on SIAM J. Control and Optimization.

[F] M. Falcone, *A numerical approach to infinite horizon problem*, Appl. Math. Optim. 15 (1987), pp. 1–13 and 23 (1991), pp. 213–214.

[FD] M. Falcone and A. Digrisolo, *An approximation scheme for optimal control problems with state constraints*, preprint 1992.

[FF1] M. Falcone and R. Ferretti, *Discrete-time high-order schemes for viscosity solutions of Hamilton-Jacobi-Bellman equations*, Numerische Mathematik, 67 (1994), pp. 315–344.

[FF2] M. Falcone and R. Ferretti, *Fully discrete high-order schemes for viscosity solutions of Hamilton-Jacobi-Bellman equations*, preprint 1992.

[FLS] M. Falcone, P. Lanucara and A. Seghini, *A splitting algorithm for Hamilton–Jacobi–Bellman equations*, Applied Numerical Mathematics, 15 (1994), pp. 207–218.

[GR] R. Gonzales and E. Rofman, *On deterministic control problems: an approximation procedure for the optimal cost (part I and II)*, SIAM J. Control and Optimization 23 (1985), pp. 242–285.

[IK] H. Ishii and S. Koike, *A new formulation of state constraints problems for first-order PDEs*, preprint 1993.

[Lo] P. Loreti, *Some properties of constrained viscosity solutions of Hamilton-Jacobi-Bellman equations*, SIAM J. Control and Optimization 25 (1987).

[LT] P. Loreti and M. E. Tessitore, *Approximation and regularity results on constrained viscosity solutions of Hamilton-Jacobi-Bellman equations*, J. of Mathematical Systems, Estimation and Control, Vol 4, No. 4 (1994), pp. 467–483.

[S] H. M. Soner, *Optimal control problem with state-space constraint*, SIAM J. Control and Optimization 24 (1986), pp. 552–562.

[Se] S.P. Sethi, *Dynamic optimal control models in advertising: a survey*, SIAM Review 19 (1977), pp. 685–725.

[Su] M. Sun, *Domain decomposition algorithms for solving Hamilton–Jacobi–Bellman equations*, Numerical Functional Analysis and Optimization 14 (1993), pp. 145–166.

[VW] M.L. Vidale and H.B. Wolfe, *An operations research study of sales response to advertising*, Operations Res. 5 (1957), pp. 370–381.

DISCRETE APPROXIMATIONS IN OPTIMAL CONTROL*

A.L. DONTCHEV†

Abstract. In this paper we present two techniques for analysis of discrete approximations in optimal control. In Section 2 we study convergence properties of the optimal value and optimal solutions. In Section 3 we obtain an estimate for the optimal control error in the case when the Euler discretization scheme is used for solving the first-order optimality conditions. Section 4 contains a survey on related results.

1. Introduction. When solving an optimal control problem we deal with functions which, except in very special cases, are to be replaced by numerically tractable approximations. Also, during the solution process we usually solve differential equations and compute integrals by employing appropriate finite-difference approximations. Discrete approximations can be applied directly to the problem at hand or to auxiliary problems used in the solution procedure. By an *a priori* discretization we obtain a discrete-time optimal control problem which can be regarded as a finite-dimensional mathematical program with a special structure. When we reduce the initial optimal control problem to (a sequence of) other infinite-dimensional problems, e.g. differential equations, boundary value problems, variational inequalities, we apply an *a posteriori* discretization for solving these problems.

In convergence analysis of algorithms for solving infinite-dimensional mimimum problems one of the difficulties is to find a compromise between the requirement for continuity or differentiability of the functionals involved and the need to use compactness arguments. The same difficulty occurs when one proves convergence of discrete approximations. In this paper, instead of imposing conditions for the problem and then deriving convergence of an approximating sequence, we suppose that the approximating sequence, generated by the algorithm, satisfies certain conditions, and then prove that this sequence converges to a solution. In such a way we avoid assumptions which are impossible to check and are able to consider broader classes of problems.

In Section 2 we study convergence properties of the optimal value and optimal solutions obtained by an a priori discretization. This problem fits in a natural way in the general framework of variational convergences. We consider an optimal control problem with control constraints and final state constraints and give conditions for epi-convergence of the Euler approximation of the problem.

In Section 3 we consider a nonlinear optimal control problem with

* This research was supported in part by the Institute for Mathematics and its Applications with funds provided by the National Science Foundation.

† On leave from the Institute of Mathematics, Bulgarian Academy of Sciences, Sofia, Bulgaria. Now at Mathematical Reviews, 416 Fourth Street, Ann Arbor, MI 48107.

convex control constraints. We obtain an estimate for the optimal control error in the case when an approximate Euler scheme is applied to the first-order optimality conditions. Using an extension of the Dahlquist theorem we prove that if a sequence of approximate solutions satisfies a coercivity-type condition, then for a sufficiently small step size h there exists a solution of the original problem which is at distance $O(h)$ from the approximating sequence.

The discretization of optimal control problems is an approximation procedure whose accuracy, as is typical in numerical analysis, depends on the regularity properties of the solution sought. In this paper we show that the Riemann integrability of optimal controls plays an important role in convergence analysis of discrete approximations.

There are a number of results scattered in the literature on discrete approximations that are very often closely related, although apparently independent. In Section 4 we present a survey on some of the topics in the area.

2. Convergence. Consider the following optimal control problem:

$$\text{minimize } J(x,u) = \int_0^1 g(x(t),u(t))dt \tag{2.1}$$

subject to

$$\dot{x}(t) = f(x(t),u(t)) \text{ for a.e. } t \in [0,1], \tag{2.2}$$

$$u(t) \in U \text{ for a.e. } t \in [0,1], \tag{2.3}$$

$$x(0) = a, \quad x(1) \in K, \tag{2.4}$$

where $x(t) \in \mathbf{R}^n$, $U \subset \mathbf{R}^m$, $f : \mathbf{R}^{n+m} \to \mathbf{R}^n, g : \mathbf{R}^{n+m} \to \mathbf{R}, a$ is a fixed initial state, a.e. means almost everywhere. The admissible controls are measurable and bounded functions in time t and the admissible state trajectories are absolutely continuous. We apply an a priori discretization of problem (2.1) by using the Euler scheme:

$$\text{minimize } J_N(x,u) = \sum_{i=0}^{N-1} hg(x_i,u_i) \tag{2.5}$$

subject to

$$\begin{aligned} &x_{i+1} = x_i + hf(x_i,u_i), i = 0,1,\cdots,N-1, \quad x_0 = a, \\ &u_i \in U, i = 0,1,\cdots,N-1, \quad x_N \in K, \end{aligned} \tag{2.6}$$

where N is a natural number and $h = 1/N$ is the step size.

Suppose that we apply a minimization algorithm (e.g. an infinite-dimensional version of a gradient method) to problem (2.1), obtaining a sequence of controls $u_1, u_2, \cdots, u_i, \cdots$. Under appropriate conditions the sequence u_i converges to an optimal control u^* of (2.1). Let us apply the same algorithm to the approximating problem (2.5) with fixed step size $h = 1/N$, obtaining a sequence of controls $u_1^N, u_2^N, \cdots, u_i^N, \cdots$. Assume that the discrete controls u_i^N are piecewise constant functions across the grid points $t_i = ih$. The algorithm converges uniformly with respect to the discretization if (for sufficiently small h) the corresponding controls u_i^N converge as $i \to \infty$ to a solution u^{N*} of the N-th discrete problem. The discretization is convergent if the optimal controls u^{N*} converges to u^* as $N \to \infty$. The corresponding sequences are presented in the following array:

$$
\begin{array}{cccccccc}
u_1^1 & u_2^1 & u_3^1 & \ldots & u_i^1 & u_{i+1}^1 & \ldots & u^{1*} \\
u_1^2 & u_2^2 & u_3^2 & \ldots & u_i^2 & u_{i+1}^2 & \ldots & u^{2*} \\
\cdots & & & & & & & \\
u_1^N & u_2^N & u_3^N & \ldots & u_i^N & u_{i+1}^N & \ldots & u^{N*} \\
- & - & - & - & - & - & \ldots & - \\
u_1 & u_2 & u_3 & \ldots & u_i & u_{i+1} & \ldots & u^*
\end{array}
$$

Simultaneously applying the algorithm and refining the discretization, we obtain a *diagonal* sequence of successive approximations. Any diagonal sequence u_i^N can be viewed as a sequence of ϵ_N-solutions of the N-th problem, for an appropriate choice of the sequence ϵ_N of positive numbers. In the following subsection we show that, for approximations generating the so-called epi-convergence, the limit of any convergent diagonal sequence u_i^N is a solution of the original problem if and only if any u_i^N is an ϵ_N-solution of the N-th problem with $\epsilon_N \to 0$.

2.1. An abstract theorem. Let X be a space equipped with two convergences τ_* and τ^* (not necessarily compatible) and let $I_N : X \to [-\infty, +\infty], N = 1, 2, \cdots$, be a sequence of extended-real-valued functionals.

DEFINITION 2.1. The sequence I_N is (τ_*, τ^*)-epi-convergent to the functional I iff the following two conditions hold:

$$x_N \xrightarrow{\tau_*} x \Rightarrow \liminf_{N\to\infty} I_N(x_N) \geq I(x); \tag{2.7}$$

$$\forall x \in X \ \exists x_N \xrightarrow{\tau^*} x \text{ such that } \limsup_{N\to\infty} I_N(x_N) \leq I(x). \tag{2.8}$$

The difference between the standard definition of epi-convergence and Definition 2.1 is that we employ two different convergences τ_* and τ^* in (2.7) and (2.8). As we shall see later, this setting is suitable in optimal control.

Consider the sequence of minimum problems denoted by (X, I_N):

$$\text{minimize } I_N(x) \text{ subject to } x \in X,$$

and let (X, I) be the "original" or "limit" problem which is approximated by the sequence (X, I_N). We use the following notation: the optimal value is defined as

$$\mathrm{val}(X, I) = \inf\{I(x) : x \in X\},$$

while the set of ϵ-optimal solutions is

$$\epsilon\text{-argmin}(X, I) = \{x \in X : I(x) \leq \mathrm{val}(X, I) + \epsilon\},$$

where ϵ is a real number.

The following theorem shows that, under epi-convergence, a convergent sequence goes to a solution of the limit problem if and only if it is approximately minimizing.

THEOREM 2.2. *Let $I_N \rightarrow I$ in the sense of (τ_*, τ^*)-epi-convergence and let the sequence x_N be τ_*-convergent to some x. Then the following are equivalent:*

(i) *there exists $\epsilon_N \geq 0, \epsilon_N \rightarrow 0$ such that $x_N \in \epsilon_N$-argmin(X, I_N) and δ-argmin$(X, I) \neq \emptyset$ for every $\delta > 0$;*

(ii) $x \in \mathrm{argmin}(X, I)$, $\lim I_N(x_N) = \mathrm{val}(X, I)$, *and* $\lim \mathrm{val}(X, I_N) = \mathrm{val}(X, I)$.

Proof. (i) $\Rightarrow$ (ii). Since $x_N \in \epsilon_N$-argmin(X, I_N), we have

$$I_N(x_N) \leq \mathrm{val}(X, I_N) + \epsilon_N,$$

and from (2.7),

$$I(x) \leq \liminf I_N(x_N) \leq \liminf \mathrm{val}(X, I_N). \tag{2.9}$$

Hence

$$\mathrm{val}(X, I) \leq \liminf \mathrm{val}(X, I_N). \tag{2.10}$$

Take $y \in \delta$-argmin(X, I_N) for an arbitrary $\delta > 0$. Then

$$I(y) \leq \mathrm{val}(X, I) + \delta.$$

By (2.8) there exists $y_N \rightarrow y$ such that

$$\limsup I_N(y_N) \leq I(y).$$

Then

$$\limsup \mathrm{val}(X, I_N) \leq \mathrm{val}(X, I) + \delta.$$

Since δ is arbitrary, using (2.10) we conclude that

$$\lim \mathrm{val}(X, I_N) = \mathrm{val}\,(X, I). \tag{2.11}$$

Moreover, from (2.9) and (2.11),

$$\lim I_N(x_N) = I(x) = \text{val}(X, I). \tag{2.12}$$

That is, $x \in \text{argmin}(X, I)$.

(ii) $\Rightarrow$ (i). Suppose that the sequence ϵ_N satisfies $\epsilon_N \geq |I_N(x_N) - I(x)|$ and $\epsilon_N \to 0$. Then

$$\begin{aligned} I_N(x_N) &= I_N(x_N) - I(x) + I(x) \leq |I_N(x_N) - I(x)| + \text{val}(X, I) \\ &\leq \epsilon_N + \text{val}(X, I_N). \end{aligned}$$

Hence $x_N \in \epsilon_N\text{-argmin}(X, I_N)$. □

2.2. Epi-convergence and discrete approximation. Let us apply Theorem 2.1 to the Euler approximation (2.5) of problem (2.1). An appropriate candidate for a space X is the set of feasible pairs (x, u); that is, the product of the set of measurable and essentially bounded controls u and the set of absolutely continuous states x. The functional $I(x, u) = J(x, u)$ if the pair $(x, u) \in X$ satisfies the constraints (2.2), (2.3), (2.4), and $+\infty$ otherwise. We assume that the discrete state x is piecewise linear and continuous across the grid points $t_i = ih$ and the discrete control u is piecewise constant across t_i. Then we can rewrite problem (2.5) in the following way:

$$\text{minimize } J_N(x, u) = \sum_{i=0}^{N-1} \int_{t_i}^{t_{i+1}} g(x(t_i), u(t))dt \tag{2.13}$$

subject to

$$\begin{aligned} &\dot{x}(t) = f(x(t_i), u(t)) \text{ and } u(t) \in U \text{ for all } t \in [t_i, t_{i+1}), i = 1, \\ &\qquad \cdots, N-1, \\ &x - \text{ piecewise linear, } u - \text{ piecewise constant,} \\ &x(0) = a, \quad x(1) \in K. \end{aligned} \tag{2.14}$$

We take $I_N(x, u) = J_N(x, u)$ if the pair (x, u) satisfies the constraints (2.14), (2.14), (2.14), and $+\infty$ otherwise.

Suppose that we apply a diagonalization procedure, obtaining a sequence of ϵ_N-optimal solutions to problem (2.13), for some sequence $\epsilon_N \to 0$. An important step in our analysis is to choose suitable convergences τ_* and τ^*. First, note that Theorem 2.1 assumes that this sequence (or a subsequence) is τ_*-convergent. This requirement will be automatically satisfied if we choose τ_* such that any sequence of ϵ_N-optimal solutions is sequentially compact. For instance, if τ_* is the product of the weak L^p convergence, $1 < p < +\infty$, for the controls and the C convergence (with the supremum norm) for the states, then any feasible sequence is sequentially compact provided that U is compact and convex and f is continuous and has a linear growth in x (the boundedness of U can be replaced by a coercivity condition for g if we consider a sequence of ϵ_N-optimal solutions).

However, with $\tau_* =$ weak-$L^p \times C$ condition (2.7) becomes very restrictive; essentially, it requires convexity of $g(x,\cdot)$ and linearity of $f(x,\cdot)$. Taking a stronger τ_* will increase our chances of staying with nonlinear problems. We present two simple lemmas which give sufficient conditions for epi-convergence of discrete approximations.

LEMMA 2.3. *Let f be continuous and g be lower semicontinuous and suppose that the sets U and K are closed. Then condition (2.7) is fulfilled with τ_* induced by the L^∞ norm.*

Proof. Let (x_N, u_N) be a sequence of admissible pairs for the discrete problem (2.13), i.e. satisfying (2.14), (2.14), (2.14), which converges in L^∞ to some (x,u). Since f is continuous, the sequence of derivatives $\dot{x}_N$ is bounded in L^∞. Hence $\dot{x}_N$ has a subsequence $\dot{x}_{N_i}$ which is converging L^1-weakly to some y. Then for each $t \in [0,1]$,

$$x(t) = \lim x_{N_i}(t) = \lim \int_0^t \dot{x}_{N_i}(s)ds = \int_0^t y(s)ds;$$

that is, $y = \dot{x}$. By the continuity of f and the closedness of K and U, (x,u) is feasible for the continuous problem. Since g is lower semicontinuous, $\liminf J_N(x_N, u_N) \geq J(x,u)$. □

If we take a weaker convergence τ_*, then there are more convergent sequences; however, condition (2.7) may require stronger assumptions for the problem.

Consider now condition (2.8) with τ^* induced by the $L^1 \times C$ norm. Suppose that we are given a measurable and essentially bounded control u and we would like to find a sequence of piecewise constant functions u_N which converges to u in the L^1 norm. By the density such a sequence exists, but this sequence may be not feasible, i.e. u_N may have values outside U or the corresponding $x_N(1) \notin K$. The easiest way to satisfy the control constraint is to take $u_N(t) = u(t_i), t \in [t_i, t_{i+1}), i = 0, 1, \cdots, N-1$; then we need a regularity property for u. For instance, if u is Riemann integrable[1], then the piecewise constant approximation u_N converges L^1-strongly to u. Another difficulty may occur in connection with the state constraints. Consider the following example:

Example. Minimize $x(1)$ subject to $\dot{x} = u, x(0) = 0, x(1) = -\pi/4, u \in \{-1,1\}$. The reachable set is $R = [-1,1]$ and the optimal value is $-\pi/4$. With the Euler scheme we obtain $x_{i+1} = x_i + hu_i, i = 1, 2, \cdots, N-1, x_0 = 0, x_N = -\pi/4, u_i \in \{-1,1\}$, and the reachable set of the discrete problem contains rational numbers only, more precisely $R_N = \{i/N, i = -N, \cdots, -2, -1, 0, 1, 2, \cdots, N\}$. There are no feasible controls driving the discrete system to $-\pi/4$. Condition (2) is not satisfied and there is no value convergence; $\mathrm{val}_N = +\infty$ for all N.

[1] If we require convergence for every partition of the interval [0, 1], then u must be Riemann integrable.

In this example the reason for nonconvergence of the optimal values is that the set of admissible controls is not convex but for the continuous problem it acts as if it coincides with its convex hull, by the Lyapunov theorem. One of the ways to avoid this difficulty is to enlarge the target set K in the discrete problem. If one supposes that the final state condition of the discrete problem in the above example is $x(1) \in \{[\pi/4-2h, \pi/4+2h]\}$, then the value convergence follows. The following lemma gives conditions under which the discretization (2.13) satisfies (2.8):

LEMMA 2.4. *Let g be upper semicontinuous in $\mathbf{R}^n \times U$ and f be locally Lipschitz continuous in $\mathbf{R}^n \times U$ and satisfy the following growth condition: there exist positive constants α and β such that $\| f(x,u) \| \leq \alpha \| x \| + \beta$ for every $x \in \mathbf{R}^n$ and every $u \in U$. Denote by (X, τ^*) the product of the space of Riemann integrable functions equipped with the L^1 norm and the space of absolutely continuous functions equipped with the C norm. Then for every $\delta > 0$, if the final state constraint (2.4) in the discrete problem is replaced by*

$$x_N \in K + \delta B, \tag{2.15}$$

where B is the unit ball, condition (2.8) holds.

Proof. Let (x, u) be a feasible state-control pair for the continuous problem (2.1). Let $u_N(t) = u(t_i), t \in [t_i, t_{i+1}), i = 0, 1, \cdots, N-1$, and let x_N be a piecewise linear function satisfying (2.14) with $u = u_N$. Then

$$|x_N(t_{i+1})| \leq (1 + \alpha h)|x_N(t_i)| + \beta h, i = 0, 1, \cdots, N-1.$$

This implies that x_N and $\dot{x}_N$ are uniformly bounded. Let $\Delta_i = |x_N(t_i) - x(t_i)|$. Then

$$\Delta_{i+1} \leq \Delta_i + \int_{t_i}^{t_{i+1}} [|f(x_N(t_i), u(t_i)) - f(x(t_i), u(t_i))|$$

$$+|f(x(t_i), u(t_i)) - f(x(t), u(t_i))| + |f(x(t), u(t_i)) - f(x(t), u(t))|]dt.$$

Since f is Lipschitz continuous on bounded sets and x is absolutely continuous on [0, 1] we have

$$\Delta_{i+1} \leq (1 + ch)\Delta_i + c(\int_{t_i}^{t_{i+1}} |u(t_i) - u(t)|dt + h^2).$$

Here and below c denotes a generic constant which is independent of N and t. We obtain

$$\max_{0 \leq i \leq N-1} |\Delta_i| \leq c(\sum_{i=0}^{N-1} \int_{t_i}^{t_{i+1}} |u(t_i) - u(t)|dt + h).$$

Since u_N is Riemann integrable it follows that $x_N \to x$ in C. For sufficiently large N we have $x_N \in K + \delta B$; that is, (x_N, u_N) is feasible and converges to (x, u). The convergence of $J_N(x_N, u_N)$ to $J(x, u)$ follows from the upper semicontinuity of g. □

Summarizing, suppose that we apply a diagonalization procedure to the discrete problem (2.5) with an enlarged target set as in (2.15), obtaining a sequence of ϵ_N-optimal solutions with $\epsilon_N \to 0$. Since we have access to this sequence during computations it is not unrealistic to suppose that we are able to determine whether this sequence, or its subsequence, is convergent, say in L^∞ (or we could try to modify the procedure to obtain such a convergence). From Theorem 2.1 we obtain that if the assumptions of both Lemma 2.1 and Lemma 2.2 hold and every optimal control is Riemann integrable, then the limit of the sequence at hand is an optimal solution. If the set of piecewise constant functions with values in U is L^1 dense in the set of L^∞ functions with values in U, then the Riemann integrability condition can be removed.

3. Error estimates. In this section we obtain error estimates for the optimal state and control for a discretization of the first-order optimality conditions. We consider the nonlinear optimal control problem (2.1) without final state constraints; that is:

$$\text{minimize} \int_0^1 g(x(t), u(t))dt \tag{3.1}$$

subject to

$$\dot{x}(t) = f(x(t), u(t)) \text{ for a.e. } t \in [0,1], x(0) = a,$$

$$u(t) \in U \text{ for a.e. } t \in [0,1], u \in L^\infty, x \in W^{1,\infty},$$

where U is a closed and convex set in $\mathbf{R}^m$. Assuming that the functions f and g are continuously differentiable, we can write the first-order conditions (Pontryagin maximum principle) as a variational inequality:

$$\dot{x}(t) = f(x(t), u(t)), \quad x(0) = a, \tag{3.2}$$

$$\dot{\lambda}(t) = -\nabla_x H(x(t), u(t), \lambda(t)), \quad \lambda(1) = 0, \tag{3.3}$$

$$\nabla_u H(x(t), u(t), \lambda(t)) \in \partial U(u(t)), \tag{3.4}$$

for a.e. $t \in [0,1]$, where λ is the adjoint variable, H is the Hamiltonian, $H(x, u, \lambda) = g(x, u) + \lambda^T f(x, u)$, and $\partial U(u)$ is the normal cone to the set U at the point u.

Suppose that we approximately solve the problem (3.2)–(3.4) by applying the Euler scheme, obtaining vectors $(x^N, u^N, \lambda^N) \in \mathbf{R}^{Nn} \times \mathbf{R}^{Nm} \times \mathbf{R}^{Nn}$;

that is, (x^N, u^N, λ^N) satisfy

$$x_{i+1} = x_i + hf(x_i, u_i) + h\delta_i^N, \quad x_0 = a, \tag{3.5}$$

$$\lambda_i = \lambda_{i+1} + h\nabla_x H(x_i, u_i, \lambda_{i+1}) + h\eta_i^N, \quad \lambda_N = 0, \tag{3.6}$$

$$\nabla_u H(x_i, u_i, \lambda_{i+1}) + \kappa_i^N \in \partial U(u_i), \tag{3.7}$$

for $i = 0, 1, \cdots, N-1$. Here the vector $\epsilon^N = (\delta^N, \eta^N, \kappa^N)$ represents the error of the method used for solving (3.2)–(3.4). Assume that the sequence of vectors $(x_i^N, u_i^N), i = 0, 1, \cdots, N$, is contained in a compact set $\mathcal{X} \subset \mathbf{R}^n \times \mathbf{R}^m$ for all N and the functions f and g are twice continuously differentiable in an open set containing $\mathcal{X}$. Let $A_N = \nabla_x f(x^N, u^N), B_N = \nabla_u f(x^N, u^N), R_N = \nabla^2_{uu} H(x^N, u^N, \lambda^N), S_N = \nabla^2_{xu} H(x^N, u^N, \lambda^N), Q_N = \nabla^2_{xx} H(x^N, u^N, \lambda^N)$.

THEOREM 3.1. *Suppose that the error* ϵ^N *satisfies*

$$\lim_{N\to\infty} \max_{0\le i\le N-1} |\epsilon_i^N| = 0, \tag{3.8}$$

and let there exist a constant $\alpha > 0$ *such that for every sufficiently large* N,

$$\int_0^1 [x^T Q_N x + 2x^T S_N u + u^T R_N u] dt \ge \alpha \int_0^1 |u|^2 dt \tag{3.9}$$

for all $x \in W^{1,2}, u \in L^2, x(0) = 0, \dot{x} = A_N x + B_N u, u(t) \in U - U$. *Then there exists a constant c such that for every sufficiently large N there exist a local minimizer* (x^{*N}, u^{*N}) *of the continuous problem and a corresponding adjoint variable* λ^{*N} *such that*

$$\begin{aligned} &\| x^{*N} - x^N \|_{W^{1,\infty}} + \| \lambda^{*N} - \lambda^N \|_{W^{1,\infty}} + \| u^{*N} - u^N \|_{L^\infty} \\ &\le c(h + \max_{0\le i\le N-1} |\epsilon_i^N|). \end{aligned} \tag{3.10}$$

A proof of this theorem is given in Subsection 3.2. It is based on a general inverse-function-type result for a sequence of set-valued maps presented in the following subsection.

3.1. An extension of the Dahlquist theorem. Let (X, ρ) be a metric space and let Y be a linear normed space. We denote by $B_a(x)$ the closed ball centered at x with radius a.

DEFINITION 3.2. Let ξ_N be a sequence of points in X. The sequence of functions $\phi_N : X \to Y$ is *strictly stationary at* ξ_N *uniformly in* N if for every $\epsilon > 0$ there exists $\delta > 0$ such that for every N and for every $u, v \in B_\delta(\xi_N)$,

$$\| \phi_N(u) - \phi_N(v) \| \le \epsilon \rho(u, v).$$

A sequence of functions ϕ_N from a Banach space X to Y is strictly stationary at ξ_N uniformly in N if there exists a constant $\alpha > 0$ such that every ϕ_N is continuously differentiable in $B_\alpha(\xi_N)$ and the derivatives $\nabla\phi_N(\xi_N) = 0$ for every N.

The following theorem can be viewed as an extended version of a basic observation in numerical analysis, sometimes called the Dahlquist theorem or Lax principle: *consistency plus stability implies convergence.*

THEOREM 3.3. *Let (X, ρ) be complete, let ξ_N be a sequence in X, let the sequence of functions $\phi_N : X \to Y$ be strictly stationary at ξ_N uniformly in N, and let Θ_N be a sequence of set-valued maps from X to Y with $0 \in \Theta_N(\xi_N)$ for all N. Suppose that the following two conditions hold:*

- *Consistency.* $\lim_{N\to\infty} \| \phi_N(\xi_N) \| = 0.$
- *Stability. There exist constants β and γ such that for every sufficiently large N there exists a function $\Psi_N : Y \to X$ with the following properties:*

$$\xi_N = \Psi_N(0);$$

$$\Psi_N(y) \in \Theta_N^{-1}(y) \text{ for every } y \in B_\beta(0);$$

$$\Psi_N \text{ is Lipschitz in } B_\beta(0) \text{ with constant } \gamma.$$

Then for every $\gamma^+ > \gamma$ there exists N^+ such that for every $N \geq N^+$ there exists $\xi_N^* \in (\Theta_N - \phi_N)^{-1}(0)$ satisfying

$$\rho(\xi_N^*, \xi_N) \leq \gamma^+ \| \phi_N(\xi_N) \| .$$

Moreover, if Θ_N^{-1} is single-valued near 0 then there exists exactly one ξ_N^* with the above properties.

Proof. Let $\gamma^+ > \gamma$ and let ϵ be such that

$$0 < \gamma\epsilon < 1 \text{ and } \frac{\gamma}{1-\epsilon\gamma} \leq \gamma^+. \tag{3.11}$$

From the stability condition there exist constants β and γ such that for N sufficiently large there exists Ψ_N satisfying

$$\rho(\Psi_N(y_1), \Psi_N(y_2)) \leq \gamma \| y_1 - y_2 \|, \tag{3.12}$$

whenever $\| y_i \| \leq \beta, i = 1, 2$. Since ϕ is strictly stationary at ξ_N there exists $\alpha > 0$ such that

$$\| \phi_N(u) - \phi_N(v) \| \leq \epsilon\rho(u, v) \tag{3.13}$$

for every $u, v \in B_\alpha(\xi_N)$. Let

$$r_N = \gamma^+ \parallel \phi_N(\xi_N) \parallel .$$

Now choose N^+ so large that for $N \geq N^+$,

$$r_N \leq \max\{\frac{\beta}{\epsilon + 1/\gamma^+}, \alpha\}. \tag{3.14}$$

Fix $N \geq N^+$ and define

$$\Phi(x) = \Psi_N(\phi_N(x)).$$

Let $x \in B_{r_N}(\xi_N)$. Then, from (3.13) and (3.14),

$$\parallel \phi_N(x) \parallel \leq \parallel \phi_N(x) - \phi(\xi_N) \parallel + \parallel \phi(\xi_N) \parallel \leq \epsilon r_N + r_N/\gamma^+ \leq \beta,$$

and, using (3.11),

$$\rho(\xi_N, \Phi(x)) = \rho(\Psi_N(0), \Psi_N(\phi_N(x))) \leq \gamma \parallel \phi_N(x) \parallel$$

$$\leq \gamma \parallel \phi_N(x) - \phi(\xi_N) \parallel + \gamma \parallel \phi(\xi_N) \parallel \leq \gamma\epsilon r_N + r_N\gamma/\gamma^+ \leq r_N.$$

Hence Φ maps $B_N(x_N)$ into itself. Moreover, if $x', x'' \in B_{r_N}(\xi_N)$, then

$$\begin{aligned}\rho(\Phi(x'), \Phi(x'')) &= \rho(\Psi_N(\phi_N(x')), \Psi_N(\phi_N(x''))) \\ &\leq \gamma \parallel \phi_N(x') - \phi_N(x'') \parallel \leq \gamma\epsilon\rho(x', x'') < \rho(x', x'').\end{aligned}$$

Thus Φ is a contracting mapping from $B_{r_N}(\xi_N)$ to $B_{r_{\bar{N}}}(\xi_N)$. By the contracting mapping principle, for every $N > N^+$ there exists $\xi_N^* \in B_{r_N}(\xi_N)$ such that $\xi_N^* = \Phi(\xi_N^*)$. From the definition of Φ we conclude that $\xi_N^* \in (\Theta_N - \phi_N)^{-1}(0)$. If Θ_N^{-1} is single-valued near 0, then ξ_N^* is the unique fixed point of Φ in $B_{r_N}(\xi_N)$. This completes the proof. □

3.2. Proof of Theorem 3.1. Consider x^N and λ^N as piecewise linear and continuous functions and u_N as a piecewise constant function across the grid points. We apply Theorem 3.1 with the following specifications:

$$X = \{\xi = (x, \lambda, u) \in W^{1,\infty} \times W^{1,\infty} \times L^\infty, x(0) = a, \lambda(1) = 0\}, \quad Y = L^\infty,$$

$$\phi_N(\xi) = - \begin{pmatrix} \dot{x} - f(x,u) - (\dot{x} - \dot{x}^N) + A_N(x - x^N) \\ + B_N(u - u^N) \\ \dot{\lambda} + \nabla_x H(x,u,\lambda) - (\dot{\lambda} - \dot{\lambda}^N) - A_N^T(\lambda - \lambda^N) \\ -Q_N(x - x^N) - S_N(u - u^N) \\ -\nabla_u H(x,u,\lambda) + \nabla_u H^N + R_N(u - u^N) + S_N^T(x - x^N) \\ + B_N^T(\lambda - \lambda^N) - \kappa^N \end{pmatrix},$$

$$\Theta_N(\xi) = \begin{pmatrix} (\dot{x} - \dot{x}^N) - A_N(x - x^N) - B_N(u - u^N) \\ (\dot{\lambda} - \dot{\lambda}^N) + A_N^T(\lambda - \lambda^N) + Q_N(x - x^N) \\ +S_N(u - u^N) \\ -\nabla_u H^N - R_N(u - u^N) - S_N^T(x - x^N) - B_N^T(\lambda - \lambda^N) \\ +\kappa^N + \partial U(u) \end{pmatrix},$$

where $\nabla_u H^N(t) = \nabla_u H(x^N(t_i), u^N(t), \lambda^N(t_{i+1}))$ for $t \in [t_i, t_{i+1})$, δ^N, η^N, κ^N are assumed piecewise constant across the grid. Clearly, ϕ_N is strictly stationary at $\xi_N = (x^N, \lambda^N, u^N)$ uniformly in N as a function from $W^{1,\infty} \times W^{1,\infty} \times L^\infty$ to L^∞, and $0 \in \Theta(\xi_N)$.

Proof of Consistency. The relations (3.5)–(3.7) imply that ξ_N satisfies: $\xi_N \in X$ and

$$\begin{aligned} &\dot{x}(t) = f(x(t_i), u(t)) + \delta_i^N, \\ &\dot{\lambda}(t) = -\nabla_x H(x(t_i), u(t), \lambda(t_{i+1})) + \eta_i^N, \\ &\nabla_u H(x(t_i), u(t), \lambda(t_{i+1})) + \kappa_i^N \in \partial U(u(t)) \end{aligned}$$

for $t \in [t_i, t_{i+1}), i = 0, 1, \cdots, N-1$. Since x^N, u^N are bounded and f, g and their derivatives are continuous, it is not difficult to show that λ^N and the derivatives $\dot{x}^N$ and $\dot{\lambda}^N$ are bounded in L^∞. Then from (3.8),

$$\begin{aligned} \| \phi_N(\xi_N) \|_{L^\infty} \le \; & \max\nolimits_{0 \le i \le N-1} \sup\nolimits_{t_i \le t \le t_{i+1}} \\ & [|f(x^N(t_i), u^N(t)) - f(x^N(t), u^N(t))| \\ & + |\nabla_x H(x^N(t_i), u^N(t), \lambda^N(t_{i+1})) \\ & \quad -\nabla_x H(x^N(t), u^N(t), \lambda^N(t))| \\ & + |\nabla_u H(x^N(t_i), u^N(t), \lambda^N(t_{i+1})) \\ & \quad -\nabla_u H^N(x(t), u^N(t), \lambda^N(t))|] + |\epsilon_i^N|\} \\ = \; & O(h) + \max\nolimits_{0 \le i \le N-1} |\epsilon_i^N| \to 0 \end{aligned}$$

when $N \to \infty$; that is, consistency holds.

Proof of Stability. Let $y = (p, q, r) \in L^\infty$. Then $\xi \in \Theta_N^{-1}(y)$ iff $\xi = (x, \lambda, u)$ satisfies the following relations:

(3.15) $$\dot{x} = A_N x + B_N u + p + \alpha_N, \quad x(0) = a,$$

(3.16) $$\dot{\lambda} = -A_N^T \lambda - Q_N x - S_N u + q + \beta_N, \quad \lambda(1) = 0,$$

(3.17) $$R_N u + S_N^T x + B_N^T \lambda + r + \gamma_N \in \partial U(u),$$

where

$$\alpha_N = \dot{x}^N - A_N x^N - B_N u^N,$$

$$\beta_N = \dot{\lambda}^N + A_N^T \lambda^N + Q_N x^N + S_N u^N,$$

$$\gamma_N = \nabla_u H^N - R_N u^N - S_N^T x^N - B_N^T \lambda^N - \kappa^N.$$

Because of (3.9) the system (3.15)–(3.17) is equivalent to the following linear-quadratic problem:

$$(3.18)\qquad \begin{aligned} &\text{minimize } 0.5 \int_0^1 [x^T Q_N x + u^T R_N u + 2x^T S_N u \\ &-(q+\beta_N)^T x + (r+\gamma_N)^T u]dt \end{aligned}$$

subject to

$$(3.19)\qquad \begin{aligned} &\dot{x} = A_N x + B_N u + p + \alpha_N, u(t) \in U \text{ for a.e. } t \in [0,1], \\ &x(0) = a. \end{aligned}$$

Let $w_N(p)$ be the solution of

$$\dot{w} = A_N w + p + \alpha_N, w(0) = a.$$

Changing the state variable $x = z + w_N(p)$ we obtain the following problem, which is equivalent to (3.18): minimize

$$(3.20)\qquad \begin{aligned} &0.5 \int_0^1 [z^T Q_N z + u^T R_N u + 2z^T S_N u - (q + \beta_N - 2Q_N w_N(p))^T z \\ &+(r + \gamma_N + 2S_N^T w_N(p))^T u]dt \end{aligned}$$

subject to

$$(3.21)\qquad \dot{z} = A_N z + B_N u, \quad u(t) \in U, \text{ a.e. } t \in [0,1], \quad z(0) = 0.$$

Since the components of A_N and B_N are bounded in L^∞ there exists a constant c independent of N such that

$$\| \dot{z} \|_{L^2}^2 + \| z \|_{L^2}^2 \le c \| u \|_{L^2}^2,$$

whenever z satisfies (3.21). Then the coercivity condition (3.9) is equivalent to

$$(3.22)\qquad \int_0^1 [z^T Q_N z + 2z^T S_N u + u^T R_N u]dt \ge \alpha_1 (\| u \|_{L^2}^2 + \| z \|_{W^{1,2}}^2)$$

for all $z \in W^{1,2}, u \in L^2, z(0) = 0, \dot{z} = A_N z + B_N u, u(t) \in U - U$, and for some $\alpha_1 > 0$ independent of N. Let

$$\mathcal{R}_N = \begin{pmatrix} Q_N & S_N \\ S_N^T & R_N \end{pmatrix}$$

and

$$\nu_N(y) = \begin{pmatrix} -q - \beta_N + 2Q_N w_N(p) \\ r + \gamma_N + 2S_N^T w_N(p) \end{pmatrix},$$

and let $\Omega_N = \{(z,u) \in W^{1,2} \times L^2 : (z,u) \text{ satisfies } (3.21)\}$. Then (3.20) can be rewritten in the following way:

$$\text{(3.23)} \quad \text{minimize } 0.5 < \chi, \mathcal{R}_N \chi > + < \nu_N(y), \chi > \text{ subject to } \chi \in \Omega_N,$$

where $\chi = (x, u)$ and $< \cdot, \cdot >$ is the L^2 scalar product. The set Ω_N is a closed and convex subset of the Hilbert space $H = W^{1,2} \times L^2$ and from (3.22) the following coercivity condition holds:

$$\text{(3.24)} \qquad < \chi, \mathcal{R}_N \chi > \geq \alpha_1 \| \chi \|_H^2 \text{ for all } \chi \in \Omega_N - \Omega_N.$$

It is a simple observation (see [25], Lemma 4) that, under (3.24), for every y there exists a unique solution $\chi_N(y)$ of the problem (3.23) and, given y', y'', the corresponding solutions $\chi_N(y'), \chi_N(y'')$ satisfy

$$\| \chi_N(y') - \chi_N(y'') \|_H \leq \frac{1}{\alpha_1} \| \nu_N(y') - \nu_N(y'') \|_{L^2}.$$

This result means that there exists an unique optimal solution $(x_N(y), u_N(y))$ of the problem (3.18) which is Lipschitz in y from L^∞ to $W^{1,2} \times L^2$ with a Lipschitz constant independent of N. By the adjoint equation we conclude that the optimal adjoint variable $\lambda_N(y)$ is Lipschitz with respect to y from L^∞ to $W^{1,2}$ uniformly in N. Furthermore, the coercivity condition (3.9) implies that for some $\alpha_2 > 0$,

$$\text{(3.25)} \qquad u^T R_N(t) u \geq \alpha_2 |u|^2 \text{ for a.e. } t \in [0,1],$$

whenever $u \in U - U$. Applying again Lemma 4 from [25] to the optimality condition (3.17) with the coercivity condition (3.25), we obtain that the optimal control $u_N(y)$ is Lipschitz in y from L^∞ to L^∞ uniformly in N, hence $x_N(y)$ and $\lambda_N(y)$ are Lipschitz from L^∞ to $W^{1,\infty}$ uniformly in N. This proves Stability.

Theorem 3.1 implies that for sufficiently large N there exist x^{*N}, u^{*N} and λ^{*N} satisfying the estimate (3.10) and the first-order conditions (3.2)–(3.4). The last step of the proof if to show that (x^{*N}, u^{*N}) is a local solution of (3.1). This follows from the observation that the coercivity condition (3.9) is stable under perturbations; that is, it holds at $(x^{*N}, u^{*N}, \lambda^{*N})$, and, together with the maximum principle, it is a second-order sufficient condition.

In the above theorem we use the uniform grid in [0, 1], i.e. with a constant step size h. Clearly, an analogous result holds for any regular partition of [0, 1] in which the maximal step size goes to zero. From Theorem 3.1 we obtain the following corollaries:

COROLLARY 3.4. *Suppose that problem (3.1) has no more than one optimal solution and for every regular partition of [0, 1] in N intervals the sequence (x^N, u^N, λ^N) obtained from (3.5)–(3.7) satisfies coercivity condition (3.9). Then there exists a (unique) optimal control for (3.1) which is Riemann integrable.*

Proof. From Theorem 3.1 problem (3.1) has a solution (x^*, u^*) and by assumption it is unique and satisfies (3.10). Then for every regular partition $\{t_i\}_{i=0}^N$,

$$\max_{0 \le i \le N-1} \sup_{t_i \le t < t_{i+1}} |u^*(t) - u(t_i)| \to 0 \text{ as } N \to \infty,$$

which implies Riemann integrability of u^*. □

4. Bibliographical remarks. Although there is an extensive literature on computational optimal control — see the monographs [34], [44], [51], [58], [59] and the references therein — relatively few results have been published on error analysis of discrete approximations.

4.1. Convergence. Epi-convergence was first introduced in finite-dimensional spaces by Wijsman [65] and has been extensively used in the analysis of variational problems: for references see [30]. Theorem 2.1 is closely related to a result due to Attouch and Wets [2]; see also [30], Chapter 4, Section 3. Two convergences in the definition of epi-convergence were used recently by Polak [53] to prove convergence of an algorithm involving successive discretizations.

Relying on unpublished work by J.-P. Aubin and J.-L. Lions, J. W. Daniel [18] (see also his papers [16], [17]) developed an abstract framework for proving convergence of discrete approximations of optimal control problems, extending the direct method of the calculus of variations in a way parallel to the epi-convergence approach (but formally independent of it; see also [60]). In a series of papers Cullum [13], [14], [15] showed value and solution convergence of Euler approximations, applying virtually the same idea. In [13] the problem is linear-convex, [14] considers problems of Mayer type that are linear in control, and [15] is an extension of [14] for a state and control constrained problem. For a survey of earlier works on computational optimal control, including discrete approximations, see [52]. There is also a body of work in Russian focusing mainly on convergence of discrete approximations combined with Tikhonov's regularizations; see [3], [7], [8], [9], [32], [43], [59]. For convergence analysis of algorithms involving discrete approximations see [31], [42], [54], [55].

4.2. Differential inclusions. Consider the problem of minimizing $J(x, u) = \phi(x(1))$ subject to (2.2)–(2.4). In this case one can eliminate the control, considering the differential inclusion

$$\dot{x}(t) \in F(x(t)), \quad x(0) = a, \quad x(1) \in K, \tag{4.1}$$

where the set-valued map $F(x) = f(x, U)$. The Euler approximation to this problem consists in minimizing $\phi(x_N)$ subject to

$$x_{i+1} \in x_i + hF(x_i), i = 0, 1, \cdots, N-1, x_0 = a, x_N \in K. \tag{4.2}$$

Under some standard conditions (e.g. a growth condition and a Lipschitz condition for F) any sequence of discrete trajectories is compact in C.

Every limit point of such a sequence is from the closure in C of the set of feasible trajectories. This closure is described by the solution set of the so-called relaxed problem (introduced by Warga [64]), in which F is replaced by the closed and convex hull of F. The continuous problem is relaxable if its value is equal to the value of the relaxed problem. Enlarging the target set K as in Lemma 2.2 and assuming continuity of ϕ, one can show epi-convergence to the relaxed problem with both τ_* and τ^* induced by the C norm. Hence, by Theorem 2.1, the optimal values of the discretized problems will converge to the optimal value of the relaxed problem. This implies that the value convergence is equivalent to the relaxability of the continuous problem. This result was established first by Mordukhovich [49]; for a more general setting see [21]. Virtually the same result is given in [11]. We note that in an earlier paper Cullum [15] essentially proved that a sequence of discrete solutions converges to a solution of the relaxed problem; instead of relaxability she uses necessary optimality conditions for the relaxed problem.

If the map F is convex- and closed-valued and Lipschitz continuous, the convergence of the Euler method can be estimated by the Hausdorff distance between the set of solutions of the difference inclusion (4.2) and the set of solutions of the differential inclusion (4.1). In our recent paper [27] we consider the following problem: find a pair of functions (x, u), x absolutely continuous in [0,1] with values in $\mathbf{R}^n$, u bounded and measurable with values in $\mathbf{R}^m$, such that

$$\begin{aligned} &\dot{x}(t) = f(x(t), u(t)) \text{ and } u(t) \in U \text{ for all } t \in [0,1], \\ &x(0) = a, x(1) \in K, \end{aligned} \tag{4.3}$$

where $f : \mathbf{R}^n \times \mathbf{R}^m \to \mathbf{R}^n$, $U \subset \mathbf{R}^m, K \subset \mathbf{R}^n$, and $a \in \mathbf{R}^n$ is fixed. Let us apply the Euler method, i.e.

$$\begin{aligned} &x_{i+1} = x_i + hf(x_i, u_i), u_i \in U, i = 0, 1, \cdots, N-1, \\ &x_0 = a, x_N \in K. \end{aligned} \tag{4.4}$$

We estimate the error of the discretization, measured by a "distance" between the set of solutions of the finite-dimensional problem (4.4) and the set of solutions of the original problem (4.3). To avoid assumptions concerning smoothness of the right-hand side of the differential equation we employ the so-called averaged modulus of continuity introduced and studied in [57]. Let v be a bounded function defined on [0, 1]. The local modulus of continuity of v is defined by

$$\omega(v; t, h) = \sup \{|v(s) - v(q)| : s, q \in [t - h/2, t + h/2] \cap [0, 1]\},$$

while the averaged modulus of continuity of v is

$$\tau(v; h) = \int_0^1 \omega(v; t, h) dt.$$

The following properties of the modulus τ can be found in [57]:

1) $\tau(v;h) \to 0$ if and only if v is Riemann integrable.

2) If v is of bounded variation, then $\tau(v;h) = O(h)$.

Let (x^*, u^*) be a feasible pair of (4.3). We assume that the sets U and K are closed and convex and f is continuously differentiable in a convex and open set in $\mathbf{R}^{n+m}$ containing $(x^*(t), u^*(t))$ for all $t \in [0,1]$. Consider the linear problem

$$(4.5)\ \dot{x}(t) = A(t)x(t) + B(t)u(t), x(0) = a, u(t) \in U \text{ for a.e. } t \in [0,1],$$

where $A = \nabla_x f, B = \nabla_u f$, both computed at (x^*, u^*). Denote by $\mathcal{R}$ the reachable set at $t = 1$ of (4.5); that is, the set of all points in $\mathbf{R}^n$ that are values at $t = 1$ of solutions of (4.5) and correspond to some feasible control u, i.e. u bounded and measurable, $u(t) \in U$ for a.e. $t \in [0,1]$.

THEOREM 4.1. [27] *Let $\dot{x}^*, A$ and B be Riemann integrable and suppose that*

$$(\text{int}\mathcal{R} \cap K) \cup (\text{int}K \cap \mathcal{R}) \neq \emptyset.$$

Then for N sufficiently large there exists a feasible pair (x^N, u^N) for the discretized problem (4.4) satisfying

$$\max_{0 \le i \le N-1}(|x^*(t_i) - x_i^N| + |u^*(t_i) - u_i^N|) = O(\tau(\dot{x}^*; h)).$$

In [27] we also obtain a similar estimate for the $C \times L^\infty$ distance from a sequence of solutions (x^N, u^N) of the discrete problem (4.4) to the set of solutions of (4.3). An error bound for the Euler method applied to an initial value problem for a differential inclusion is found in [24]. For further results in this direction, including higher-order schemes and error estimates for the reachable set, see [29], [61], [62], [63], [66].

4.3. Error estimates for the optimal solution. There are very few results in the literature providing error estimates for discrete approximations in optimal control. Hager [37] considered higher-order schemes applied to unconstrained problems, obtaining error estimates under appropriate smoothness assumptions; see also [56]. Error estimates for the Euler scheme applied to state and control constrained convex optimal control problems were derived in [19]. Related results, employing arguments from sensitivity analysis in optimization, are given in [1], [45], [46], [47]. In our recent paper [25] we consider a nonlinear problem of the form (3.1) and prove the following theorem:

THEOREM 4.2. [25] *Let (x^*, u^*) be a local solution for the continuous problem (3.1), let f and g be twice continuously differentiable in an open set containing $(x^*(t), u^*(t))$ for all $t \in [0,1]$. Suppose that u^* is Riemann integrable and there exists a constant $\alpha > 0$ such that*

$$\int_0^1 [x^T Q x + 2x^T S u + u^T R u] dt \geq \alpha \int_0^1 |u|^2 dt$$

for all $x \in W^{1,2}, u \in L^2, x(0) = 0, \dot{x} = Ax + Bu, u(t) \in U - U$, *where* $A = \nabla_x f(x,u), B = \nabla_u f(x,u), R = \nabla^2_{uu} H(x,u,\psi,), S = \nabla^2_{xu} H(x,u,\psi,), Q = \nabla^2_{xx} H(x,u,\psi)$, *all computed at* x^*, u^*, ψ^*. *Then for* h *sufficiently small there exists a local minimizer* (x^h, u^h) *of the corresponding discrete problem obtained by the Euler scheme such that*

$$\max_{0 \leq i \leq N}(|x^*(t_i) - x_i^h| + |u^*(t_i) - u_i^h|) = O(h + \tau(u^*; h)).$$

Note that the only assumption for the regularity of the optimal control u^* is the Riemann integrability, i.e. the optimal control may have infinitely many points of discontinuity (but must be almost everywhere continuous). Since u^* is Riemann integrable, $\tau(u^*; h) \to 0$ as $h \to 0$; thus, from the above estimate we immediately conclude that the error is convergent to zero. If the optimal control u^* is of bounded variation, then $\tau(u^*; h) = O(h)$, and we automatically obtain that the error is $O(h)$. An abstract inverse-function theorem for set-valued maps related to Theorem 3.2 is presented in [26].

4.4. Dual methods. In a series of papers Hager [36], [38], [39], [40] studies approximations to dual problems in optimal control. Dual problems can be defined in various ways, depending on the assumed spaces for the variables and the constraints. Choosing appropriate spaces one can obtain dual problems which are better suited to discrete approximations and numerical computations.

The dual approach is exceptionally efficient when one deals with entropy-like minimization problems; see [5], [6]. In [22], [23], [28] we applied dual methods to a class of problems from computer-aided geometric design. Given $0 = t_1 < t_2 < \cdots < t_n = 1$ and reals $y_i, i = 1, \cdots, n$ the *convex best interpolation problem* consists of finding a convex function f such that $f(t_i) = y_i$ for $i = 1, 2, \cdots, n$, and the second derivative of f has a minimal L^2 norm. Hornung [41] proved that the solution of this problem is a cubic spline whose second derivative is the positive part of a piecewise linear function; for further results see [48]. In [22] we show that Hornung's result can be easily obtained by a duality argument. Let $f = x_1, f' = x_2, f'' = u$; then the convex best interpolation problem can be rewritten as an optimal control problem:

$$\text{minimize } \| u \|_{L^2}$$

subject to

$$\dot{x}_1 = x_2, \quad \dot{x}_2 = u, \quad u \geq 0,$$

$$\int_0^1 B_i^2 u dt = d_i, i = 1, 2, \cdots, m = n - 2, \tag{4.6}$$

where the interpolation conditions are represented by the integral control constraints (4.6), B_i^2 are the normalized B-splines of order 2 and support $[t_i, t_{i+2}]$ and d_i are the second divided differences of the data (t_i, y_i). Define a dual functional

$$\mathcal{L}(\lambda) = \inf \{0.5 \parallel u \parallel_{L^2}^2 - \sum_{i=1}^{m} \lambda_i (\int_0^1 B_i^2 u dt - d_i) : u \in L^2, u \geq 0\}.$$

Then the corresponding dual problem is an unconstrained nonlinear program of the following form:

$$\text{maximize } -0.5 \int_0^1 (\sum_{i=1}^{m} \lambda_i^* B_i^2(t))_+^2 dt + \sum_{i=1}^{m} \lambda_i d_i \text{ subject to } \lambda \in \mathbf{R}^m.$$

In [28] we proved convergence of a class of gradient methods for solving the dual problem. A constrained interpolation problem, where the function sought is required to satisfy bilateral constraints, is studied by duality methods in [23].

Another approach related to duality is based on the solution of the Hamilton-Jacobi equation. The Hamilton-Jacobi equation has typically a nonsmooth (viscosity) solution whose computation requires special procedures. For a number of results in this direction see [4], [10], [12], [33], [35].

4.5. Some open problems. Although discrete approximations are primarily intended for computational purposes, they are also a useful tool for obtaining purely theoretical results. The idea of using broken lines to prove existence of solutions to differential equations goes back to Euler. Apparently one can use various approximations, depending on what property of the continuous problem one would like to study. For instance, one can derive necessary optimality conditions from necessary conditions for a discretized problem by passing with the step size to zero; see [50]. It seems that this approach has not been fully exploited in the literature of optimal control. Depending on the approximation used one may obtain various necessary (or sufficient) conditions.

The presence of state constraints considerably complicates the numerical analysis in optimal control. An estimate for the L^2 norm of the optimal control error was obtained in [19] for a convex problem with linear inequality state and control constraints. It is not yet clear whether this estimate holds with a stronger norm and for more general nonlinear problems.

The implementation and rigorous error analysis of higher-order approximations (e.g. Runge-Kutta schemes) applied to optimal control problems constitute another area for future research. Some interesting results in this direction have been recently obtained by Veliov [61], [62].

REFERENCES

[1] W. ALT, *On the approximation of infinite optimization problems with an application to optimal control problems*, Appl. Math. Optim. **12** (1984), 15–27.

[2] H. ATTOUCH, R.J.-B. WETS, *Approximation and convergence in nonlinear optimization*, Nonlinear Programming 4, (edited by) O. MANGASARIAN, R. MEYER, S. ROBINSON, Academic Press, New York, 1981, pp. 367–394.

[3] E.R. AVAKOV, F.P. VASIL'EV, *Difference approximation of a maximin problem of optimal control with phase constraints*, Vestnik Moscov. Univ., Ser XV, Vychisl. Mat. Kibernet. (*Russian*) **2** (1982), 11–17.

[4] M. BARDI, M. FALCONE, *An approximation scheme for the minimum time function*, SIAM J. Control Optim. **28** (1990), 950–965.

[5] J.M. BORWEIN, A.S. LEWIS, *Duality relationship for entropy-like minimization problems*, SIAM J. Control Optim. **29** (1991), 325–338.

[6] J.M. BORWEIN, A.S. LEWIS, *Partially finite convex programming, Part I: quasi relative interiors and duality theory*, Mathematical Programming **57** (1992), 15–48.

[7] B.M. BUDAK, E.M. BERKOVICH, E.N. SOLOV'EVA, *Difference approximations in optimal control problems*, SIAM J. Control **7** (1969), 18–31. (*originally published in Russian in* Vestnik Mosk. Univ. Seriya I: Matemat. Mekhanika **2** (1968), 41–55).

[8] B.M. BUDAK, E.M. BERKOVICH, E.N. SOLOV'EVA, *The convergence of difference approximations in optimal control*, Zh. Vychisl. Mat. i Math. Fiz. (*Russian*) **9** (1969), 533–547.

[9] B.M. BUDAK, F.P. VASIL'EV, Some Computational Aspects of Optimal Control Problems, Moscow University Press, Moscow (*Russian*), 1975.

[10] I. CAPUZZO DOLCETTA, *On a discrete approximation of the Hamilton-Jacobi equation of dynamic programming*, Appl. Math. Optim. **10** (1983), 367–377.

[11] I. CHRYSSOVERGHI, A. BACOPOULOS, *Discrete approximation of relaxed optimal control problems*, J. Optim. Theory Appl. **65** (1990), 395–407.

[12] M.G. CRANDALL, P.L. LIONS, *Two approximations of solutions of Hamilton-Jacobi equations*, Math. of Comp. **43** (1984), 1–19.

[13] J. CULLUM, *Discrete approximations to continuous optimal control problems*, SIAM J. Control **7** (1969), 32–49.

[14] J. CULLUM, *An explicit procedure for discretizing continuous, optimal control problems*, Journ. of Optim. Theory and Appl. **8** (1971), 15–34.

[15] J. CULLUM, *Finite-dimensional approximations of state constrained continuous optimal control problems*, SIAM J. Control **10** (1972), 649–670.

[16] J.W. DANIEL, *On the approximate minimization of functionals*, Math. Computation **23** (1969), 573–581.

[17] J.W. DANIEL, *On the convergence of a numerical method in optimal control*, J. Optim. Theory and Appl. **4** (1969), 330–342.

[18] J.W. DANIEL, The Approximate Minimization of Functionals, Wiley-Interscience, New York, 1983.

[19] A.L. DONTCHEV, *Error estimates for a discrete approximation to constrained control problems*, SIAM J. Numer. Anal. **18** (1981), 500–514.

[20] A.L. DONTCHEV, Perturbations, Approximations and Sensitivity Analysis of Optimal Control Systems, Lecture Notes in Contr. Inf. Sci. **52**, Springer, Berlin, 1983.

[21] A.L. DONTCHEV, *Equivalent perturbations and approximations in optimal control*, International Series Numer. Math. **84**, Birkhäuser, 1988, pp. 43–54.

[22] A.L. DONTCHEV, *Duality methods in constrained best interpolation*, Mathematica Balkanica **1** (1987), 96–105.

[23] A.L. DONTCHEV, *Best interpolation in a strip*, J. Approx. Theory, **73** (1993), 334–342.

[24] A.L. DONTCHEV, E.M. FARHI, *Error estimates for discretized differential inclusions*,

Computing **41** (1989), 349–358.
[25] A.L. DONTCHEV, W.W. HAGER, *Lipschitz stability in nonlinear control and optimization*, SIAM J. Control Optim. **31** (1993), 569–603.
[26] A.L. DONTCHEV, W.W. HAGER, *An inverse mapping theorem for set-valued maps*, Proc. Amer. Math. Soc. **121** (1994), 481–489.
[27] A.L. DONTCHEV, W.W. HAGER, *Euler approximation to the feasible set*, Num. Funct. Anal. Optim. **15** (1994), 245–262.
[28] A.L. DONTCHEV, BL. KALCHEV, *Duality and well-posedness in convex interpolation*, Num. Funct. Anal. Optim. **10** (1989), 673–690.
[29] A.L. DONTCHEV, F. LEMPIO, *Difference methods for differential inclusions—a survey*, SIAM Review **34** (1992), 263–294.
[30] A.L. DONTCHEV, T. ZOLEZZI, Well-posed Optimization Problems, Lecture Notes in Math. **1543** Springer, Berlin, 1993.
[31] J.C. DUNN, *Diagonally modified conditional gradient method for input constrained optimal control problems*, SIAM J. Control Optim. **24** (1986), 1177–1191.
[32] Y.M. ERMOL'EV, V.P. GULENKO, T.I. TSARENKO, Finite Element Methods in Optimal Control, Naukova Dumka, Kiev (*Russian*), 1978.
[33] M. FALCONE, *A numerical approach to the infinite horizon problem of deterministic control theory*, Appl. Math. Optim. **15** (1987), 1–13. Corrigenda: Appl. Math. Optim. **23** (1991), 213–214.
[34] R.P. FEDORENKO, Approximate Solution of Optimal Control Problems, Nauka, Moscow (*Russian*), 1971.
[35] R. GONZALES, E. ROFMAN, *On deterministic control problems: an approximation procedure for the optimal cost. Part I: the stationary problem. Part II: the nonstationary case.*, SIAM J. Control Optim. **23** (1985), 242–265, 267–285.
[36] W.W. HAGER, *The Ritz-Trefftz method for state and control constrained optimal control problems*, SIAM J. Numer. Anal. **12** (1975), 854–867.
[37] W.W. HAGER, *Rate of convergence for discrete approximations to unconstrained control problems*, SIAM J. Numer. Anal. **13** (1976), 449–471.
[38] W.W. HAGER, *Convex control and dual approximations. Part 1. Part 2*, Control & Cybern. **8** (1979), 5–12, 321–338.
[39] W.W. HAGER, *Multiplier method for nonlinear optimal control*, SIAM J. Numer. Anal. **27** (1990), 1061–1080.
[40] W.W. HAGER, G.D. IANCULESCU, *Dual approximations in optimal control*, SIAM J. Control Optim. **22** (1984), 423–465.
[41] U. HORNUNG, *Interpolations by smooth functions under restrictions in the derivatives*, J. Approx. Theory **28** (1980), 227–237.
[42] C.T. KELLEY, E.W. SACHS, *Mesh independence of the gradient projection method for optimal control problems*, SIAM J. Control Optim. **30** (1992), 477–493.
[43] A.A. LEVIKOV, *Error estimate for solution of the linear optimal control problem*, Avtomat. i Telemekh. **3** (*Russian*) (1982), 71–78.
[44] K.C.P. MACHIELSEN, *Numerical solution of optimal control problems with state constraints by sequential quadratic programming in function space.* (dissertation) Technische Hogeschool Eindhoven, Eindhoven, 1987.
[45] K. MALANOWSKI, *On convergence of finite-difference approximations to control and state constrained optimal control problems*, Archiwum Aut. i Telem. **24** (1979), 319–337.
[46] K. MALANOWSKI, *On convergence of finite-difference approximations to optimal control problems for systems with control appearing linearly*, Archiwum Aut. i Telem. **24** (1979), 319–337.
[47] K. MALANOWSKI, *Convergence of approximations vs. regularity of solutions for convex control-constrained optimal control problems*, Appl. Math. Optim. **8** (1981), 69–95.
[48] C.A. MICHELLI, P.W. SMITH, J. SWETITS, J.D. WARD, *Constrained L_p approximation*, Constr. Approx. **1** (1985), 93–102.
[49] B. MORDUKHOVICH, *On difference approximations of optimal control systems*,

Appl. Math. Mech. **42** (1978), 452–461.

[50] B. Mordukhovich, *Discrete approximations and refined Euler-Lagrange conditions for nonconvex differential inclusions*, IMA preprint Series **1115**, March 1993. (also in) SIAM J. Control Optim. **33** (1995), 882–915.

[51] E. Polak, Computational Methods in Optimization: A Unified Approach, Academic Press, New York, 1971.

[52] E. Polak, *An historical survey of computations methods in optimal control*, SIAM Review **15** (1973), 548–553.

[53] E. Polak, *On the use of consistent approximations in the solution of semi-infinite optimization and optimal control problems*, Dept. of Electr. Eng. and Comp. Sc., University of California, Berkeley (preprint).

[54] E. Polak, Limin He, *Rate-preserving strategies for semi-infinite programming and optimal control*, SIAM J. Control Optim. **30** (1992), 543–572.

[55] E. Polak, T.H. Yang, D.Q. Mayne, *A method of centers based on barrier functions for solving optimal control problems with continuous state and control constraints*, SIAM J. Control Optim. **31** (1993), 159–179.

[56] G.W. Reddien, *Collocation at gauss points as a discretization in optimal control*, SIAM J. Control Optim. **17** (1979), 298–316.

[57] Bl. Sendov, V.A. Popov, The Averaged Moduli of Smoothness, J. Wiley & Sons, New York 1988.

[58] K.L. Teo, C.J. Goh, K.H. Wang, A Unified Computational Approach to Optimal Control Problems, Longman Sci. Tech., Harlow, 1991.

[59] F.P. Vasil'ev, Methods for Solving Extremum Problems, Nauka, Moscow (*Russian*) 1981.

[60] V.V. Vasin, *Discrete approximation and stability in extremal problems*, Zh. Vychisl. Mat. i Mat. Fiz. **22** (1982), 824–839.

[61] V.M. Veliov, *Second order discrete approximations to strongly convex differential inclusions*, Systems & Control Letters, **13** (1989), 263–269.

[62] V.M. Veliov, *Second order discrete approximations to linear differential inclusions*, SIAM J. Numer. Anal. **29** (1992), 439–451.

[63] V.M. Veliov, *Best approximations of control/uncertain differential systems by means of discrete-time systems*, IIASA WP-91-45, November 1991.

[64] J. Warga, *Relaxed variational problems*, J. Math. Anal. Appl. **4** (1962), 111–145.

[65] R.A. Wijsman, *Convergence of sequences of convex sets, cones and functions*, Bull. Amer. Math. Soc. **70** (1964), 186–188.

[66] P.R. Wolenski, *The exponential formula for the reachable set of a Lipschitz differential inclusion*, SIAM J. Control Optim. **28** (1990), 1148–1161.

THE MAXIMUM PRINCIPLE IN OPTIMAL CONTROL OF SYSTEMS GOVERNED BY SEMILINEAR EQUATIONS

BORIS GINSBURG* AND ALEXANDER IOFFE†

Abstract. We prove a maximum principle for an abstract semilinear nonsmooth optimal control problem under a fairly weak set of assumptions that typically do not imply the existence of solution of the equation. The result is further applied to obtain maximum principles in a problem with a distributed time delay and in a problem for a system of semilinear elliptic equations with domain and boundary controls.

1. Introduction. The class of controllable systems covered by the main theorem (Theorem 3.3) of this article can be loosely described by the equation

$$Lx = \phi(x,u)\ , \tag{1.1}$$

where x is from a Banach space X , $L : X \to Y$ is a generally unbounded (e.g. differential) operator , u belongs to the set of admissible controls $\mathcal{U}$ and $\phi : X \times \mathcal{U} \to Y$ is a control mapping.

Needless to say that overwhelming majority of optimal control (OC) problems for which the basic necessary condition for a minimum, the maximum principle, has been established, fall into this category.

Our main theorem is an "abstract" maximum principle, applicable to all OC problems for (1.1)-like equations, provided, of course, that L and ϕ satisfy certain (suprisingly mild, as we shall see) conditions. The first basic requirement is that L must have a compact inverse G . The second is that $G \circ \phi$ must possess a certain property in the solution called in the paper "relaxability at a point", which is rather typical for controllable systems (in particular, it is automatic if the set $\phi(x,\mathcal{U})$ is convex). If these properties hold, the theorem immediately allows to write the maximum principle (modulo some technical work involving description of G^* and/or L^* and calculation all necessary subdifferentials).

The theorem (and its proof) leads us to a rather surprising observation that the properties of the nonlinear equation (1.1) and even its linearization play no role in the maximum principle. Namely, the question of whether or not (1.1) has a solution, given a control u other than the optimal control $\bar{u}$ turns out to be irrelevant to the maximum principle, its validity, and specific form. The same can be said about the question of existence and the form of solution of the "linearized" equation. (The last question, however, becomes important when we want to prove that the problem is normal). To emphasize the really unorthodox nature of these observations we refer

* Both authors at: Department of Mathematics, The Technion, Haifa 32000, Israel.
† The second author's research was supported by the US-Israel Binational Science Foundation grant 90-00455.

to many known proofs of maximum principles ([9], [24], [22], [6], [11], [10], [5]).[1]

We believe that the theorem may help to save much effort in specific situations. Two examples are given in §4 and §5. In §4 we consider optimal control of a system of ODE with delays of fairly general structure, state constraints, etc. The type of functional relations defining the delay was suggested by the very recent study by Clarke and Wolenski [8]. Actually, our relation is somewhat less general than that in [8] but in other respects the problem we consider is more general (we include, in particular, end point constraints without adding any convexity assumptions, and state constraints as well). What we wish to emphasize here, in the introduction, is the extreme simplicity of obtaining the maximum principle for the fairly complex problem considered in §4 , from the main theorem.

The OC problems considered in §5 includes a system of semilinear elliptic equations with both domain and boundary controls. The maximum principle obtained in §5 contains in particular a recent result of Casas [5] (and some of his earlier results) as a very specific case. We assume in §5 that f is an affine function of u in order to trivialize the relaxability requirement. This does not allow us to claim that Theorem 5.1 contains those of Bonnans and Casas [2] in which this combination of the domain and boundary controls was considered for the first time. But in all other respects Theorem 5.1 is noticeably stronger and more general—in particular it uses no crude stability (or calmness) requirements.

Returning to the main theorem and the basic problem to which the theorem is applied, we observe that the philosophy of the approach we use in this paper goes back to the "principle of Lagrange" for problems with mixed smooth-convex structure proved 20 years ago in the monograph of Ioffe-Tihomirov [17]. The principle was modified in [15] to cover nonsmooth OC problems using the machinery of generalized gradients of Clarke and the theory of fans. A more subtle modification of the proof introduced in [15] opens gates to the mentioned "expulsion" of even linearized equation (1.1) from the proof of maximal principle here in this paper.

Finally, we mention one more step forward made here in comparison with [15]. As in that work we do not impose any differentiability assumptions on f. But we use here the different nonsmooth technique. The apparatus of approximate subdifferentials in Banach spaces developed in [16], allows to eventually get more precise transversality conditions in-

[1] It has to be noted, however, that there are proofs of the Pontryagin's maximum principle for ordinary differential equation which are free from references to any existence theorems (e.g. [21]); the same seems to apply to [18] as well), but under the conditions which do imply existence and uniqueness of solutions of the corresponding nonlinear equations. It is worth mentioning in this connection a personal communication of H. Sussmann who (using an earlier results of Lojasiewicz) succeded to weaken the Lipschitz condition (responsible for uniqueness in ODE) in a proof substantially employing the existence theorem for ODE!

volving Mordukhovich's generalized derivatives instead of the generalized gradients of Clarke.

2. Analytical background. In what follows, X and Y are separable Banach spaces with duals X^* and Y^* and canonical pairings $\langle x^*, x\rangle$, $\langle y^*, y\rangle$ etc.

2.1. Subdifferentials. Unless otherwise specified, we always assume that functions and maps are Lipschitz continuous around points of interest. Definitions are given only to cover this particular case. For more details and more general settings see [15], [16] and [4]. For given function f, we define

the *Clarke directional derivative*

$$d^o f(x;h) = \limsup_{\substack{u \longrightarrow x \\ t \longrightarrow +0}} t^{-1}(f(u+th) - f(u)) \ ;$$

the *generalized gradient of Clarke*

$$\partial_c f(x) = \{x^* : \langle x^*, h\rangle \leq d^o f(x;h), \forall h\} \ ;$$

the *lower Dini directional derivative*

$$d^- f(x;h) = \liminf_{t \longrightarrow +0} t^{-1}(f(x+th) - f(x)) \ ;$$

and the *Dini subdifferential*

$$\partial^- f(x) = \{x^* : \langle x^*, h\rangle \leq d^- f(x;h), \forall h\} \ .$$

The set

$$\partial_a f(x) = \limsup_{u \longrightarrow x} \partial^- f(u)$$

is called the *approximate subdifferential or a-subdifferential* of f at x. Here the upper limit is defined with respect to the weak* topology in X^*. For Lipschitz functions $\partial_c f(x) = \mathrm{cl}^* \mathrm{conv}\ \partial_a f(x)$.

We denote by $\rho(S, x)$ the distance from x to S. Obviously this is a Lipschitz function with constant one. Define the *G-normal cone* to S at $x \in S$

$$N_G(S,x) := cl^* \bigcup_{\lambda > 0} \lambda \partial_a \rho(S,x) \ .$$

If dim $X < \infty$ (more generally, if X has a Frechet differentiable renorm), then $N_G(S,x)$ coincides with the weak-star closure of the limiting proximal normal or Mordukhovich normal cone (see [20], [15], [21] and [18].) *This cone does not depend on the specific norm used in the definition of* ρ. The set

$$\partial_G f(x) = \{x^* : (x^*, -1) \in N_G(\mathrm{epi}\ f, (x, f(x)))\}$$

is called the *approximate (geometric) subdifferential* or *G-subdifferential*. It turns out that $\partial_a f(x) = \partial_G f(x)$ if f is Lipschitz continuous at x.

For a function $f(x, y, z, ..)$ of several variables we write $\partial f(x, \cdot, z, ..)(\bar{y})$ to denote the subdifferential of the function $y \rightarrow f(x, y, z, ..)$ with respect to y at $\bar{y}$.

More generally, consider a map $F : X \rightarrow Y$, which is continuous around x. The set-valued map

$$y^* \rightarrow D_G^* F(x)(y^*) = \{x^* : (x^*, -y^*) \in N_G(\text{graph } F, (x, F(x)))$$

is called the *G-coderivative* of F. The calculation of G-coderivatives may be difficult unless F is Lipschitz continuous. In the last case we shall use a special norm in $X \times Y$ for calculation of coderivatives.

PROPOSITION 2.1. *Let F be Lipschitz continuous around x with constant K. There is an equivalent norm in $X \times Y$, namely $||(x, y)|| := K||x|| + ||y||$ such that in this norm*

$$\rho(\text{graph } F, (u, v)) = ||F(u) - v||$$

for all (u, v) sufficiently close to $(x, F(x))$.

We shall use the notation $\rho_F(u, v) := ||F(u) - v||$.

Proof. We have

$$\rho(\text{graph } F, (u, v)) = \inf_w (K||u - w|| + ||F(w) - v||) \leq ||F(u) - v||$$

On the other hand, for any w

$$\begin{aligned} ||(u, v) - (w, F(w))|| &= K||u - w|| + ||F(w) - v|| \geq \\ K||u - w|| + ||F(u) - v|| &- ||F(u) - F(w)|| \geq ||F(u) - v|| \end{aligned}$$

□

Consider now a special class of Lipschitz continuous maps.

DEFINITION 2.2. We say that G has a *compact strict prederivative* at x if there is a norm compact set $R \subset Y$ such that

$$G(u + h) - G(u) \in ||h||R + r(x, h)||h||B,$$

where $r(x, h) \rightarrow 0$ as $u \rightarrow x$, $h \rightarrow 0$.

The fundamental fact about G-coderivatives playing a crucial role in derivation of necessary optimality conditions is the following *scalarization formula*:

THEOREM 2.3. *[16] Assume that $F = A + G$, where A is a linear bounded operator and G has a compact strict prederivative at x. Then*

$$D_G^* F(x)(y^*) = \partial_a (y^* \circ F)(x) \ , \ \forall y^* \in Y^* \ .$$

2.2. Regularity . Let F be a map continuous in a neighborhood of x and $S \in X$ be a closed set containing x. Set $M_x := \{u \in S : F(u) = F(x)\}$.

DEFINITION 2.4. We say that F is *regular* with respect to S at x if there is a constant K, such that

$$\rho(M_x, u) \leq K \|F(u) - F(x)\|$$

for all $u \in S$ of a neighborhood of x.

PROPOSITION 2.5. *Let F be a Lipschitz map with constant L. If F is not regular at x, then there are sequences x_n, x_n^*, y_n^*, , u_n^* such that*

$$\begin{aligned} &x_n \in S \ , \ x_n \to x \ , \ \|y_n^*\| = 1 \ , \\ &x_n^* \in D_G^* F(x_n)(-y_n^*) \ , \ u_n^* \in (L+1)\partial_a \rho(S, x_n) \ , \\ &x_n^* + u_n^* \to 0 \end{aligned}$$

Proof. By definition, there is a sequence of $u_n \to x$ such that

$$\rho(M_x, u_n) \geq 2\, n \, \|F(u_n) - F(x)\| \ . \tag{2.1}$$

Set $\alpha_n := (1/2)\rho(M_x, u_n)$ and let $\phi(u) := \|F(u) - F(x)\|$. Then (2.1) implies that $\phi(u_n) \geq \inf_S \phi + (\alpha_n/n)$. By the Ekeland variational principle for any n there is an $x_n \in S$ such that $\|x_n - u_n\| < \alpha_n$ and the function $\phi(u) + (\|u - x_n\|/n)$ attains a local minimum on S at x_n. This function is Lipschitz with constant $L + 1$ where L is the Lipschitz constant of F. Therefore the function

$$\phi(u) + \|u - x_n\|/n + (L+1)\rho(S, u)$$

has an unconditional local minimum at x_n.

Finally, as $\phi(u) \leq \|F(u) - y\| + \|y - F(x)\|$ for any y, we conclude that $(x_n, F(x_n))$ is a local minimum of the function

$$f(u, y) := \|F(u) - y\| + \|y - F(x)\| + \|u - x_n\|/n + (L+1)\rho(S, u)$$

This means that $(0, 0) \in \partial_a f(x_n, F(x_n))$. The formula for subdifferential of the sum ([16], Thm 5.6) implies the existence of

$$\begin{aligned} &(x_n^*, y_n^*) \in \partial_a \rho_F(x_n, F(x_n)) \ , \\ &v_n^* \in \partial_a \|.\|(F(x_n) - F(x)) \ , \\ &w_n^* \in (1/n)\partial_a \|u - x_n\|(x_n) \ , \\ &u_n^* \in (L+1)\partial_a \rho(S, u)(x_n) \ , \end{aligned}$$

such that $\quad x_n^* + w_n^* + u_n^* = 0 \;$ and $\; y_n^* + v_n^* = 0$.

It is clear, that $x_n \to x$. Since $\|x_n - u_n\| \leq (1/2)\rho(M_x, u_n)$ we have $x_n \notin M_x$ and $F(x_n) \neq F(x)$. Therefore $\|y_n^*\| = \|v_n^*\| = 1$. Furthermore $\|w_n^*\| \leq (1/n)$, so $x_n^* + u_n^* \to 0$. Finally, the inclusion

$$x_n^* \in D_G^* F(x_n)(-y_n^*) \tag{2.2}$$

follows from the first inclusion by definition. □

Remark 2.6. Assume that $F = A + G$, where A is a linear bounded operator and G has a compact strict prederivative at x. Then we can use the scalarization formula of Theorem 2.3. Replacing $-y^*$ by y^* write instead of (2.2) the condition

$$x_n^* \in \partial_a(y_n^* \circ F)(x_n) \ .$$

The sequences $\{x_n\}, \{x_n^*\}, \{y_n^*\}, \{u_n^*\}$ obtained in the proposition are bounded, so there is weak* limit point y^*, such that

$$0 \in D_G^* F(x)(-y^*) + (K+1)\partial_a \rho(S,u)(x)$$

We cannot, however, exclude the possibility that $y^* = 0$. In this case $x^* = u^* = 0$. To derive the necessary conditions we should have the natural conditions which would guarantee that $y^* \neq 0$.

DEFINITION 2.7. We say F has the *finite codimension property* with respect to S at x if there is a weak* closed subspace $V^* \subset Y^*$ of finite codimension and constants $\epsilon > 0, c > 0$ such that if

- $u \in S$ and $\|u - x\| < \epsilon$;
- $(x^*, y^*) \in \partial_a \rho_F(u, F(u))$;
- $u^* \in N_G(S, u)$;
- $\|y^*\| = 1$; $\rho(V^*, y^*) \leq \epsilon$,

then $\|x^* + u^*\| \geq c$. This is a modification of a similar condition introduced in [14] in terms of Clarke generalized gradients and fans. If F is a linear bounded map it means exactly that the range of F has finite codimension. Let us agree to call the pair (V^*, c) the *cosupport* of F at x.

THEOREM 2.8. *Let F has the finite codimension property with respect to S at x. If F is not regular at x, then there is $y^* \neq 0$ such that*

$$0 \in D_G^* F(x)(y^*) + N_G(S, x) \ . \tag{2.3}$$

Proof. Let V^*, ϵ, c be given by the finite codimension property of F. Denote by W^* a complementary to V^* finite dimensional subspace of Y^*. Every y^* has the unique representation $y^* = v^* + w^*$, where $v^* \in V^*$ and $w^* \in W^*$. There is a constant $0 < m \leq 1$ such that

$$m(\|v^*\| + \|w^*\|) \leq \|y^*\| \leq \|v^*\| + \|w^*\| \tag{2.4}$$

for all y^*. Let $\{x_n\}, \{x_n^*\}, \{y_n^*\}, \{u_n^*\}$ be sequences from Prop.2.5 and x^*, y^*, u^* be weak* limit points. We have $y_n^* = v_n^* + w_n^*$. Since $\dim W^* < \infty$ we may assume that $w_n^* \to w^* \in W^*$. If $w^* = 0$ then $\rho(V^*, y_n^*) \to 0$. It implies that for large n

$$1 = ||y_n^*|| \leq c||x_n^* + u_n^*|| \to 0 .$$

Therefore $w^* \neq 0$, so is not y^*. □

2.3. Fredholm maps

DEFINITION 2.9. F is called the *Fredholm map* at x if it can be represented as a sum $F = A + G$ where A is a bounded linear operator whose image is a closed subspace of finite codimension, and G is a map with compact strict prederivative at x.

Such Fredholm maps naturally arise in connection with optimal control problem involving differential equation with nonsmooth right-hand part.

DEFINITION 2.10. We say that closed set S is a *cylinder of finite codimension* if there is a closed subspace $X_1 \subset X$ of finite codimension such that $x + X_1 \subset S$ for any $x \in S$.

THEOREM 2.11. *Let F be a Fredholm map at x. Assume that S is a closed cylinder of finite codimension. Then F has the finite codimension property with respect to S at x.*

To prove the theorem we need three lemmas.

LEMMA 2.12. *Let $V \subset Y$ be a subspace complementary to* Im A *and let $V^\perp$ be the annihilator of V. Then there is a $\mu > 0$ such that for any $y^* \in V^\perp$ there exists $h \in X$ with $||h|| = 1$ for which*

$$\mu||y^*|| \leq \langle y^*, Ah \rangle .$$

Proof. Take a $y \in Y$ such that $||y|| = 1$ and $\langle y^*, y \rangle \geq ||y^*||/2$. Then $y = w + v$, where $w \in \mathrm{Im} A$ and $v \in V$. There is k, such that $||w|| \leq k$. Let $w = Ah$. By the Banach open mapping theorem there is an r, not depending on w, such that $||h|| \leq r\ ||w||$. We arrive at the desired conclusion, taking $\mu = 1/(2kr)$.

Denote by $s_R(v^*)$ the support function of R : $s_R(v^*) := \sup_{y \in R} \langle v^*, y \rangle$. □

LEMMA 2.13. *Let $R \subset Y$ be a compact strict prederivative of G. Then for any $\sigma > 0$ there is a $\delta > 0$ such that the inequality*

$$\langle u^*, h \rangle + \langle v^*, Ah \rangle \leq (s_R(-v^*) + \sigma)||h||$$

holds for any $h \in X$, whenever $(u^*, v^*) \in \partial_a \rho_F(u, F(u))$ and $||u - x|| < \delta$.

Proof. As R is norm compact, the function $v^* \to s_R(v^*)$ is weak* continuous. On the other hand, $\partial_a \rho_F(u, F(u))$ is a weak* upper limit of $\partial^- \rho_F(w, F(w))$ as $w \to u$ (see [4], Corollary 2). Therefore it is sufficient to prove the inequality only for $(u^*, v^*) \in \partial^- \rho_F(u, v)$. For such (u^*, v^*) we have by definition

$$\langle u^*, h\rangle + \langle v^*, v\rangle \le \liminf_{t\to 0+} \|t^{-1}(F(u+th) - F(u)) \ - v\|, \ \forall h, v \ .$$

By the assumption we can choose $\delta > 0$ such that

$$G(u+h) - G(u) \in \|h\|R + \sigma\|h\|B$$

if $\|u - x\| < \delta$ and $\|h\| < \delta$. Therefore, if t is sufficiently small, we have

$$t^{-1}(F(u+th) - F(u)) = Ah + \|h\|e(t) + g(t)$$

where $e(t) \in R$ and $\|g(t)\| \le \sigma\|h\|$. Let e be any limit point of $e(t)$ as $t \to \infty$. Then we get

$$\langle u^*, h\rangle + \langle v^*, v\rangle \le \|(Ah + \|h\|e - v)\| + \sigma\|h\|, \ \forall h, v \ .$$

Taking $v = Ah + \|h\|e$ we conclude with

$$\langle u^*, h\rangle + \langle v^*, Ah\rangle \le \|h\|\langle -v^*, e\rangle + \sigma\|h\| \le (s_R(-v^*) + \sigma)\|h\| \ .$$

□

LEMMA 2.14. *Let $Q = \{y_1, .., y_n\}$ be a finite collection of elements of Y and $Q^\perp \subset Y^*$ be the annihilator of Q. Then there are constants $0 < m < M$ such that for all $y^* \in Y^*$*

$$m \cdot \max_i |\langle y^*, y_i\rangle| \le \rho(Q^\perp, y^*) \le M \cdot \max_i |\langle y^*, y_i\rangle| \ .$$

Proof. Let π be the canonical projection $Y^* \to Y^* \backslash Q^\perp$. Then both $\rho(Q^\perp, y^*)$ and $\max_i |\langle y^*, y_i\rangle|$ depend only on πy^* and are norms on the factor space. Since it is finite dimensional both norms are equivalent.

Proof of the theorem. We have to find $V^*, \epsilon > 0$ and $c > 0$ such that $c\|x^* + u^*\| \ge 1$ if $\|u - x\| < \epsilon, u \in S$, $(x^*, y^*) \in \partial_G \rho_F(u, F(u))$, $\|y^*\| = 1$, $u^* \in N_G(S, u)$ and $\rho(V^*, y^*) \le \epsilon\|y^*\|$. Consider the operator A_1 which is the restriction of A to X_1. Then Im A_1 has finite codimension. Denote by V_1 the subspace complementary to Im A_1. By lemma 2.12 there is a $\mu > 0$ such that for all $y^* \in V_1^\perp$ one can find $h_1 \in X_1$ with $\|h_1\| = 1$ and

$$\mu\|y^*\| \le \langle y^*, Ah_1\rangle \tag{2.5}$$

Take a $\sigma < \mu/4$ and a corresponding $\delta > 0$ from lemma 2.13 and choose a finite σ-net R_1 for R. Set $V^* := V_1^\perp \bigcap R_1^\perp$. Clearly V^* is weak* closed subspace of finite codimension. By Lemma 2.14 there is a constant M such that

$$s_{R_1}(y^*) \le M\rho(R_1^\perp, y^*) \le M\rho(V^*, y^*)$$

Let furthermore $W^* \subset Y^*$ be a complementary subspace to V^*. Then any y^* has a unique representation as a sum $y^* = v^* + w^*$. It follows from the inequality (2.4) that there is $0 < m$ such that for any $z^* \in V^*$

$$m(\|v^* - z^*\| + \|w^*\|) \le \|y^* - z^*\|$$

and, consequently, $m\|w^*\| \le \rho(V^*, y^*)$.

Take now $\epsilon \in (0, \delta)$ such that $\epsilon M_1 < \sigma$ and $\epsilon < m/8$. If now $u \in S$, $\|u - x\| < \epsilon < \delta_1$, $(x^*, y^*) \in \partial_G \rho_F(u, F(u))$, $\|y^*\| = 1$, $u^* \in N_G(S, u)$ and $\rho(V^*, y^*) \le \epsilon\|y^*\|$ then by lemma 2.14 for any $h \in X$ we have

$$\begin{aligned} \langle y^*, Ah\rangle &\le -\langle x^*, h\rangle + [s_R(-y^*) + \sigma]\|h\| \\ &\le -\langle x^*, h\rangle + [s_{R_1}(-y^*) + 2\sigma]\|h\| \\ &\le -\langle x^*, h\rangle + [M\rho(V^*, y^*) + 2\sigma]\|h\| \\ &\le -\langle x^*, h\rangle + 3\sigma\|h\| \,. \end{aligned}$$

Note further that if $y^* = v^* + w^*$ then

$$\|w^*\| \le \frac{1}{m}\rho(V^*, y^*) \le \frac{\epsilon}{m}\|y^*\| \le \frac{1}{8}\|y^*\| \,.$$

It implies that $\|v^*\| \ge (7/8)\|y^*\|$. Take now $h = h_1$ from (2.5). We get

$$\mu\|v^*\| \le \langle v^*, Ah_1\rangle = \langle y^*, Ah_1\rangle \le -\langle x^*, h_1\rangle + 3\sigma\|h_1\| \,.$$

It only remains to observe that any $u^* \in N_G(S, x)$ belongs to $X_1^\perp$, so $\langle u^*, h_1\rangle = 0$ and

$$(7/8)\mu\|y^*\| \le \mu\|v^*\| \le -\langle x^* + u^*, h_1\rangle + 3\sigma\|h_1\| \le \|x^* + u^*\| + 3\sigma\|h_1\| \,.$$

Finally because $\|y^*\| = \|h_1\| = 1$ and $\sigma < \mu/4$

$$\frac{\mu}{8}\|y^*\| \le \|x^* + u^*\|.$$

□

3. An abstract maximum principle

3.1. The Lagrange multiplier rule. Let X and Y be Banach spaces. Suppose f_i are functions on X and F is a mapping from X to Y. We consider the problem

(P) minimize $f_0(x)$
s.t. $F(x) = 0$; $f_i(x) \le 0$, $i = 1, .., m$; $x \in S$

under the assumptions that

(A_1) F is Fredholm at $\bar{x}$;
(A_2) f_i, $i = 0,..,m$, are Lipschitz continuous around a given point $\bar{x}$;
(A_3) S is a closed cylinder of finite codimension (that is to say, there is a closed subspace $X_1 \subset X$ of finite codimension such that $x + X_1 \subset S$ for any $x \in S$).

Before stating the results, we note that, as follows from Theorem 2.11, F has the finite codimension property at $\bar{x}$. This means that we can associate with F a weak* closed subspace $V^* \subset Y^*$ of finite codimension and a number $c > 0$ such that the corresponding relations (see Definition 2.2) are satisfied. We denote by W^* a complementary to V^* space and by π the projection of Y^* onto W^*.

Define *the Lagrangian* of the problem

$$\mathcal{L}(\lambda_0, .., \lambda_m, y^*, x) := \sum_{i=0}^{m} \lambda_i f_i(x) + (y^* \circ F)(x) \ .$$

It will be mainly considered as a function of x, especially when calculating subdifferentials etc.

THEOREM 3.1 (LAGRANGE MULTIPLIER RULE). *If $\bar{x}$ is a local solution of* (**P**), *then there exist multipliers $(\lambda_0 \geq 0, .., \lambda_m \geq 0, y^*)$ not all equal to zero, such that $\lambda_i f_i(\bar{x}) = 0$ for $i = 1..m$ and for some $r > 0$*

$$0 \in \partial_a \mathcal{L}(\lambda_0, .., \lambda_m, y^*, \bar{x}) + r \partial_a \rho(S, \bar{x}) \tag{3.1}$$

The collection of multipliers $(\lambda_0, .., \lambda_m, y^*)$ satisfying the additional normalization condition

$$\sum \lambda_i + \|\pi y^*\| = 1 \tag{3.2}$$

is nonempty and weak compact.*

Proof. First we notice (as in [13]) that $\bar{x}$ is a local minimum in the problem

(P') minimize $f(x)$ under conditions $F(x) = 0 \ ; \ x \in S$.

where $f(x) := \max\{f_0(x) - f_0(\bar{x}), f_1(x), .., f_m(x)\}$.

If F is *regular* with respect to S then standard reduction argument (see [13]) implies that $\bar{x}$ is an unconditional local minimum of the function

$$f(x) + M\|F(x)\| + r\rho(S, x)$$

where M is the product of the regularity constant of F and a common Lipschitz constant for f_i and r is the Lipschitz constant of

$$\phi(x) := f(x) + M\|F(x)\| \, .$$

By the formula for subdifferential of a sum, we have

$$0 \in \partial_a \phi(\bar{x}) + r\partial_a \rho(S, \bar{x}) \; . \tag{3.3}$$

$\phi(x)$ can be interpreted as a composition of the mapping $x \rightarrow (f(x), F(x))$ (obviously Fredholm at $\bar{x}$) and the function $(\alpha, y) \rightarrow (\alpha + M\|y\|)$. Applying the chain rule (see Corollary 7.8.1 of [16]), we conclude that there is y^* with $\|y^*\| \leq M$ such that

$$\partial_a \phi(\bar{x}) \subset \partial_a (f + y^* \circ F)(\bar{x}) \; .$$

Finally, observe that

$$(f + y^* \circ F)(x) = \max\{f_0(x) - f_0(\bar{x}) + y^* \circ F(x), .., f_m(x) + y^* \circ F(x)\} \, .$$

The formula for the subdifferential of a maximum (Corollary 5.4.6 of [16]) ensures that there exist $\lambda_0 \geq 0, .., \lambda_m \geq 0$, with $\sum \lambda_i = 1$ and $\lambda_i = 0$ if $f_i(\bar{x}) < 0$ for $i = 1, .., m$ such that for any $x^* \in \partial_a (f + y^* \circ F)(\bar{x})$ we have

$$x^* \in \partial_a (\lambda_0 (f_0 + y^* \circ F) + \lambda_m (f_m + y^* \circ F))(\bar{x}) = \partial_a \mathcal{L}(\lambda_0, .., \lambda_m, y^*)(\bar{x})$$

which, together with (3.3) gives (3.1). We also observe that in this case

$$\sum \lambda_i + \|\pi y^*\| \geq \sum \lambda_i = 1 \geq (1/M)\|y^*\| \; . \tag{3.4}$$

On the other hand, if F *is not regular* with respect to S at $\bar{x}$, then by Theorem 2.8 there are $y^* \neq 0$, $r > 0$ and $c > 0$ such that

$$0 \in \partial_a (y^* \circ F)(\bar{x}) + r\partial_a \rho(S, \bar{x})$$

and $\|\pi y^*\| \geq c\|y^*\|$. The inclusion gives (3.1) with $\lambda_0 = ... = \lambda_m = 0$ and the inequality implies (3.4) with c instead of M.

As Lagrange multipliers are defined up to a multiplicative constant, this means that the set of multipliers normalized by (3.2) is nonempty and bounded. As the function $(\lambda_0, .., \lambda_m, y^*) \rightarrow \sum \lambda_i + \|\pi y^*\|$ is weak* continuous and π is a projection onto a finite dimensional subspace, it is weak* compact. □

3.2. An abstract optimal control problem. Consider a generalization of **(P)** which involves a "control" variable u along with the "state" variable x.

(C) minimize $f_0(x)$

s.t. $F(x, u) = 0 \;\; ; f_i(x) \leq 0 \, , \;\; i = 1, .., m \;\; ; x \in S \, , \, u \in \mathcal{U} \; .$

Here $\mathcal{U}$ is the set of admissible controls.

We shall say that $(\bar{x}, \bar{u})$ is a *local solution* of **(C)** if for any other admissible pair (x, u) with x sufficiently close to $\bar{x}$ we have $f_0(x) \geq f_0(\bar{x})$.

Let us set $R^k_+ := \{a = (\alpha_1, .., \alpha_k); \alpha_i \geq 0\}$, $\alpha^+ := \max\{0, \alpha\}$ and $a^+ := (\alpha_1^+, .., \alpha_k^+)$. We define the norm $\|a\| := \sum |\alpha_i|$.

DEFINITION 3.2. Given finite collection of controls, we set

$$\Phi(x, a) := F(x, \bar{u}) + \sum \alpha_j (F(x, u_j) - F(x, \bar{u})) .$$

with all $\alpha_j \in R_+$. We say that F is *relaxable* at $(\bar{x}, \bar{u})$ if for any $\delta > 0$ there are a $\sigma > 0$ and a map $v_\delta(x, a) : X \times R^k_+ \to U$ such that

$$\|F(x, v_\delta(x, a)) - F(x', v_\delta(x', a')) - \Phi(x, a) + \Phi(x', a')\| \leq \delta(\|x - x'\| + \|a - a'\|)$$

provided $\|x - \bar{x}\|, \|x' - \bar{x}\|, \|a\|, \|a'\| < \sigma$.

Observe that F trivially satisfies the relaxability property if $F(x, U)$ is a convex set for all x of a neighborhood of $\bar{x}$.

We are now able to formulate the basic assumptions on the components of **(C)**.

As far as f_i and S are concerned, we shall adopt the same assumptions (A_2) and (A_3) as in **(P)**.

About F we assume that it has the form

$$F(x, u) := Ax + (G \circ \phi)(x, u) , \tag{3.5}$$

where A, G and ϕ satisfy the following hypotheses:

(A_4) $A : X \to Y$ is a bounded linear operator onto a closed subspace of finite codimension;

(A_5) G is a compact linear operator $Z \to Y$ (Z being a certain intermediate space);

(A_6) for any $u \in \mathcal{U}$ the mapping $\phi : X \to Z$ defined by $x \to \phi(x, u)$ is Lipschitz continuous near $\bar{x}$ and sends a neighborhood of $\bar{x}$ onto a (relatively) weak compact set in Z;

(A_7) the map $G \circ \phi$ is relaxable at $(\bar{x}, \bar{u})$.

Define the *Lagrangian* of **(C)**

$$\mathcal{L}(\lambda_0, .., \lambda_m, y^*, x, u) := \sum_{i=0}^{m} \lambda_i f_i(x) + (y^* \circ F)(x, u) .$$

THEOREM 3.3 (THE LAGRANGIAN MINIMUM PRINCIPLE). *We posit* (A_2)–(A_7). *If* $(\bar{x}, \bar{u})$ *is a local solution of* **(C)**, *then there are Lagrange multipliers* $(\lambda_0 \geq 0, .., \lambda_m \geq 0, y^*)$ *not all equal to zero, such that* $\lambda_i f_i(\bar{x}) = 0$, $i = 1, .., m$, *and*

$$0 \in \partial_a \mathcal{L}(\lambda_0, .., \lambda_m, y^*, \cdot, \bar{u})(\bar{x}) + N_G(S, \bar{x}) , \tag{3.6}$$

$$\mathcal{L}(\lambda_0, .., \lambda_m, y^*, \bar{x}, \bar{u}) = \min_{u \in U} \mathcal{L}(\lambda_0, .., \lambda_m, y^*, \bar{x}, u) . \tag{3.7}$$

We shall first prove two auxiliary lemmas concerning the finite codimension property.

LEMMA 3.4. *Assume that $H(x)$ is a Fredholm map at $\bar{x} \in S$ having the finite codimension property with respect to S. Let (V^*, c) be a cosupport of H. Then there is a $\delta > 0$ such that any map H' which is Fredholm at $\bar{x}$ and satisfies*

$$\|H(x') - H(x) - H'(x') + H'(x)\| \leq \delta \|x - x'\| \tag{3.8}$$

also has the finite codimension property with respect to S with the cosupport $(V^*, c/2)$. We also have for any y^*

$$\partial_a(y^* \circ H')(x) \subset \partial_a(y^* \circ H)(x) + \delta B\|y^*\| \,. \tag{3.9}$$

Proof. If H' satisfies (3.8) then

$$d^-(y^* \circ H')(x; h) \leq d^-(y^* \circ H)(x; h) + \delta \|h\| \|y^*\|$$

for all x close to $\bar{x}$ and all h. This immediately implies (3.9).

By the assumptions there are $\epsilon > 0$ and $c > 0$ such that $\|x^* + u^*\| \geq c$ whenever $x^* \in \partial_a(y^* \circ H)(x)$, $u \in N_G(S, x)$, x is sufficiently close to $\bar{x}$ and y^* with $\|y^*\| = 1$, $\rho(V^*, y^*) \leq \epsilon$. Note now that if we have $w^* \in \partial_a(y^* \circ H')(x)$ then by (3.9) $\|w^* + u^*\| \geq c - \delta$ and it is sufficient to take $\delta < c/2$. □

LEMMA 3.5. *Let $H_0, .., H_k$ be mappings $X \to Y$ which are Fredholm at $\bar{x}$. If H_0 has the finite codimension property with respect to S at $\bar{x}$, and (V^*, c) is cosupport of H_0 then the map*

$$\mathcal{H}\,(x, a) = H_0(x) + \sum_{j=1}^{k} \alpha_j (H_j(x) - H_0(x))$$

from $X \times R^k$ has the finite codimension property at $(\bar{x}, 0)$ with respect to $S \times R^k_+$ with the cosupport $(V^*, c/2)$. For any y^* we have

$$\begin{aligned} \partial_a(y^* \circ \mathcal{H}\,)(\bar{x}, 0) &\subset \partial_a(y^* \circ H)(\bar{x}) \\ &\times \{\langle y^*, H_1(\bar{x}) - H_0(\bar{x})\rangle, .., \langle y^*, H_k(\bar{x}) - H_0(\bar{x})\rangle\} \end{aligned} \tag{3.10}$$

Proof. Given a y^* with $\|y^*\| = 1$, we have

$$\begin{aligned} &d^-(y^* \circ \mathcal{H}\,)((x, a); (h, b)) \leq \\ &\quad d^-(y^* \circ H_0)(x; h) + \sum \beta_j \langle y^*, H_j(x) - H_0(x)\rangle + 2K\|a\|\|h\| \end{aligned}$$

where K is a common Lipschitz constant for all H_j and $b = (\beta_1, .., \beta_k)$. This means that for any $\delta > 0$ we can choose a neighborhood of $(\bar{x}, 0)$

small enough to guarantee that

$$\partial_a(y^* \circ \mathcal{H})(\bar{x},0) \subset$$
$$(\partial_a(y^* \circ H)(\bar{x}) + \delta B_x) \times \{\langle y^*, H_1(\bar{x}) - H_0(\bar{x})\rangle, .., \langle y^*, H_k(\bar{x}) - H_0(\bar{x})\rangle\}$$

which immediately implies (3.10) and brings us to the situation similar to that in the proof of the preceding lemma.

Proof of the theorem. Let (V^*, c) be a cosupport of $F(.,\bar{u})$ at $\bar{x}$ on S and π a projection in Y^* parallel to V^*.

The theorem will be proved if we show that for any finite collection $u_1, .., u_k$ and any $\delta > 0$ the set $\Lambda(u_1, .., u_k, \delta)$ of multipliers $(\lambda_0 \geq 0, .., \lambda_m \geq 0, y^*)$, satisfying the complementary slackness condition as well as the multiplier rule (3.6) and such that

$$0 \in \partial_a \mathcal{L}(\lambda_0, .., \lambda_m, y^*, \cdot, \bar{u})(\bar{x}) + \delta B \tag{3.11}$$
$$\mathcal{L}(\lambda_0, .., \lambda_m, y^*, \bar{x}, \bar{u}) \leq \min_j \mathcal{L}(\lambda_0, .., \lambda_m, y^*, \bar{x}, u_j) + \delta \tag{3.12}$$
$$\sum \lambda_i + \|\pi y^*\| = 1 \tag{3.13}$$

is nonempty and weak* compact.

Indeed, if this true, then as, obviously,

$$\Lambda(u_1, .., u_k, u_{k+1}, .., u_r, \delta') \subset \Lambda(u_1, .., u_k, \delta)$$

for $\delta' < \delta$, the intersection of all possible $\Lambda(u_1, .., u_k, \delta)$ is nonempty and any collection of multipliers belonging to the intersection has the desired properties.

So fix a finite set $u_1, .., u_k$ and any $\delta > 0$. Set

$$\Phi_\delta(x, a) := F(x, v_\delta(x, a^+)) + \sum \alpha_j^-(F(x, u_j) - F(x, \bar{u}))$$

and consider the problem

$$(\mathbf{C}_\delta) \qquad \begin{array}{l} \text{minimize } f_0(x) \\ \text{s.t. } f_i(x) \leq 0 \ , \ i = 1, .., m \ ; \ \Phi_\delta(x, a) = 0 \ ; \ x \in S \ , \ a \geq 0. \end{array}$$

We first observe that $(\bar{x}, 0)$ is a local solution of $(\mathbf{C}_\delta)$. Indeed, if $a \geq 0$ then $\Phi_\delta(x, a) = F(x, v_\delta(x, a))$, so that (x, a) being admissible in $(\mathbf{C}_\delta)$ implies that $(x, v_\delta(x, a))$ is admissible in $(\mathbf{C})$ and consequently $f_0(x) \geq f_0(\bar{x})$.

Next we notice that $(\mathbf{C}_\delta)$ satisfies all the conditions of Theorem 3.1, namely $(A_1) - (A_3)$ if we replace x by (x, a), F by Φ_δ and S by $S \times R^k_+$. (A_2) and (A_3) are obvious.

To prove (A_1) observe that Φ and Φ_δ are Fredholm maps at $(\bar{x}, 0)$. Applying Lemma 3.5 with $F(x, \bar{u})$ as H_0 and with $F(x, u_j)$ as H_j, we see that $\Phi(x,a) = \mathcal{H}(x,a)$ satisfies the finite codimension property at $(\bar{x},0)$ with respect to $S \times R_+^k$ with the cosupport $(V^*, c/4)$ when (V^*, c) is a cosupport of $F(x,\bar{u})$. Applying further Lemma 3.4 with Φ as H and Φ_δ as H' we find that for sufficiently small $\delta > 0$ Φ_δ satisfies the finite codimension property at $(\bar{x},0)$ with respect to $S \times R_+^k$ with the cosupport $(V^*, c/4)$. In other words, there is a common cosupport $(V^*, c/4)$ at $(\bar{x},0)$ for all Φ_δ if δ is small enough.

Define the *Lagrangian* of $(\mathbf{C}_\delta)$:

$$L_\delta(\lambda_0, .., \lambda_m, y^*, x, a) := \sum_{i=0}^{m} \lambda_i f_i(x) + (y^* \circ \Phi_\delta)(x,a)$$

and set

$$L_0(\lambda_0, .., \lambda_m, y^*, x, a) := \sum_{i=0}^{m} \lambda_i f_i(x) + (y^* \circ \Phi)(x,a) \ .$$

Applying Theorem 3.1 to $(\mathbf{C}_\delta)$, we find that the collection of Lagrange multipliers $(\lambda_0 \geq 0, .., \lambda_m \geq 0, y^*)$ such that $\lambda_i f_i(\bar{x}) = 0$ for $i = 1, .., m$, $\sum \lambda_i + \|\pi y^*\| = 1$ and

$$\begin{aligned}(0,0) \in\ & \partial_a L_\delta(\lambda_0, .., \lambda_m, y^*, \cdot, \cdot)(\bar{x}, 0) \\ & + r(\partial_a \rho(R_+^k, 0) + \partial_a \rho(S, \bar{x}))\end{aligned} \tag{3.14}$$

is nonempty and weak* compact. It remains to show that conditions (3.11)–(3.12) hold.

By using Lemma 3.4 with $L_0(\lambda_0, .., \lambda_m, y^*, ., .)$ as H and $L_\delta(\lambda_0, .., \lambda_m, y^*, \cdot, \cdot)$ as H', we see that

$$\partial_a L_\delta(\lambda_0, .., \lambda_m, y^*, \cdot, \cdot)(\bar{x}, 0) \subset \partial_a L_0(\lambda_0, .., \lambda_m, y^*, \cdot, \bar{u})(\bar{x}) + \delta B$$

On the other hand, applying Lemma 3.5 with $f_0, .., f_m, F(., \bar{u})$ as H_0 and with $f_0, .., f_m, F(., u_j)$ as H_j, we see that $L(\lambda_0, .., \lambda_m, y^*, x, a)$ appears as $\mathcal{H}(x,a)$ and

$$\begin{aligned}&\partial_a L_0((\lambda_0, .., \lambda_m, y^*, \cdot, \cdot, \cdot)(\bar{x}) \subset \\ &\quad \partial_a(\sum \lambda_i f_i + (y^* \circ F)(\cdot, \bar{u})(\bar{x})) \times \\ &\quad\quad \{\langle y^*, F(\bar{x}, u_1) - F(\bar{x}, \bar{u})\rangle, .., \langle y^*, F(\bar{x}, u_k) - F(\bar{x}, \bar{u})\rangle\}\end{aligned}$$

(as H is a linear function of α_i). Combining these two inclusion with (3.14) we conclude that

$$\begin{aligned}&0 \in \partial_a \mathcal{L}((\lambda_0, .., \lambda_m, y^*, \cdot, \bar{u})(\bar{x}) + r\partial_a \rho(S, \bar{x}) + \delta B \\ &\langle y^*, F(\bar{x}, u_j) - F(\bar{x}, \bar{u})\rangle \geq -\delta \quad j = 1, .., k\end{aligned}$$

which implies (3.11)–(3.12). This completes the proof of the theorem. □

3.3. The maximum principle. We finally turn to the problem

$$\textbf{(OC)} \qquad \begin{array}{ll} & \text{minimize } f_0(x) \\ \text{s.t.} & Lx = \phi(x,u)\ ; \\ & f_i(x) \le 0\ ,\ i = 1,..,m\ ;\ x \in S;\ u \in \mathcal{U} \end{array}$$

Here L is an operator from X into another Banach space Z and ϕ is a mapping from $X \times \mathcal{U}$ into Z. Assume that

(A_8) L is a closed operator with domain $D_L \subset X$, the equation $Lx = y$ has a unique solution $x = G(y)$ for any $y \in Y$ and G is a compact operator.

Define the *Hamiltonian* of **(OC)**

$$H(p,x,u) := \langle p, \phi(x,u) \rangle$$

where $p \in Z^*$.

THEOREM 3.6 (THE MAXIMUM PRINCIPLE). *We posit* (A_2), (A_3), (A_6) - (A_8). *If* $(\bar{x},\bar{u})$ *is a local solution to the problem* **(OC)**, *then there are* $\lambda_0 \ge 0,..,\lambda_m \ge 0$ *and* $p \in Z^*$, *not all equal to zero, such that* $\lambda_i f_i(\bar{x}) = 0$, $i = 1,..,m$ *and*

$$L^*p \in \partial_a H(p,\cdot,\bar{u})(\bar{x}) - \sum \lambda_i \partial_a f_i(\bar{x}) - N_G(S,\bar{x})\,, \tag{3.15}$$

$$H(p,\bar{x},\bar{u}) = \max_{u \in U} H(p,\bar{x},u) \tag{3.16}$$

Proof. The proof is of course an easy reformulation of Theorem 3.3. By (A_8) we can rewrite the control equation $Lx = \phi(x,u)$ as

$$x - G \circ \phi(x,u) = 0$$

which has the Fredholm structure as in (3.5) with $A = I$ and $Y = X$. Let λ_i, x^* be chosen according to the Theorem 3.3. Setting $p = G^* x^*$, we immediately get that

$$\mathcal{L}(\lambda_0,..,\lambda_m,x^*,x,u) = \sum \lambda_i f_i(x) + \langle x^*, x \rangle - H(p,x,u)$$

so the Lagrangian minimum condition (3.7) is exactly the maximum principle (3.16). The inclusion (3.6) gives

$$x^* \in \partial_a H(p,\cdot,\bar{u})(\bar{x}) - \sum \lambda_i \partial_a f_i(\bar{x}) - N_G(S,\bar{x})$$

Consequently

$$p \in G^*[\partial_a H(p,\cdot,\bar{u})(\bar{x}) - \sum \lambda_i \partial_a f_i(\bar{x}) - N_G(S,\bar{x})]$$

Since L is a closed operator, this inclusion implies (3.15). □

Thus we see that, given an optimal control problem, we have to check exactly two things to prove a maximum principle for this problem, namely that (A_8) holds and that $G \circ \phi$ is relaxable at $(\bar{x}, \bar{u})$. The remaining is a technical work of description of G^* (or L^*) and all the subdifferentials.

4. Optimal control of a system with a distributed time-dependent delay

4.1. Statement of the problem and the main result. As the first application of the general principle, established in the previous section, we consider the following problem.

(TD) minimize $\psi_0(x(1))$ over all solutions to the equation

$$\dot{x}(t) = \phi(t, \int_{-\Delta}^{0} x(t+\tau)\, dS_t \;, u(t)) \;, \quad t \in [0,1] \;, \; x \in R^n \tag{4.1}$$

satisfying the boundary conditions

$$x(t) = c(t) \;, \; t \in [-\Delta, 0] \;, \; x(1) \in S \tag{4.2}$$

and state constraints

$$\psi_i(t, x(t)) \leq 0 \; i = 1, .., m \;, \tag{4.3}$$

and corresponding to control functions $u(t)$ which are measurable selections of a set-valued map $t \rightarrow U(t)$. To avoid any discussion on the measurability subject, we assume that $U(t)$ are subsets of a complete metric space U. Here S_t for any t is a matrix-valued function of bounded variation on $[-\Delta, 0]$. One can think of S_t as of a matrix $r \times n$ whose elements are functions of bounded variation which are constant for $\tau > 0$ and $\tau < -\Delta$. Therefore $\phi(t, y, u)$ is a function on $[0,1] \times R^r \times U$ with values in R^n.

The formulation, namely the way the delays are presented, has been essentially suggested by (although differs from) the one in a recent paper by Clarke and Wolenski [7] in which differential inclusions with delay and convex bounded values are considered. It actually embraces many traditional formulations connected with time delays. Here are several examples.

Example 1. If $r = n$ and $dS_t = \epsilon_0\, I\, d\tau$ where ϵ_0 is the unit mass at zero, then $\int_{-\Delta}^{0} x(t+\tau) dS_t = x(t)$, so (4.1) becomes the standard control equation

$$\dot{x}(t) = \phi(t, x(t), u(t)).$$

Example 2. If $r = 2n$ and $dS_t = (\epsilon_0\, I \;, \; \epsilon_{-\Delta}\, I) d\tau$ then $\int_{-\Delta}^{0} x(t+\tau) dS_t = (x(t), x(t-\Delta))$, so (4.1) becomes the standard control equation with fixed time delay

$$\dot{x}(t) = \phi(t, x(t), x(t-\Delta), u(t)).$$

Example 3. If $r = m \cdot n$ and $dS_t = (\epsilon_0 I\ ,\ \epsilon_{-\Delta_1(t)} I, .., \ \epsilon_{-\Delta_m(t)} I)d\tau$ then $\int_{-\Delta}^{0} x(t+\tau)dS_t = (x(t), x(t-\Delta_1(t)), .., x(t-\Delta_m(t))$ and we get an equation with m time dependent delays

$$\dot{x}(t) = \phi(t, x(t), x(t-\Delta_1(t)), .., x(t-\Delta_m(t)), u(t)).$$

Example 4. Let again $r = n$ but $dS_t = (\sum \alpha_i(t)\epsilon_{-\Delta_i(t)})Id\tau$. Then the equation becomes

$$\dot{x}(t) = \phi(t, \sum \alpha_i x(t - \Delta_i(t)), u(t)).$$

We can refer to [8], [1], [19] for earlier or other formulations.

Observe, to avoid confusion, that in the statement and the examples we think of x as a row vector.

Let $(\bar{x}(t), \bar{u}(t))$ be a solution of (TD) in the sense that for any other admissible pair (x, u) with x sufficiently close to $\bar{x}$ (in the uniform topology) we have $\psi(x(1)) \geq \psi(\bar{x}(1))$. Set

$$\bar{y}(t) = \int_{-\Delta}^{0} \bar{x}(t+\tau)dS_t.$$

We shall assume the following hypotheses:

(H_1) $t{\rightarrow}S_t$ is a bounded measurable map into the space of matrix-valued functions of bounded variations; this means that there is a is a $K > 0$ such that the variation of all elements of S_t are not greater than K for all $t \in [0,1]$ and the map $t{\rightarrow} \int_{-\Delta}^{0} x(t+\tau)dS_t$ is measurable for any continuous vector function x on $[-\Delta, 1]$;

(H_2) for any measurable selections $u(t)$ of $U(t)$ and any bounded measurable $y(t)$ the mapping $t{\rightarrow}\phi(t, y(t), u(t))$ is measurable and there are measurable functions $\epsilon(t) > 0$ and $K(t) \geq 0$ such that

$$|\phi(t, y, u(t)) - \phi(t, y', u(t))| \leq K(t)\|y - y'\|$$

if $\|y - \bar{y}\| \leq \epsilon(t)$ and $\|y' - \bar{y}\| \leq \epsilon(t)$;

(H_3) ψ_0 is Lipschitz continuous near $\bar{x}(t)$ and S is a closed set;

(H_4) $\psi_i(t, x)$ are continuous and there is a summable function $\kappa(t)$ on $[0, 1]$, such that

$$|\psi_i(t, x) - \psi_i(t, x')| \leq \kappa(t)\|x - x'\|$$

if $\|x, x' - \bar{x}(t)\| \leq \epsilon$;

(H_5) We shall also assume for technical convenience that $S_t(0) \equiv 0$ and $\Delta \geq 1$ (which, of course is no restriction at all).

We need a few more notation before stating the theorem. First we set, as usual,

$$H(t, y, p, u) = p\phi(t, y, u).$$

Denote $T_i = \{t \in [0,1] : \psi_i(t, \bar{x}(t) = 0\}$. Let us write

$$\overline{\partial}\psi_i(t,x) = \limsup_{\tau \longrightarrow t, w \longrightarrow x} \partial_c \psi_i(\tau, w).$$

THEOREM 4.1. *Let* ($\bar{x}$,$\bar{u}$) *be a local solution of* **(TD)**. *Then there are a number* $\lambda \geq 0$, *nonnegative measures* μ_i, *supported on* T_i, $i = 1, .., k$ *and a vector-valued function* $p(t)$ *of bounded variation such that*

$$(4.4) \quad \begin{aligned} &p(t) \in -\lambda\partial_a\psi(\bar{x}(1)) - N_G(S, x(1)) \\ &\quad - \textstyle\int_t^1 [S_\xi(t-\xi)\partial_c H(\xi, \bar{x}(\xi), p(\xi), \bar{u}(\xi))d \sum \overline{\partial}\psi_i(t, \bar{x}(t))d\mu_i] \end{aligned}$$

and

$$(4.5) \quad \int_0^1 H(t, \bar{y}(t), p(t), \bar{u}(t))dt = \max_{\mathcal{U}} \int_0^1 H(t, \bar{y}(t), p(t), u(t))dt$$

Moreover, if U *is a complete metric space,* $U(t)$ *is a measurable set-valued map and* $\psi(t,y,u)$ *is jointly continuous in* y, u *then* (4.5) *can be replaced by the standard pointwise maximum principle*

$$H(t, \bar{y}(t), p(t), \bar{u}(t)) = \max_{U(t)} H(t, \bar{y}(t), p(t), u)$$

4.2. Proof. It can be easily shown that the just formulated problem belongs to the class of problems covered by Theorem 3.6. But we choose in this section another line of proof involving reduction to the Lagrangian minimum principle (Theorem 3.3)— mainly with the purpose of demonstration that this theorem is as well applicable for specific cases. In the next section we shall use the Theorem 3.6 to get maximum principle for another problem.

4.2.1. Reformulation of the problem. We begin by rewriting (4.1) as

$$\begin{aligned} \dot{x}(t) &= \phi(t, y(t), u(t)) \\ y(t) &= \int_{-\Delta}^0 x(t+\tau)dS_t \end{aligned}$$

If we set t^+ for $\max\{0,t\}$ and t^- for $\min\{0,t\}$, then the first equation together with the initial condition (4.2) can be written as

$$(4.6) \quad x(t) - \int_0^{t^+} \phi(s, y(s), u(s))ds = c(t^-) \quad t \in [-\Delta, 1];$$

$$(4.7) \quad y(t) - \int_{-\Delta}^0 x(t+\tau)dS_t = 0 \qquad t \in [0,1],$$

and the right end point constraints (4.2) can be written as

$$z - x(1) = 0 \tag{4.8}$$

$$z \in S. \tag{4.9}$$

Consider the spaces $X = C^n([-\Delta, 1])$ and $Y = L^n_\infty([0,1])$ (the space of continuous functions and the space of bounded measurable functions with values in R^n). Then we can formalize (4.6)–(4.9) as the equation

$$F(x, y, z, u) = c_0$$

where for any fixed u F is the operator from $X \times Y \times R^n$ into itself defined by the left-hand sides of (4.6)–(4.9) and $c_0(t) := (c(t^-), 0, 0)$. We see that F has the structure

$$F = A + G \circ P ,$$

where A is the linear operator defined on $X \times Y \times R^n$:

$$(x(t), y(t), z) \rightarrow (x(t)\ , y(t) - \int_{-\Delta}^{0} x(t+\tau) dS_t\ , z - x(1))$$

whose image is obviously the whole of $X \times Y \times R^n$; P is the an operator from $X \times Y \times R^n$ into $L_1^n[0,1]$ which, for a fixed u, sends $y(t)$ into $\phi(t, y(t), u(t))$, and G transforms every $v \in L_1^n$ into $(\int_0^{t^+} v(s)ds, 0, 0) \in X \times Y \times R^n$.

Let us now define $\mathcal{U}$ as the collection of all measurable selections of $U(t)$ for which $\phi(t, \bar{y}(t), u(t))$ is summable and (H_2) holds with some $K(t)$ and $\epsilon(t)$. Then the image of the ϵ-neighborhood of $\bar{y}$ under $P(\cdot, u)$ is weak compact in L_1 by the Lebesgue theorem on majorized convergence.

We further observe that the constraint $z \in S$ can be rewritten as

$$(x, y, z) \in X \times Y \times S$$

which makes (A_3) valid. The state constraints rewritten as

$$f_i(x(\cdot)) = \max_{[0,1]} \psi_i(t, x(t)) \le 0 \tag{4.10}$$

satisfy (A_2).

We refer to [17] Chapter 5, for a proof of the relaxability of operators of the form

$$(y, u) \rightarrow \int_0^t \phi(s, y(s), u(s)) ds.$$

As for compactness of G, it is obvious.

Thus, we have restated our problem in the form of minimizing

$$f_0(x) = \psi_0(x(1))$$

subject to the constraints (4.6)–(4.9). It satisfies all the assumptions of Theorem 3.3.

4.2.2. Deciphering the necessary conditions. The Lagrangian of the problem is

$$L(\lambda_0,..,\lambda_k,l,\mu,y^*;x,y,z,u) := \lambda_0\psi_0(z)+\sum_{i=1}^{k}\lambda_i\max_{[0,1]}\psi_i(t,x(t))$$
$$+\int_{-\Delta}^{1}[x(t)-\int_0^{t^+}\phi(\xi,y(\xi),u(\xi))d\xi]\,d\mu$$
$$+\langle y^*,\ y(t)-\int_{-\Delta}^{0}x(t+\tau)dS_t(\tau)\ \rangle+l(z-x(1))\,.$$

Here $\lambda_i\geq 0, i=0,..,k$, $l\in R^n$, μ is a Radon measures on $[-\Delta,1]$ with values in R^n and y^* is a bounded functional on $L^n_\infty([0,1])$.

By Theorem 3.3 these multipliers can be chosen to ensure that first, zero belongs to the sum of the $a-$subdifferential of L at $(\bar{x},\bar{y},\bar{u})$ and the G-normal cone to **S** at this point, and second, the minimum of the Lagrangian with respect to u is attained at $\bar{u}$.

As the variables x,y,z enter separately into different parts of L and **S** is defined only in terms of the $z-$components, we can compute the subdifferentials with respect to each component separately.

The easiest is the subdifferentiation with respect to z which gives

$$l\in-\lambda_0\partial_a\psi(\bar{x}(1))-N_G(S,\bar{x}(1)) \tag{4.11}$$

The part of L containing y is

$$\langle y^*,y\rangle-\int_{-\Delta}^{1}\int_0^{t^+}\phi(\xi,y(\xi),u(\xi))d\xi]d\mu$$
$$=\langle y^*,y\rangle-\int_0^1(\int_\xi^1 d\mu)\phi(\xi,y(\xi),u(\xi))d\xi$$
$$=\langle y^*,y\rangle-\int_0^1 H(\xi,y(\xi),p(\xi),u(\xi))d\xi$$

where we set $p(\xi):=\int_\xi^1 d\mu$. Denote

$$\mathcal{H}(y,p,u):=\int_0^1 H(t,y(t),p(t),u(t))dt$$

Thus, the condition that the subdifferential of the Lagrangian with respect to y contains zero leads to the conclusion that

$$y^*\in\partial_a\mathcal{H}(\cdot,p,u)(y). \tag{4.12}$$

By [6] or [16] the subdifferential $\partial_a\mathcal{H}$ is contained in the collection of all R^n-valued measurable functions $e(t)$ such that

$$e(t)\in\partial_c H(t,p(t),\cdot,u(t))(\bar{y}(t))\text{ a.e. on }[0,1]\,. \tag{4.13}$$

So there is a function $e(t)$ such that for any y

$$\langle y^*, y\rangle = \int_0^1 e(t)\cdot y(t)dt$$

It follows from (H_2) and from the boundedness of p that e is a summable function.

Now the x-part of the Lagrangian takes the form

$$\int_{-\Delta}^1 x(t)d\mu - \int_0^1 \Big(\int_{-\Delta}^0 x(t+\tau)dS_t d\tau\Big)\, e(t)dt + \sum \lambda_i \max_{[0,1]} \psi_i(t, x(t)) - l\cdot x(1)\,.$$

As total variations of components S_t are uniformly bounded, we can be sure that

$$\int_0^1\int_{-\Delta}^0 v(t,\tau)dS_t(\tau)dt = \lim_{\epsilon\to 0}\epsilon^{-1}\int_0^1\int_{-\Delta}^0 v(t,\tau)(S_t(\tau+\epsilon)-S_t(\tau))d\tau dt$$

for any continuous function $v(t,\tau)$. By introducing the new variable $\xi = t+\tau$ (instead of τ) we obtain

$$\begin{aligned}\int_0^1\Big[\int_{-\Delta}^0 x(t+\tau)dS_t(\tau)\Big]\, e(t)dt &= \int_0^1\Big[\int_{t-\Delta}^t x(\xi)dS_t(\xi-t)d\xi\Big]\, e(t)dt\\ &= \lim_{\epsilon\to 0}\int_{-\Delta}^1 x(\xi)\Big[\int_{\xi+}^{\min(1,\xi+\Delta)} \epsilon^{-1}(S_t(\xi-t+\epsilon)-S_t(\xi-t))e(t)dt\Big]\, d\xi\\ &= \int_{-\Delta}^1 x(\xi)\, dR(\xi)\,,\end{aligned}$$

where

$$R(\xi) = \int_{\xi+}^{\min(1,\xi+\Delta)} S_t(\xi-t)e(t)dt\,. \tag{4.14}$$

Note that for $t\geq 0$ we have, using (4.13) and (H_5), that

$$R(t) = \int_t^1 S_\xi(t-\xi)e(\xi)d\xi \in \int_t^1 S_\xi(t-\xi)\partial_c H(\xi,\cdot,p(\xi),\bar{u}(\xi))(\bar{y}(\xi))d\xi\,.$$

It remains to calculate the subdifferentials of functions f_i. The a–subdifferential of a Lipschitz function always belongs to its generalized gradient of Clarke. Thus we get (see [6]) that $\partial_a f_i(\bar{x})$ belongs to the set of all measures of the form $d\eta_i = a_i(t)d\tilde{\mu}_i$, where $\tilde{\mu}_i$ is a probability measure supported on T_i and $a_i(t)\in\overline{\partial}\psi_i(t,\bar{x}(t))$ a.e.. We therefore conclude that there are such $\tilde{\mu}_i$ and $a_i(\cdot)$ with which the following equality holds for any $x \in C^n[-\Delta, 1]$:

$$(4.15) \quad \int_{-\Delta}^{1} x(t)d\mu - \int_{-\Delta}^{1} x(t)dR(t) + \sum_{1}^{k} \int_{0}^{1} x(t)a_i(t)d\mu_i - l \cdot x(1) = 0$$

where we set $\mu_i = \lambda_i \tilde{\mu}_i$.

As $R(1) = 0$ by (4.14), we deduce from (4.15) that for any $t \in [0, 1]$

$$(4.16) \qquad p(t) = \int_{t}^{1} d\mu = l - R(t) - \sum_{1}^{k} \int_{t}^{1} a_i(t)d\mu_i \, .$$

Combining this with (4.11) and (4.16) we get (4.4).

The integral maximum principle (4.5) is an immediate corollary of general maximum principle and the proof of the pointwise maximum principle under the assumptions of the concluding part of the theorem is exactly the same as in [14].

5. The maximum principle for elliptic systems with distributed and boundary control

5.1. Statement of the problem and the main result. Let Ω be a bounded domain in R^n with a $C^{1,1}$ boundary Γ. We shall consider functions $x(t)$ with values in R^m and expressions like Δx or $\frac{\partial x}{\partial n}$ will denote vectors whose components are the results of application of the corresponding operators (Laplace operator, the derivative along the outward normal to Γ etc.) to components of x, that is to say

$$\Delta x(t) = (\Delta x_1(t), .., \Delta x_m(t)) \, .$$

The problem we shall consider can be formulated as follows:

$$(5.1) \quad \textbf{(EC)} \qquad \text{minimize } J(x, u) = \int_{\Omega} \phi_0(t, x)dt + \int_{\Gamma} \psi_0(\tau, x)d\tau$$

over the collection of all triples $(x(\cdot), u(\cdot), v(\cdot))$ such that x is a solution (in the weak sense) of the Neumann problem for the system of equation

$$(5.2) \qquad \begin{aligned} -\Delta x &= \phi(t, x, u) \quad \text{on } \Omega \\ \tfrac{\partial x}{\partial n} &= \psi(\tau, x, v) \quad \text{on } \Gamma \end{aligned}$$

corresponding to domain and boundary control function $u(t), v(\tau)$, taking values in the sets $U(t)$ and $V(\tau)$ respectively and satisfying the state constraints

$$(5.3) \qquad \begin{aligned} \phi_i(t, x) &\le 0 \text{ i=1,..,k} \quad \text{a.e. in } \Omega \, , \\ \psi_j(\tau, x) &\le 0 \text{ j=1,..,l} \quad \text{a.e. on } \Gamma . \end{aligned}$$

Let $(\bar{x}, \bar{u}, \bar{v})$ be a local solution of the problem in the sense that there is an $\epsilon > 0$ such that for every other triple (x, u, v) which is admissible (the full meaning of this word will be clear after we formulate all the hypotheses) and such that $|x(t) - \bar{x}(t)| \leq \epsilon$ for all $t \in \overline{\Omega}$, we have $J(x) \geq J(\bar{x})$.

Now the hypotheses:

(H_6) $U(t)$ and $V(\tau)$ are measurable convex-valued maps from Ω and Γ into R^r and R^s respectively;

(H_7) the vector-functions $\phi(t, x, u)$, $\psi(t, x, u)$ are measurable as functions of t, continuous as functions of x and affine as functions of u, that is to say

$$\phi(t, x, u) = h_0(t, x) + h_1(t, x)u \quad ; \quad \psi(t, x, u) = g_0(t, x) + g_1(t, x)u$$

where the vector functions h_0, g_0 and matrix function h_1, g_1 satisfy the Carathéodory condition;

(H_8) $\phi(t, \bar{x}(t), \bar{u}(t)) \in L^m_\omega(\Omega)$ and $\psi(\tau, \bar{x}(\tau), \bar{v}(\tau)) \in L^m_\gamma(\Gamma)$ for some $\omega > n/2$ and $\gamma > n - 1$;

(H_9) for any measurable selections $u(t)$ of $U(t)$ and $v(\tau)$ of $V(\tau)$ there are an $\epsilon > 0$ and measurable functions $\alpha(t) \geq 0$ on Ω and $\beta(\tau) \geq 0$ on Γ such that

$$\|\phi(t, x, u(t)) - \phi(t, x', u(t))\| \leq \alpha(t)\|x - x'\| \ \text{a.e. in } \Omega,$$
$$\|\psi(\tau, x, v(\tau)) - \psi(\tau, x', v(\tau))\| \leq \beta(\tau)\|x - x'\| \ \text{a.e. on } \Gamma,$$

provided $\|x, x' - \bar{x}\| \leq \epsilon$; for $(\bar{u}, \bar{v})$ we can choose $\alpha \in L_\omega(\Omega)$ and $\beta \in L_\gamma(\Gamma)$ with the same ω, γ as in (H_8);

(H_{10}) $\phi_i(t, x), \psi_j(t, x)$ are continuous functions satisfying the Lipschitz condition in x on every compact subset of R^n uniformly with respect to t;

(H_{11}) ϕ_0, ψ_0 are Carathéodory functions and there are $\bar{\epsilon} > 0$ and summable function $\bar{\alpha}(t)$ and $\bar{\beta}(\tau)$ such that

$$\|\phi_0(t, x) - \phi_0(t, x')\| \leq \bar{\alpha}(t)\|x - x'\| \ \text{a.e. in } \Omega,$$
$$\|\psi_0(\tau, x) - \psi_0(\tau, x')\| \leq \bar{\beta}(\tau)\|x - x'\| \ \text{a.e. on } \Gamma,$$

provided $\|x, x' - \bar{x}\| \leq \bar{\epsilon}$.

In what follows we fix ω, γ for which $(H_8), (H_9)$ holds.

To state the theorem we introduce the "domain" and "boundary" Hamiltonians

$$H(t, x, u; p) := p(t)\phi(t, x, u) \quad t \in \Omega \tag{5.4}$$
$$H_\Gamma(\tau, x, v; p) := p(\tau)\psi(\tau, x, v) \quad \tau \in \Gamma \tag{5.5}$$

and sets

$$T_i := \{t \in \overline{\Omega} : \ \phi_i(t, \bar{x}(t)) = 0\}, \quad , i = 1, .., k$$
$$T_j := \{\tau \in \Gamma : \ \psi_j(\tau, \bar{x}(\tau)) = 0\}, \quad , j = k + 1, .., k + l$$

We finally agree to call a function p a solution to the boundary value inclusion problem

$$\begin{aligned} -\Delta p(t) &\in A(t)\mu \quad t\in\Omega \\ \tfrac{\partial p}{\partial n}(\tau) &\in B(\tau)\nu \quad \tau\in\Gamma \end{aligned}$$

where $A(t), B(\tau)$ are set-valued maps and μ,ν are (scalar) Radon measures, if for any y of class C^∞ on $\overline{\Omega}$

$$\int_\Omega p(t)\Delta y(t)dt + \int_\Gamma p(\tau)\frac{\partial y}{\partial n}(\tau)d\tau \in \int_\Omega y(t)A(t)d\mu + \int_\Gamma y(\tau)B(\tau)d\nu.$$

Here by an integral of a set-valued map we, as usual, understand the collection of integrals of its measurable selections.

THEOREM 5.1. *Let $(\bar{x},\bar{u},\bar{v})$ be a local solution of (EC). Then there are a number $\lambda\geq 0$, a function $p\in W^{1,\sigma}(\overline{\Omega})$ for any $\sigma\in[1,n/(n-1)]$ and nonnegative R^m Radon measures ν_i, $i=1,..k$ and η_j, $j=1,..l$, supported on T_i and T_{k+j} respectively such that $p(x)$ is a solution of boundary value inclusion problem*

$$(5.6)\quad \begin{aligned} &-\Delta p(t)\in\partial_c H(t,\cdot,\bar{u};p)(\bar{x}(t)) - \lambda\partial_c\phi_0(t,\bar{x}(t)) - \sum\overline{\partial}\phi_i(t,\bar{x}(t))\nu_i \\ &\qquad t\in\Omega \\ &\tfrac{\partial p}{\partial n}(\tau)\in\partial_c H_\Gamma(\tau,\cdot,\bar{v};p)(\bar{x}(\tau)) - \lambda\partial_c\psi_0(\tau,\bar{x}(\tau)) \\ &\qquad -\sum\overline{\partial}\phi_i(\tau,\bar{x}(\tau))\nu_i|_\Gamma - \sum\overline{\partial}\psi_j(\tau,\bar{x}(\tau))\eta_j \quad \tau\in\Gamma \end{aligned}$$

and

$$(5.7)\quad \begin{aligned} H(t,\bar{x}(t),\bar{u}(t);p(t)) &= \max_{u\in U(t)} H(t,\bar{x}(t),u;p(t)) \text{ a.e. on } \Omega \\ H_\Gamma(\tau,\bar{x}(\tau),\bar{v}(\tau);p(\tau)) &= \max_{v\in V(\tau)} H_\Gamma(\tau,\bar{x}(\tau),v;p(\tau)) \text{ a.e. on } \Gamma \end{aligned}$$

We use the notation $\nu|_\Gamma$ for the restriction of ν to Γ.

5.2. Proof

5.2.1. Reformulation of the problem. Consider the following scalar linear boundary problem

$$(5.8)\quad \begin{aligned} -\Delta x(t)+x(t) &= \xi(t) \; t\in\Omega \\ \tfrac{\partial y}{\partial n}(\tau) &= \zeta(\tau) \; \tau\in\Gamma \end{aligned}$$

LEMMA 5.2. *([5],[12]) Let $\xi\in L_\omega(\Omega)$, $\zeta\in L_\gamma(\Gamma)$ for $\omega>n/2$ and $\gamma>n-1$. Then the boundary value problem has a unique solution $y=E(\xi,\zeta)$, belonging to $H^1(\Omega)\bigcap C(\overline{\Omega})$ and the Green operator*

$$E: L_\omega(\Omega)\times L_\gamma(\Gamma)\to C(\overline{\Omega})$$

is compact.

This is the crucial fact which enables us to easily reformulate the problem in desired terms, actually together with Lemma 5.4 below, the only information from the theory of PDE we need to prove the theorem. It follows from Lemma 5.2 and (H_8) that $\bar{x}(t)$ is continuous on $\overline{\Omega}$.

Denote by $\mathcal{U}$ the collection of all measurable selection $(u(\cdot), v(\cdot))$ of $U(\cdot)$ and $V(\cdot)$ respectively having the properties that:

1. $\phi(t, \bar{x}(t), u(t)) \in L^m_\omega(\Omega)$ and $\psi(t, \bar{x}(t), v(t)) \in L^m_\gamma(\Gamma)$;
2. there are nonnegative functions $\alpha \in L_\omega(\Omega)$ and $\beta \in L_\gamma(\Gamma)$ and $\epsilon > 0$ such that

$$\text{(5.9)} \quad \begin{aligned} &\|\phi(t,x,u(t)) - \phi(t,x',u(t))\| \le \alpha(t)\|x - x'\| \quad \text{a.e. in } \Omega \\ &\|\psi(\tau,x,v(\tau)) - \psi(\tau,x',v(\tau))\| \le \beta(\tau)\|x - x'\| \quad \text{a.e. on } \Gamma \end{aligned}$$

provided x, x' lie within ϵ-neighborhood of $\bar{x}(t)$ and $\bar{x}(\tau)$ respectively.

LEMMA 5.3. *Let $u(\cdot)$ and $v(\cdot)$ be measurable selections of $U(\cdot)$ and $V(\cdot)$. Then for any $\delta > 0$ there are subsets $\Omega' \subset \Omega$ and $\Gamma' \subset \Gamma$ whose corresponding measures are smaller than δ such that functions*

$$u'(t) = \begin{cases} \bar{u}(t), & \text{if } t \in \Omega' \\ u(t), & \text{if } t \in \Omega \backslash \Omega' \end{cases} \qquad v'(\tau) = \begin{cases} \bar{v}(\tau), & \text{if } \tau \in \Gamma' \\ v(\tau), & \text{if } \tau \in \Gamma \backslash \Gamma' \end{cases}$$

belong to $\mathcal{U}$.

Proof. Obvious from (H_8) and (H_9).

Let we now consider the spaces $X = C^m(\overline{\Omega})$, $Z = L^m_\omega(\Omega) \times L^m_\gamma(\Gamma)$ and the operator $L : X \to Z$ defined in the following way.

Let N be the closed operator from $C(\overline{\Omega})$ into $L_\omega(\Omega) \times L_\gamma(\Gamma)$ determined by (5.8). Then

$$L = \begin{pmatrix} N & 0 & 0 & \dots & 0 \\ 0 & N & 0 & \dots & 0 \\ \dots & & & & \\ 0 & \dots & & & N \end{pmatrix}$$

with the identity $m \times m$ matrix on the right.

Let further the map $\Psi : X \times \mathcal{U} \to Z$ be defined by

$$\Psi(x,u,v) = (\phi(t,x(t),u(t)) + x(t) \ , \ \psi(\tau,x(\tau),v(\tau)) \ t \in \Omega \ , \ \tau \in \Gamma.$$

Comparing this with (5.8), we see that every admissible triple (x, u, v) in our problem satisfies the equation

$$\text{(5.10)} \qquad Lx = \Psi(x,u,v) \ , \ (u,v) \in \mathcal{U}$$

We finally define the function $f_0, ..., f_{k+l}$ by

$$f_0(x) = \int_\Omega \phi_0(t, x(t))dt + \int_\Gamma \psi_0(\tau, x(\tau))d\tau$$
$$f_i(x) = \max_{t\in\Omega} \phi_i(t, x(t)) \qquad i = 1, .., k$$
$$f_{k+j}(x) = \max_{\tau\in\Gamma} \psi_j(\tau, x(\tau)) \quad j = 1, .., l$$

We therefore conclude that $(\bar{x}, \bar{u}, \bar{v})$ is a local solution to the problem of minimizing $f_0(x)$ subject to (5.10) and $f_i(x) \le 0$, $i = 1, .., k + l$. We need to check that all the assumptions of Theorem 3.6 are satisfied for this new problem.

First of all, we notice that (A_2) immediate from $(H_{10}), (H_{11})$ and (A_3) trivially holds. We have further by Lemma 5.2 that the equation $Lx = z$ has unique solution $x = Gz$ for every $z \in Z$ and

$$G = \begin{pmatrix} E & 0 & 0 & \dots & 0 \\ 0 & E & 0 & \dots & 0 \\ \dots & & & & \\ 0 & \dots & & & E \end{pmatrix}$$

is a compact operator. Thus (A_8) holds. (A_6) is an easy consequence of the definition of $\mathcal{U}$ (in view of the Lebegue theorem on majorized convergence). We note finally that $\Psi(x, U)$ is a convex set, owing to (H_6), (H_7). So (A_7) is also valid. Thus we can apply Theorem 3.6 to our problem.

5.2.2. Deciphering the necessary conditions. By the abstract maximum principle there are $(p, q) \in Z^* = L^m_{w'}(\Omega) \times L_{\gamma'}(\Gamma)$, (where as usual $1/w' + 1/w = 1$, $1/\gamma' + 1/\gamma = 1$) and numbers $\lambda_i \ge 0$, $i = 0, .., k + l$, not all equal to zero such that $\lambda_i\ f_i = 0$, $i = 1, .., k + l$, and for

$$\mathcal{H}\ (p, q; x, u, v) := \langle (p, q), \Psi(x, u, v) \rangle$$

we have

$$\mathcal{H}\ (p, q; \bar{x}, \bar{u}, \bar{v}) = \max_{(u,v)\in\mathcal{U}} \mathcal{H}\ (p, q; \bar{x}, u, v) \tag{5.11}$$

and

$$(p, q) \in G^*[\partial_a\ \mathcal{H}\ (p, q; \cdot, \bar{u}, \bar{v})(\bar{x}) + \sum \lambda_i \partial_a f_i(\bar{x})]. \tag{5.12}$$

As follows from the definition of Ψ

$$\mathcal{H}\ (p, q; x, u, v) = \int_\Omega [H(p; t, x, u) + p \cdot x]dt + \int_\Gamma H_\Gamma(q; \tau, x, v)d\tau\,. \tag{5.13}$$

As H and H_Γ are Lipschitz continuous function of x near $\bar{x}$, the a-subdifferential of $\mathcal{H}$ with respect to x (i.e. as a function on $C^m(\overline{\Omega})$) is contained in the collection of all R^m-valued Radom measures μ on $\overline{\Omega}$ that can be represented in the form

$$\int_{\overline{\Omega}} h d\mu = \int_{\Omega} h(\alpha + p)dt + \int_{\Gamma} h\beta d\tau$$

where $\alpha(t)$ and $\beta(\tau)$ are measurable selections of the set-valued maps $t \to \partial_c H(p;\cdot,\bar{u})(\bar{x}(t))$ and $\tau \to \partial_c H_\Gamma(q;\cdot,\bar{v})(\bar{x}(\tau))$ respectively ([6], [16]). Likewise, $\partial_a f_0(\bar{x})$ is defined by measurable selections of the set-valued maps $t \to \partial_c \phi_0(t, \bar{x}(t))$, $t \in \Omega$ and $\tau \to \partial_c \psi_0(\tau, \bar{x}(\tau))$, $\tau \in \Gamma$.

For the subdifferentials of f_i we also have representations analogous to what we had for state constraints in the previous section: $\partial_a f_i(\bar{x})$ belongs to the collection of Radon R^m-valued measures of the form $\gamma_i\, d\nu_i$, where ν_i is a probability measure on T_i and γ_i is a measurable selection of the set-valued maps $t \to \partial_c \phi_i(t, \bar{x}(t))$, $t \in \Omega$, when $i = 1, .., k$, and $\tau \to \partial_c \psi_{i-k}(\tau, \bar{x}(\tau))$, $\tau \in \Gamma$, when $i = k+1, .., k+l$.

It remains to describe the structure of G^* or, which is the same, of E^*. Observe that E^* is an operator from the space $M(\overline{\Omega})$ of Radon measures on $\overline{\Omega}$ into $L^m_{w'}(\Omega) \times L_{\gamma'}(\Gamma)$, (recall that $1/w' + 1/w = 1$, $1/\gamma' + 1/\gamma = 1$). Let μ be a Radon measure on $\overline{\Omega}$. Denote by μ_Γ the restriction of μ on the boundary. The following representation for $E^*\mu$ was proved in [5] and (for a slightly different problem) [12]:

LEMMA 5.4. *Let $(\xi, \eta) = E^*(\mu)$. Then there is the function $p \in W^{1,\sigma}(\overline{\Omega})$, for any $1 < \sigma < n/(n-1)$, which is a unique solution of the adjoint boundary value problem*

$$\begin{aligned} -\Delta p + p &= \mu \quad \text{in } \Omega \\ \frac{\partial p}{\partial n} &= \mu_\Gamma \quad \text{on } \Gamma \end{aligned} \tag{5.14}$$

in the sense that

$$\int_{\Omega} p(t)(\Delta y + y)dt + \int_{\Gamma} p(\tau)\frac{\partial y}{\partial n} = \int_{\overline{\Omega}} y d\mu \tag{5.15}$$

for any $y \in C^\infty(\overline{\Omega})$, such that (ξ, η) are the traces of p on Ω and Γ respectively.

Combining this with (5.12) and the description of the subdifferentials we conclude that the function $p = G^*\mu$ is a solution of the system of inclusions

$$\begin{aligned} -\Delta p(t) + p \;&\in\; \partial_c H(t,\cdot,\bar{u};p)(\bar{x}(t)) + p - \lambda_0 \partial_c \phi_0(t, \bar{x}(t)) \\ &\quad - \textstyle\sum \overline{\partial}\phi_i(t,\bar{x}(t))\nu_i; \quad \forall t \in \Omega \\ \frac{\partial p}{\partial n}(\tau) \;&\in\; \partial_c H_\Gamma(\tau,\cdot,\bar{v};p)(\bar{x}(\tau)) - \lambda_0 \partial_c \psi_0(\tau,\bar{x}(\tau)) \\ &\quad - \textstyle\sum \overline{\partial}\phi_i(\tau,\bar{x}(\tau))\nu_i|_\Gamma - \sum \overline{\partial}\psi_j(\tau,\bar{x}(\tau))\eta_j \perp; \quad \forall \tau \in \Gamma \end{aligned}$$

where we set $\eta_j = \nu_{k+j}$ and denoted by $\nu_i|_\Gamma$ the restriction of ν_i on Γ.

It remains to cancel p on both parts of the "domain" inclusion to get (5.6).

As to (5.7), it follows from (5.11) and (5.13) by means of the standard measurable selection argument.

This completes the proof of the theorem. □

REFERENCES

[1] H.T. Banks, *Necessary condition for problems with variable time lags*, SIAM J. Control **6** (1968), 9–47.

[2] F. Bonnans, E. Casas, *A boundary Pontriagin's principle for state constrained elliptic systems*, Intern. Ser. Num. Math. **107**, Birkhauser, Boston, 1992, pp. 241–249.

[3] F. Bonnans, E. Casas, *An extension of Pontriagin's principle for state constrained optimal control of semilinear elliptic equations and variational inequalities* (preprint), 1993.

[4] J.M. Borwein, A.D. Ioffe, *Proximal analysis in smooth spaces* (preprint), October 1993.

[5] E. Casas, *Boundary control of semilinear elliptic equations with pointwise state constraints*, SIAM J. Control Optimization **31** (1993), 993–1006.

[6] F.H. Clarke, Optimization and Nonsmooth Analysis, Wiley, New York, 1983.

[7] F.H. Clarke, G.G. Watkins, *Necessary conditions, controllability and the value function for differential-difference inclusions*, Nonlinear Analysis, Theory, Methods and Appl. **10** (1986), 1155–1179.

[8] F.H. Clarke, P.R. Wolenski, *Necessary conditions for functional differential inclusions*, CRM-1760, March 1992.

[9] A.Ya. Dubovitzkii, A.A. Miljutin, *Extremum problems in presence of restrictions*, USSR Comput. Math. and Math. Phys. **5** (1965), 1–80.

[10] H.O. Fattorini, H. Frankowska, *Necessary conditions for infinite dimensional control problems*, Math. Contr. Signal, Systems **4** (1991), 41–67.

[11] H. Frankowska, *The maximum principle for an optimal solution to a differential inclusion with endpoint constraints*, SIAM J. Control Optimization **25** (1987), 145–157.

[12] B.E. Ginsburg, *General scheme for maximum principle in optimal control with applications to elliptic systems and systems with delay*, (M.Sc. thesis) Technion, Haifa, 1993.

[13] A.D. Ioffe, *Necessary and sufficient conditions for a local minimum*, SIAM J. Control Optimization **17** (1979), 245–288.

[14] A.D. Ioffe, *Necessary conditions for nonsmooth optimization*, Math. Oper. Res. **9** (1984), 159–189.

[15] A.D. Ioffe, *Approximate subdifferentials and applications 1. The finite dimensional theory*, Trans. Amer. Math. Soc. **281** (1984), 289–316.

[16] A.D. Ioffe, *Approximate subdifferentials and applications 3: the metric theory*, Mathematika **36** (1989), 1–38.

[17] A.D. Ioffe, V.M. Tihomirov, Theory of Extremal Problems, Nauka, Moscow (Russian), 1974. (English translation) North Holland, Amsterdam, 1979.

[18] P.D. Loewen, Optimal Control and Nonsmooth Analysis, CRM Proceedings Lecture Notes **0.2**, American Mathematical Society, Providence, 1993.

[19] A. Manitius, *Optimal control of hereditary systems*, Control Theory and Topics in Functional Analysis, Inter. Center for Th. Physics, Trieste, Italy, 1974.

[20] B.S. Mordukhovich, *The maximum principle in the problems of time optimal control with nonsmooth constraints*, J. Appl. Math. Mech **40** (1976), 960–969.

[21] B.S. Mordukhovich, Approximation Methods in Problems of Optimization and

Control, Nauka, Moscow, 1988.
[22] L.W. NEUSTADT, Optimization, Princeton University Press, Princeton, 1976.
[23] D. TIBA, Optimal Control of Nonsmooth Distributed Parameter Systems, Lecture Notes Math. **1459**, Springer, Berlin, 1990.
[24] J. WARGA, Optimal Control of Differential and Functional Equations, Academic Press, New York, 1972.

ON CONTROLLED INVARIANCE FOR A SIMPLE CLASS OF DISTRIBUTIONS WITH SINGULARITIES*

KEVIN A. GRASSE†

Abstract. Invariant distributions play a key role in the geometric theory of nonlinear control systems. It has long been recognized that distributions that are not invariant under a given control system can sometimes be rendered invariant under a feedback-adjusted form of the system, in which case the distribution is called controlled invariant for the system. The standard necessary and sufficient condition for controlled invariance is known as the "Quaker Lemma" and is usually proved under rather stringent regularity assumptions on both the distribution and the input vector fields of the control system. We will present an alternative proof of the Quaker Lemma that allows a significant weakening of the standard regularity assumptions, and in particular applies to the controlled invariance of a certain class of singular distributions.

1. Introduction. The concept of an invariant distribution is a central theme in the geometric theory of linear and nonlinear control systems, and finds numerous applications in such areas as disturbance decoupling, controllability, and system decomposition (see, e. g., [7], [8], [11]–[14], [16], [19], [21], [22], [25], [29]). To briefly summarize the main ideas, let M and Ω be smooth manifolds (perhaps of differing dimensions), let $\dot{x} = f(x,\omega)$ be a smooth control system with state variable x in M and control variable ω in Ω, and let D be a smooth, regular, involutive distribution on M. We recall that D is *f-invariant* if $[f^\omega, \mathrm{vf}^\infty[\mathrm{D}]] \subseteq \mathrm{vf}^\infty[\mathrm{D}]$ for every $\omega \in \Omega$. Here f^ω is the vector field on M defined by $f^\omega(x) = f(x,\omega)$, $\mathrm{vf}^\infty[\mathrm{D}]$ denotes the family of all smooth vector fields taking values in the distribution D, and $[,]$ is the usual Lie bracket (precise definitions are given in the next section). The importance of invariant distributions stems from the well known fact that the flow of the control system preserves the leaves of such distributions ([8], [10], [13], [16]). More precisely, let $u: \mathbb{R} \to \Omega$ be a suitably "regular" control for f and let $t \mapsto \mu_t^u(x)$ denote the trajectory of f corresponding to u that goes through x at time zero. Since D is regular and involutive, the Frobenius theorem implies that the state space M is a pairwise disjoint union of the "leaves" (i. e., maximal, connected integral submanifolds) of D. If in addition D is f−invariant, then for each fixed $t \in \mathbb{R}$ and each connected subset P of a D−leaf on which μ_t^u is defined it will be the case that $\mu_t^u(P)$ is also contained in a single D−leaf.

In the study of invariant distributions of control systems, it was recognized early on that, while a given distribution might not be invariant

---

* The results of this paper were presented in preliminary form at the workshop on *Nonsmooth Analysis and Geometric Methods in Deterministic Optimal Control*, which was held at the Institute for Mathematics and Its Applications, University of Minnesota, on February 8–17, 1993.

† The author gratefully acknowledges the financial support of the IMA and the National Science Foundation, which facilitated the completion of this work.
Department of Mathematics, University of Oklahoma, Norman, OK 73019.

for a given control system, one could in some cases render the distribution invariant for a "feedback adjusted" version of the system, a phenomenon that has since come to be known as *controlled invariance* ([18],[20]). Specifically, one seeks a feedback mapping $\omega = \alpha(x,\nu)$ with the property that D is invariant under the transformed system $\tilde{f}(x,\nu) = f(x,\alpha(x,\nu))$. Investigations into finding computable necessary and sufficient conditions for *local* controlled invariance have proved quite successful. In the case of affine control systems

$$\dot{x} = f_0(x) + \sum_{k=1}^{m} \omega_k g_k(x),$$

where $f_0, g_1, \ldots, g_m$ are smooth vector fields, if we let G denote the distribution spanned by the vector fields $g_1, \ldots, g_m$, then it is well known ([4],[13],[15],[18]) that local controlled invariance is equivalent to the bracket invariance conditions

$$[f_0, \mathrm{vf}^\infty[\mathrm{D}]] \subseteq \mathrm{vf}^\infty[\mathrm{D}] + \mathrm{vf}[\mathrm{G}]$$

and

$$[g_k, \mathrm{vf}^\infty[\mathrm{D}]] \subseteq \mathrm{vf}^\infty[\mathrm{D}] + \mathrm{vf}[\mathrm{G}], \quad k = 1, \ldots, m,$$

provided that certain "nonsingularity" conditions are also satisfied. Specifically, it is also assumed that the distributions D, G and (G + D)/D are of constant rank in a neighborhood of the point at which the invariant feedback is to be constructed. Corresponding results for fully nonlinear systems are given in [20],[22] under analogous nonsingularity conditions.

These nonsingularity assumptions are not particularly severe when one is searching for locally defined invariant feedbacks, since in important special cases (e. g., real-analytic systems) they can be expected to hold on an open dense subset of the state space. However, the nonsingularity assumptions can pose more of a problem when one is searching for *globally* defined invariant feedbacks, because an obvious necessary condition for the existence of a globally defined invariant feedback is that a locally defined invariant feedback exists in a neighborhood of each point of the state space, including those points where the distributions D, G and (G + D)/D may have singularities. We should point out that the presence of such singularities is by no means the only obstruction to the existence of global invariant feedbacks; there are also topological obstructions ([1],[2],[5],[6]).

Our primary objective in this paper is to explore situations in which the aforementioned nonsingularity assumptions can be weakened. Some progess in this direction has already been made. Specifically, in the paper of Cheng and Tarn [4] it was shown that for systems affine in the control the standard nonsingularity assumptions on G and (G+D)/D can be weakened for affine systems, though they still assume that D is regular. The author generalized this result to fully nonlinear control systems and obtained the following theorem. For simplicity of notation it is stated in the case where the control space is the Euclidean space $\mathbb{R}^m$ with coordinates $\omega = (\omega_1, \ldots, \omega_m)$; the dimension of the state manifold M is n.

THEOREM 1.1. *Let D be a smooth, regular, involutive distribution on M of dimension $n - p > 0$ and let $\dot{x} = f(x, \omega)$ be a smooth control system on M with control space $\mathbb{R}^m$. Then D is locally controlled invariant for f at $(x_0, \omega_0) \in M \times \mathbb{R}^m$ if and only if there exist an open neighborhood $W_1 = U_1 \times V_1$ of (x_0, ω_0) in $M \times \mathbb{R}^m$ with U_1 the domain of a Frobenius chart $(\phi_1, \ldots, \phi_n)$ at x_0 and smooth real-valued functions $\{\gamma_{ij} \mid p+1 \leq i, j \leq n\}$ and $\{\beta_{kj} \mid 1 \leq k \leq m,\ p+1 \leq j \leq n\}$ on W_1 such that for every $\omega \in V_1$*

$$\left[f^\omega, \frac{\partial}{\partial \phi_j}\right](x) = \sum_{i=p+1}^{n} \gamma_{ij}(x,\omega) \frac{\partial}{\partial \phi_i}\bigg|_x + \sum_{k=1}^{m} \beta_{kj}(x,\omega) \frac{\partial f}{\partial \omega_k}(x,\omega))$$

for every $x \in U_1$.

The key improvement here (and also the novelty of the Cheng-Tarn result for affine systems) is that no assumption is required about constant dimensionality of the intersection of D and the distribution spanned by the vector fields $\left\{ \frac{\partial f}{\partial \omega_k} \mid 1 \leq k \leq m \right\}$. The proof of this theorem is given in [9] and is somewhat different than either of the proofs in [4] or [20]. We have since noticed that it can be extended to deal with integrable, but not necessarily regular, distributions D of a very specific type. This extension will be the main topic of this paper, and is offered as a modest first step in the study of controlled invariance for singular distributions. Other investigations into this problem have been carried out by V. Ramakrishna in [23] and [24] using a different approach that makes use of the theory of solutions of singular, overdetermined systems of PDEs.

2. Preliminaries on integrability and invariance of distributions. Since the main topic of this paper deals with the controlled invariance of distributions that may be singular, it will be worthwhile to review some of basic facts about singular distributions.

NOTATION 2.1.

(a) M denotes an n-dimensional, connected, second-countable, C^∞ (smooth) manifold with tangent bundle TM and canonical projection $\pi: TM \to M$. Occasionally, we may formulate results for C^ω (real-analytic) manifolds.

(b) $C^\infty(M)$ denotes the set of all smooth functions on M, while $\mathfrak{X}^\infty(M)$ denotes the set of all smooth vector fields on M (with analogous notations for their real-analytic counterparts).

REMARK 2.1. Recall that $\mathfrak{X}^\infty(M)$ is a real Lie algebra under the usual Lie bracket multiplication and a module over the ring $C^\infty(M)$.

DEFINITIONS/NOTATION 2.2.

(a) A *distribution* D on M is an assignment of a vector subspace $D(x)$ of T_xM to each $x \in M$.

(b) We denote the family of smooth vector fields in D by

$$\mathrm{vf}^\infty[D] = \{Y \in \mathfrak{X}^\infty(M) \mid Y(x) \in D(x) \ \ \forall x \in M\};$$

note that $\mathrm{vf}^\infty[D]$ is a $C^\infty(M)$ submodule of $\mathfrak{X}^\infty(M)$.

(c) D is *involutive* if $\mathrm{vf}^\infty[D]$ is also a Lie subalgebra of $\mathfrak{X}^\infty(M)$.

(d) D is *smooth* if $\mathrm{vf}^\infty[D](x) = D(x)$ for every $x \in M$.

(e) D is *regular* if the function $x \mapsto \dim D(x)$ is constant on M; otherwise, D is *singular* (if D is smooth, then $x \mapsto \dim D(x)$ is always at least lower semicontinuous).

(f) D is *integrable* if for every $x \in M$ there exists an integral submanifold $\iota: N \to M$ of D with $x \in \iota(N)$ (so $\iota: N \to M$ is an smoothly immersed submanifold of M and $d\iota_y(T_yN) = D(\iota(y))$ for every $y \in N$).

(g) If D is integrable, then D is involutive.

The integral submanifolds of an integrable distribution fit together in a very special way to form what is called a *foliation* of M. This concept is described in the following definitions.

DEFINITION 2.1. A smooth immersed submanifold $\iota: N \to M$ is called a *leaf* of M if given any locally connected topological space Λ and any continuous mapping $g: \Lambda \to M$ with $g(\Lambda) \subseteq \iota(N)$ the induced mapping $\iota^{-1} \circ g: \Lambda \to N$ is continuous in the submanifold topology of N. If a leaf $\iota: N \to M$ is such that $k = \dim N$, then we sometimes call it a k–leaf. A *piece* of a leaf $\iota: N \to M$ is a subset P of N that is open and connected in the submanifold topology of N.

Convention. We henceforth will identify an immersed submanifold $\iota: N \to M$ with its image $\iota(N)$ and view ι as the inclusion map.

DEFINITION 2.2. A smooth (possibly singular) *foliation* of M is a partition F of M into leaves with the additional property that for every $x \in M$ there exists a chart (ϕ, U) of M such that:

(a) $\phi(x) = 0 \in \mathbb{R}^n$;

(b) $\phi(U) = V \times W \subseteq \mathbb{R}^n$, where V is a connected neighborhood of $0 \in \mathbb{R}^k$, W is a connected neighborhood of $0 \in \mathbb{R}^{n-k}$, and $k \in \{0, 1, \ldots, n\}$ is the dimension of the F leaf through x (by convention, we set $\mathbb{R}^0 = \{0\}$, a one-point set);

(c) if L is any leaf of F, then there exists a subset W_L of W such that

$$\phi(U \cap L) = \bigcup_{w \in W_L} V \times \{w\}.$$

The foliation is called *regular* if the all leaves in F have the same dimension.

The definition of integrability is a local condition as stated, but it is more or less well known that the existence of (local) integral submanifolds through each point of M implies the existence of *maximal* integral submanifolds through each point of M. This is proved explicitly in the thesis of Stefan ([26]), but is also present in the earlier work of Hermann ([10]) and Chevalley ([3]).

THEOREM 2.3. *Let D be an integrable C^∞ distribution on M. Then each $x \in M$ is contained in a unique maximal (with respect to inclusion) integral submanifold of D. Moreover, the set of all maximal integral submanifolds of D partitions M and forms a C^∞ foliation with singularities.*

Another common way of defining distributions is by way of families of vector fields.

DEFINITION 2.3. If $\mathrm{S} \subseteq \mathfrak{X}^\infty(M)$ is an $\mathbb{R}$ vector subspace, then the assignment $x \mapsto \mathrm{S}(x)$ defines a C^∞ distribution on M, which we denote by dist [S].

We thus see that there is a correspondence between distributions and families of vector fields, and that integrable distributions induce a foliation of the ambient manifold M. However, this is not of much use unless one has computable criteria for integrability. For regular distributions the classical criterion is given by the Frobenius theorem, which says that integrability is equivalent to involutivity. For possibly singular distributions the next theorem, due to Stefan [26] and Sussmann [27], gives several criteria for integrability.

NOTATION 2.4. For $X \in \mathfrak{X}^\infty(M)$ we let $\mu^X(t,x) = \mu_t^X(x)$ denote the global flow of X; thus $t \mapsto \mu_t^X(x)$ is the unique maximally defined integral curve of X that passes through x at time $t = 0$. Recall that for $t \in \mathbb{R}$ the map $x \mapsto \mu_t^X(x)$ is a diffeomorphism between open subsets of M.

THEOREM 2.5. *For a C^∞ (possibly singular) distribution D on M the following statements are equivalent.*

(1) *D is involutive and for every $X \in vf^\infty[D]$ and for every $x \in M$ the function $t \mapsto \dim D(\mu^X(t,x))$ is constant on its interval of definition.*

(2) *There exists a Lie subalgebra $S \subseteq \mathfrak{X}^\infty(M)$ such that $dist[S] = D$, and the function $t \mapsto \dim S(\mu^X(t,x))$ is constant on its interval of definition for every $X \in S$ and $x \in M$.*

(3) *For every $X \in vf^\infty[D]$ and $(t,x) \in domain(\mu^X)$ we have*

$$d(\mu_t^X)(x)D(x) = D(\mu^X(t,x)).$$

(4) *For every $X \in vf^\infty[D]$ and $(t,x) \in domain(\mu^X)$ we have*

$$d(\mu_t^X)(x)D(x) \subseteq D(\mu^X(t,x)).$$

(5) *For every $X \in vf^\infty[D]$ and $x \in M$ there exist $\epsilon > 0$, a finite set $\{Y_1, \ldots, Y_p\} \subseteq vf^\infty[D]$, and C^∞ functions $a_{ij}: (-\epsilon, \epsilon) \to \mathbb{R}$, $1 \leq i,j \leq p$, such that for $|t| < \epsilon$*

$$D(\mu^X(t,x)) = span\,\{Y_i(\mu^X(t,x)) \mid 1 \leq i \leq p\}$$

and for $1 \leq i \leq p$

$$[X, Y_i](\mu^X(t,x)) = \sum_{j=1}^{p} a_{ij}(t) Y_j(\mu^X(t,x)).$$

(6) *D is integrable.*

Having identified when possibly singular distributions are integrable, we next turn our attention to determining conditions under which the leaves of such distributions are preserved under the flow of a vector field. Such conditions are spelled out in the following theorem, which bears a striking similiarity to the preceding theorem on integrability.

THEOREM 2.6. *For a* C^∞ *(possibly singular) distribution* D *on* M *and a smooth vector field* $X \in \mathfrak{X}^\infty(M)$ *the following statements are equivalent.*

(1) $[X, vf^\infty[D]] \subseteq vf^\infty[D]$ *and for every* $x \in M$ *the function* $t \mapsto \dim D(\mu^X(t,x))$ *is constant on its interval of definition.*

(2) *For every* $(t,x) \in domain(\mu^X)$ *we have*

$$d(\mu_t^X)(x)D(x) = D(\mu^X(t,x)).$$

(3) *For every* $(t,x) \in domain(\mu^X)$ *we have*

$$d(\mu_t^X)(x)D(x) \subseteq D(\mu^X(t,x)).$$

(4) *For every* $x \in M$ *there exist* $\epsilon > 0$, *a finite set* $\{Y_1, \ldots, Y_p\} \subseteq vf^\infty[D]$, *and* C^∞ *functions* $a_{ij}: (-\epsilon, \epsilon) \to \mathbb{R}$, $1 \leq i, j \leq p$, *such that for* $|t| < \epsilon$

$$D(\mu^X(t,x)) = span\{Y_i(\mu^X(t,x)) \mid 1 \leq i \leq p\}$$

and for $1 \leq i \leq p$

$$[X, Y_i](\mu^X(t,x)) = \sum_{j=1}^{p} a_{ij}(t) Y_j(\mu^X(t,x)).$$

Furthermore, if D is also integrable, then the above four statements are equivalent with:

(5) μ^X *locally preserves the leaves of* D*; more precisely, if* $P \subseteq M$ *is a piece of a* D *leaf and if* $t \in \mathbb{R}$ *is such that* $P \subseteq domain(\mu^X)$, *then* $\mu_t^X(P)$ *is a piece of a* D *leaf.*

Proof. Regretably, the proof of this theorem does not seem to have found its way into print, but it is carried out using the techniques developed by Stefan [26] and Sussmann [27] for the proof of Thm. 2.9. We state as a corollary the version of this result when distributions are replaced by families of vector fields. □

COROLLARY 2.1. *If* $S \subseteq \mathfrak{X}^\infty(M)$ *is an* $\mathbb{R}$ *vector subspace and* $X \in \mathfrak{X}^\infty(M)$ *is such that* $[X, S] \subseteq S$ *and for every* $x \in M$ *the function* $t \mapsto$

$\dim S(\mu^X(t,x))$ is constant on its interval of definition, then (4) of the previous theorem holds for $D = dist[S]$; in particular,

$$[X, vf^\infty[dist[S]] \subseteq vf^\infty[dist[S]].$$

Proof. We omit the straightforward proof, but we do note that it is not entirely trivial, since $vf^\infty[dist[S]]$ could be much larger than S, so the invariance condition for $vf^\infty[dist[S]]$ is not obviously equivalent to the invariance condition for S. □

REMARK 2.2. The map $t \mapsto \dim S(\mu^X(t,x))$ is always constant if either S is locally finitely generated ([10]) or real analytic ([17]). These are the two most important situations in which Thms. 2.9 and 2.10 are applied.

3. Control systems and controlled invariance. DEFINITION 3.1. Let Ω be a smooth manifold of dimension m. A smooth *control system* on M with *control space* Ω is a smooth mapping $f: M \times \Omega \to TM$ such that $(\pi \circ f)(x,\omega) = x$ for every $(x,\omega) \in M \times \Omega$. For our purposes here it is sufficient to take the family of all piecewise-smooth maps of $\mathbb{R}$ into Ω as the family of admissible controls.

DEFINITION 3.2. Let $f: M \times \Omega \to TM$ be a smooth control system. A family of vector fields $S \subseteq \mathfrak{X}^\infty(M)$ is called f-invariant if $[f^\omega, S] \subseteq S$ for every $\omega \in \Omega$, where f^ω is the smooth vector field $f^\omega(x) = f(x,\omega)$. A smooth distribution D is f-invariant if $vf^\infty[D]$ is f-invariant.

A standard application of f-invariance is given by the following theorem. It is most conveniently stated for real-analytic systems, since in the smooth case one needs additional completeness and constant dimensionality assumptions.

THEOREM 3.1. *([12],[13],[14]). Consider the real-analytic, input-output system*

$$\begin{aligned} \dot{x} &= f(x,\omega) + d(x), \\ y &= h(x), \end{aligned}$$

where $f: M \times \Omega \to TM$, $d \in \mathfrak{X}^\omega(M)$, and $h: M \to \mathbb{R}^p$, and let S denote the smallest f-invariant Lie subalgebra of $\mathfrak{X}^\omega(M)$ that contains d. Then the output y is independent of the "disturbance" d if and only if $S \subseteq \ker dh$.

DEFINITION 3.3. A smooth distribution D is called *locally controlled invariant* for a control system $f: M \times \Omega \to TM$ at (x_0, ω_0) in $M \times \Omega$ if there exist an open neighborhood $U_0 \times V_0$ of (x_0,ω_0) in $M \times \Omega$ and a smooth map $\alpha: U_0 \times V_0 \to \Omega$ such that:

(i) $\alpha(x_0,\omega_0) = \omega_0$;

(ii) the associated mapping $\Phi: U_0 \times V_0 \to M \times \Omega$ given by

$$\Phi(x,\nu) = (x, \alpha(x,\nu))$$

is a diffeomorphism of $U_0 \times V_0$ onto an open subset of $M \times \Omega$;

(iii) the feedback adjusted system

$$\tilde{f}: U_0 \times V_0 \to TU_0, \tilde{f}(x,\nu) = f(x, \alpha(x,\nu)) \tag{3.1}$$

satisfies

$$[\tilde{f}^\nu, Y] \in \mathrm{vf}[\mathrm{D}|_{U_0}] \tag{3.2}$$

for every $\nu \in V_0$ and $Y \in \mathrm{vf}[\mathrm{D}|_{U_0}]$, where $\mathrm{D}|_{U_0}$ denotes the restriction of the distribution D to the open set U_0.
A mapping α satisfying these conditions will be called an *invariant feedback* for f relative to D.

REMARK 3.1. Condition (ii) of Def. 3.4 ensures that there is, at least locally, a one-to-one correspondence between trajectories of the original system f and the feedback adjusted system $\tilde{f}$.

DEFINITION 3.4. Let D be a smooth, regular, involutive distribution on M of dimension $n - p > 0$. A *Frobenius chart* at $x_0 \in M$ is a cubic coordinate chart $\phi: U_0 \to \mathbb{R}^n$ centered at x_0 (i. e., $\phi(U_0)$ is either all of $\mathbb{R}^n$ or an open cube in $\mathbb{R}^n$ of the form $(-\epsilon, \epsilon)^n$ for some $\epsilon > 0$, and $\phi(x_0) = 0$) with the property that for every $x \in U_0$

$$\mathrm{D}(x) = \mathrm{span}_{\mathbb{R}} \left\{ \left.\frac{\partial}{\partial \phi_j}\right|_x \; : \; p+1 \le j \le n \right\}, \tag{3.3}$$

where $\phi_1, \ldots, \phi_n$ are the coordinate functions of ϕ. With the stated assumptions on D, the existence of a Frobenius chart at every point of M is guaranteed by the local form of the Frobenius theorem ([28]).

REMARK 3.2. If U_0 is the domain of a Frobenius chart $\phi = (\phi_1, \ldots, \phi_n)$ for D, then it is easy to see that the invariance condition (2) is equivalent to the set of equations

$$\frac{\partial \tilde{f}_i^\nu}{\partial \phi_j}(x) = 0, \quad 1 \le i \le p, \quad p+1 \le j \le n, \tag{3.4}$$

for every $(x, \nu) \in U_0 \times V_0$, where $\tilde{f}_i^\nu$ is the ith component function of the vector field $\tilde{f}^\nu$ in the chart ϕ; i. e., $\tilde{f}^\nu = \sum_{i=1}^n \tilde{f}_i^\nu \frac{\partial}{\partial \phi_i}$.

4. Controlled invariance for a class of singular distributions

In this section we discuss a generalization of Thm. 1.1 in which the regularity requirement on the distribution D is weakened. To simplify the exposition, we work locally in the domain of a coordinate chart of M; a global version of these results is currently in progress and will be reported elsewhere. Thus we let $(x_1, \ldots, x_n)$ be a system of local coordinates on M that allows us to identify (an open subset of) M with $\mathbb{R}^n$. We consider a family of smooth vector fields

$$\mathrm{S} = \mathrm{span}_{C^\infty(M)} \{X_{p+1}, \ldots, X_n\} \tag{4.1}$$

where $0 \leq p \leq n-1$ and $X_{p+1}, \ldots, X_n \in \mathfrak{X}^\infty(M)$ are such that there exist functions $h_{j\ell} \in C^\infty(M), p+1 \leq \ell, j \leq n$ for which

$$X_\ell = \sum_{j=p+1}^{n} h_{j\ell} \frac{\partial}{\partial x_j}, \quad p+1 \leq \ell \leq n; \tag{4.2}$$

we could say that S is *subordinate* to the regular distribution

$$\mathrm{D} = \mathrm{span}_{\mathbb{R}} \left\{ \frac{\partial}{\partial x_{p+1}}, \ldots, \frac{\partial}{\partial x_n} \right\}. \tag{4.3}$$

We further assume that:

(A1): The $(n-p) \times (n-p)$-matrix function $H(x) = [h_{j\ell}(x)]$ has nonzero determinant on an open dense subset of M.

The singularities of dist [S] then occur on the closed, nowhere-dense subset of M where $H = 0$. The family of vector fields S is locally finitely generated by assumption, so to ensure the integrability of S it suffices to assume (see Rem. 2.12):

(A2): S is involutive, or equivalently $[X_i, X_j] \in \mathrm{S}$ for $p+1 \leq i, j \leq n$.

In terms of the functions $h_{j\ell}$ the involutivity of S is equivalent to the existence of smooth functions $\alpha_{jq\ell} \in C^\infty(M)$, $p+1 \leq j, q, \ell \leq n$ such that:

(A2)′: $$\sum_{j=p+1}^{n} \left(h_{jq} \frac{\partial h_{i\ell}}{\partial x_j} - h_{j\ell} \frac{\partial h_{iq}}{\partial x_j} \right) = \sum_{j=p+1}^{n} h_{ij} \alpha_{jq\ell} \quad p+1 \leq j, q, \ell \leq n.$$

Before discussing the invariance of S under a control system, we first examine sufficient conditions for S to be invariant under a single smooth vector field. An arbitrary vector field $f \in \mathfrak{X}^\infty(M)$ has the local representation

$$f = \sum_{i=1}^{n} f_i \frac{\partial}{\partial x_i},$$

where $f_i \in C^\infty(M), 1 \leq i \leq n$, so for $p+1 \leq \ell \leq n$ we obtain

$$\begin{aligned}
[f, X_\ell] &= \left[\sum_{i=1}^{n} f_i \frac{\partial}{\partial x_i}, \sum_{j=p+1}^{n} h_{j\ell} \frac{\partial}{\partial x_j} \right] \\
&= \sum_{i=1}^{n} \sum_{j=p+1}^{n} \left(f_i \frac{\partial h_{j\ell}}{\partial x_i} \frac{\partial}{\partial x_j} - h_{j\ell} \frac{\partial f_i}{\partial x_j} \frac{\partial}{\partial x_i} \right) \\
&= \sum_{i=p+1}^{n} \left(\sum_{j=1}^{n} f_j \frac{\partial h_{i\ell}}{\partial x_j} \right) \frac{\partial}{\partial x_i} - \sum_{i=1}^{n} \left(\sum_{j=p+1}^{n} h_{j\ell} \frac{\partial f_i}{\partial x_j} \right) \frac{\partial}{\partial x_i} \qquad (4.4)
\end{aligned}$$

Since an arbitrary element of S has the form

$$\sum_{j=p+1}^{n} \alpha_j X_j = \sum_{i=p+1}^{n} \left(\sum_{j=p+1}^{n} h_{ij}\alpha_j \right) \frac{\partial}{\partial x_i} \tag{4.5}$$

for some choice of smooth functions $\alpha_{p+1}, \ldots \alpha_n \in C^\infty(M)$, a sufficient condition for S to be f-invariant (i. e., $[f, \mathrm{S}] \subseteq \mathrm{S}$) is that

$$[f, X_\ell] \in \mathrm{S} \quad p+1 \leq \ell \leq n. \tag{4.6}$$

Comparison of the coefficients of $\partial/\partial x_i$ for $1 \leq i \leq p$ in (4.4) and (4.5) shows that we must have

$$\sum_{j=p+1}^{n} h_{j\ell} \frac{\partial f_i}{\partial x_j} = 0, \quad 1 \leq i \leq p, p+1 \leq \ell \leq n.$$

Since $H(x) = [h_{j\ell}(x)]$ is invertible on an open dense subset of M, this yields

$$\text{(I1):} \qquad \frac{\partial f_i}{\partial x_j} = 0, \quad 1 \leq i \leq p, p+1 \leq j \leq n,$$

which is the usual invariance condition for the regular distribution D. Comparison of the coefficients of $\partial/\partial x_i$ for $p+1 \leq i \leq n$ in (4.4) and (4.5) shows that we must have smooth functions $\alpha_{j\ell} \in C^\infty(M)$, $p+1 \leq j, \ell \leq n$ such that

$$\text{(I2):} \qquad \sum_{j=1}^{n} f_j \frac{\partial h_{i\ell}}{\partial x_j} - \sum_{j=p+1}^{n} h_{j\ell} \frac{\partial f_i}{\partial x_j} = \sum_{j=p+1}^{n} h_{ij}\alpha_{j\ell}$$
$$p+1 \leq i \leq n, p+1 \leq \ell \leq n.$$

Thus we have deduced the following result.

PROPOSITION 4.1. *The family of vector fields S is invariant under the vector field* $f = \sum_{i=1}^n f_i \frac{\partial}{\partial x_i} \in \mathfrak{X}^\infty(M)$ *if and only if the invariance conditions (I1) and (I2) are satisfied.*

REMARK 4.1. It is easy to see that (I2) is always satisfied in the regular case where $H(x) = [h_{j\ell}(x)]$ is invertible throughout M.

Now suppose we have a smooth control system $f: M \times \Omega \to TM$. As mentioned before, we will identify M with $\mathbb{R}^n$ via the system of coordinates $(x_1, \ldots, x_n)$, and in addition we will identify Ω with $\mathbb{R}^m$ via the system of coordinates $\omega = (\omega_1, \ldots, \omega_m)$. The control system f on $M \times \Omega$ also induces a(n uncontrolled) vector field

$$f \times 0: M \times \Omega \to TM \times T\Omega$$

defined by

$$(f \times 0)(x, \omega) = (f(x, \omega), 0_\omega),$$

where 0_ω denotes the zero vector in the tangent space $T_\omega\Omega$. Likewise, for any smooth vector field $X \in \mathfrak{X}^\infty(M)$ we can define an associated vector field $X \times 0$ on $M \times \Omega$ by

$$(X \times 0)(x, \omega) = (X(x), 0_\omega).$$

In particular, for the coordinate vector field $(\partial/\partial x_i)_M$ induced by the chart $(x_1, \ldots, x_n)$ of M and the coordinate vector field $(\partial/\partial x_i)_{M\times\Omega}$ induced by the chart $(x_1, \ldots, x_n, \omega_1, \ldots, \omega_m)$ of $M \times \Omega$, it is easy to see that

$$\left(\frac{\partial}{\partial x_i}\right)_M \times 0 = \left(\frac{\partial}{\partial x_i}\right)_{M\times\Omega}.$$

For ease of notation (and with minimal risk of confusion) we shall use the same symbol $\partial/\partial x_i$ for any one of the three vector fields

$$\left(\frac{\partial}{\partial x_i}\right)_M, \quad \left(\frac{\partial}{\partial x_i}\right)_M \times 0, \quad \left(\frac{\partial}{\partial x_i}\right)_{M\times\Omega}.$$

To formulate a sufficient condition for the family of vector fields S to be controlled invariant under the control system f we define another family of vector fields G_K by

$$\mathrm{G}_K = \mathrm{span}_{C^\infty(M\times\Omega)}\left\{\frac{\partial f}{\partial \omega_k} \times 0 \mid k \in K\right\}. \tag{4.7}$$

Here K is a nonempty subset of $\{1, \ldots, m\}$ and

$$\frac{\partial f}{\partial \omega_k} \colon M \times \Omega \to TM$$

for $1 \leq k \leq m$ is the smooth control system given in local coordinates by

$$\frac{\partial f}{\partial \omega_k} = \sum_{i=1}^{n} \frac{\partial f_i}{\partial \omega_k}\frac{\partial}{\partial x_i}.$$

Observe that in the case where f is affine in the control (see Sect. I) G is just a family of vector fields spanned by a subset of the input vector fields.

We can now state our main result on the controlled invariance of certain singular distributions. For completeness, some of the earlier notation is reviewed in the statement of the thoerem. Also the reader should recall that in the local treatment given here $M = \mathbb{R}^n$ and $\Omega = \mathbb{R}^m$.

THEOREM 4.1. *Let $0 \leq p \leq n-1$ and let $X_{p+1}, \ldots, X_n$ be smooth vector fields on M for which there exist functions $h_{j\ell} \in C^\infty(M), p+1 \leq j, \ell \leq n$ such that*

$$X_\ell = \sum_{j=p+1}^{n} h_{j\ell}\frac{\partial}{\partial x_j}, \quad p+1 \leq \ell \leq n,$$

and the $(n-p)\times(n-p)$-matrix function $H(x)=[h_{j\ell}(x)]$ has nonzero determinant on an open dense subset of M. Consider the family of vector fields

$$S=span_{C^\infty(M)}\{X_{p+1},\dots,X_n\}$$

and suppose that S is involutive (in terms of the functions $h_{j\ell}$ this assumption is equivalent to (A2)$'$). Let $f\colon M\times\Omega\to TM$ be a smooth control system and suppose that there exists a nonempty subset $K\subseteq\{1,\dots,m\}$ such that the vector fields

$$\left\{\frac{\partial f}{\partial\omega_k}(x,\omega)\mid k\in K\right\}$$

are linearly independent modulo $\{\frac{\partial}{\partial x_{p+1}},\dots,\frac{\partial}{\partial x_n}\}$ on an open dense subset of $M\times\Omega$ and such that there exist vector fields

$$Y_{p+1},\dots,Y_n\in G_K=span_{C^\infty(M\times\Omega)}\left\{\frac{\partial f}{\partial\omega_k}\times 0\mid k\in K\right\}$$

for which the following invariance condition is satisfied:

$$\text{(INV)}\qquad [f\times 0,X_\ell\times 0]\in S\times 0+\sum_{j=p+1}^{n}h_{j\ell}(Y_j\times 0),\quad p+1\le\ell\le n.$$

Then S is locally controlled invariant for f at $(0,0)\in M\times\Omega$; i. e., there exists a smooth locally defined feedback $\omega=\alpha(x,\nu)$ with the property that S is invariant under the control system $\tilde f(x,\nu)=f(x,\alpha(x,\nu))$.

Proof. The assumed invariance condition implies that there exist $\alpha_{j\ell}\in C^\infty(M\times\Omega)$ for $p+1\le j,\ell\le n$ such that

$$[f\times 0,X_\ell\times 0]=\sum_{j=p+1}^{n}\alpha_{j\ell}X_j+\sum_{j=p+1}^{n}h_{j\ell}(Y_j\times 0). \tag{4.8}$$

Since $Y_j\in G_K$ there exist smooth functions $\beta_{kj}\in C^\infty(M\times\Omega)$, $k\in K, p+1\le j\le n$ such that

$$Y_j=\sum_{k\in K}\beta_{kj}\frac{\partial f}{\partial\omega_k}=\sum_{k\in K}\beta_{kj}\left(\sum_{i=1}^{n}\frac{\partial f_i}{\partial\omega_k}\frac{\partial}{\partial x_i}\right).$$

Using this and the form of the vector fields X_ℓ given by (4.2), we obtain

$$\begin{aligned}[f\times 0,X_\ell\times 0] &= \sum_{i=p+1}^{n}\left(\sum_{j=p+1}^{n}h_{ij}\alpha_{j\ell}\right)\frac{\partial}{\partial x_i}\\ &\quad+\sum_{i=1}^{n}\left(\sum_{j=p+1}^{n}\left(\sum_{k\in K}\beta_{kj}\frac{\partial f_i}{\partial\omega_k}\right)\right)\frac{\partial}{\partial x_i}.\end{aligned}\tag{4.9}$$

Comparison of the coefficients of $\partial/\partial x_i$ for $1 \le i \le p$ in (4.4) and (4.9) shows that we must have

$$\sum_{j=p+1}^{n} h_{j\ell}\left(\frac{\partial f_i}{\partial x_j} + \sum_{k\in K}\beta_{kj}\frac{\partial f_i}{\partial \omega_k}\right) = 0, \quad 1 \le i \le p, p+1 \le \ell \le n.$$

Since $H(x) = [h_{j\ell}(x)]$ is invertible on an open dense subset of M, this yields

$$\frac{\partial f_i}{\partial x_j} + \sum_{k\in K}\beta_{kj}\frac{\partial f_i}{\partial \omega_k} = 0, \quad 1 \le i \le p, p+1 \le j \le n. \tag{4.10}$$

Similarly, comparison of the coefficients of $\partial/\partial x_i$ for $p+1 \le i \le n$ in (4.4) and (4.9) shows that

$$\sum_{j=1}^{n} f_j \frac{\partial h_{i\ell}}{\partial x_j} - \sum_{j=p+1}^{n} h_{j\ell}\left(\frac{\partial f_i}{\partial x_j} + \sum_{k\in K}\beta_{kj}\frac{\partial f_i}{\partial \omega_k}\right) = \sum_{j=p+1}^{n} h_{ij}\alpha_{j\ell} \qquad p+1 \le i \le n, p+1 \le \ell \le n. \tag{4.11}$$

We next define vector fields $Z_j \in \mathfrak{X}^\infty(M \times \Omega)$ by

$$Z_j = \frac{\partial}{\partial x_j} + \sum_{k\in K}\beta_{kj}\frac{\partial}{\partial \omega_k}, \quad p+1 \le j \le n, \tag{4.12}$$

and we let $\mu_t^j(x,\nu)$ denote the flow of Z_j (thus for each $(x,\nu) \in M \times \Omega$ the map $t \mapsto \mu_t^j(x,\nu)$ denotes the unique, maximally defined integral curve of Z_j that passes through (x,ν) at time zero). Then for $1 \le i \le p, p+1 \le j \le n$ (4.10) yields

$$\begin{aligned} 0 &= \frac{\partial f_i}{\partial x_j} + \sum_{k\in K}\beta_{kj}\frac{\partial f_i}{\partial \omega_k} \\ &= df_i\left(\frac{\partial}{\partial x_j} + \sum_{k\in K}\beta_{kj}\frac{\partial}{\partial \omega_k}\right) \\ &= df_i Z_j, \end{aligned}$$

so each f_i, $1 \le i \le p$ is constant along each the flow of each Z_j, $p+1 \le j \le n$.

Since the flows μ^j of the vector fields Z_j are defined and smooth on open subsets of $\mathbb{R} \times M \times \Omega$, the mapping Φ given by

$$\Phi(x,\nu) = \mu_{x_n}^n \circ \cdots \circ \mu_{x_{p+1}}^{p+1}((x_1,\ldots,x_p,0,\ldots,0),\nu)$$

is defined and smooth on an open neighborhood of $(0,0) \in M \times \Omega$ of the form

$$U_0 \times V_0 = \{(x,\nu) \in M \times \Omega \mid |x| < \epsilon, |\nu| < \epsilon\},$$

for some $\epsilon > 0$. It is easy to see that Φ is a diffeomorphism of $U_0 \times V_0$ onto an open subset of $M \times \Omega$ and has the form

$$\Phi(x,\nu) = (x, \alpha(x,\nu))$$

for some smooth map $\alpha: U_0 \times V_0 \to \Omega$ (see [9] for a detailed argument). The proof will be completed by showing that α is the desired invariant feedback.

To this end, we first note that for $1 \le i \le p$ we have

$$\begin{aligned} f_i(\Phi(x,\nu)) &= f_i(x,\alpha(x,\nu)) \\ &= f_i(\mu^n_{x_n} \circ \cdots \circ \mu^{p+1}_{x_{p+1}}((x_1,\ldots,x_p,0,\ldots,0),\nu)) \\ &= f_i((x_1,\ldots,x_p,0,\ldots,0),\nu), \end{aligned}$$

since the functions $f_i, 1 \le i \le p$ are constant along the integral curves $t \mapsto \mu^j_t(x,\nu), p+1 \le j \le n$. Hence we obtain

$$\frac{\partial}{\partial x_j}(f_i \circ \Phi) = 0 \quad 1 \le i \le p, p+1 \le j \le n,$$

which is precisely the invariance condition (I1) for $f \circ \Phi$. Thus if S is regular, then α is the desired invariant feedback. However, in the nonregular case the invariance condition (I2) for $f \circ \Phi$ must also be verified, which in turn requires the following fact.

Claim. The vector fields $Z_{p+1}, \ldots, Z_n$ commute pairwise; i. e., $[Z_j, Z_\ell] = 0$ for $p+1 \le j, \ell \le n$.

We have already established that for $p+1 \le j, \ell \le n$ and $1 \le i \le p$ we have $Z_j f_i = Z_\ell f_i = 0$, from which it follows that

$$[Z_j, Z_\ell] f_i = Z_j(Z_\ell f_i) - Z_\ell(Z_j f_i) = 0. \tag{4.13}$$

From the form of the vector fields $Z_{p+1}, \ldots, Z_n$ (see (4.12)) we obtain

$$[Z_j, Z_\ell] = \sum_{k \in K} \lambda_{kj\ell} \frac{\partial}{\partial \omega_k}, \tag{4.14}$$

for appropriate functions $\lambda_{kj\ell} \in C^\infty(M \times \Omega)$, $k \in K, p+1 \le j, \ell \le n$. Equations (4.13) and (4.14) combine to yield

$$\sum_{k \in K} \lambda_{kj\ell} \frac{\partial f_i}{\partial \omega_k} = 0, \quad 1 \le i \le p, k \in K, p+1 \le j, \ell \le n.$$

By assumption, the vector fields

$$\frac{\partial f}{\partial \omega_k}, k \in K$$

are linearly independent modulo $\{\frac{\partial}{\partial x_{p+1}}, \ldots, \frac{\partial}{\partial x_n}\}$ on an open dense subset of $M \times \Omega$, so we infer that the functions $\lambda_{kj\ell}$ are zero on an open dense subset of $M \times \Omega$, and hence are zero everywhere on $M \times \Omega$ by continuity. Thus (4.14) implies that $[Z_j, Z_\ell] = 0$ for $p+1 \le j, \ell \le n$, and the claim is proved.

An immediate corollary of the claim is that the flows of the vector fields $Z_j, p+1 \le j \le n$ also commute; i. e.,

$$\mu_t^j \circ \mu_s^\ell(x,\nu) = \mu_s^\ell \circ \mu_t^j(x,\nu) \tag{4.15}$$

for $p+1 \le j, \ell \le n$ and for every $t, s \in \mathbb{R}$ and $(x,\nu) \in M \times \Omega$ for which both sides of (4.15) are defined. Thus for $p+1 \le j \le n$ and $p+1 \le i \le n$ we have

$$\begin{aligned}
&\frac{\partial}{\partial x_j}(f_i \circ \Phi)(x,\nu) \\
&= \frac{\partial}{\partial x_j} f_i(\mu_{x_n}^n \circ \cdots \circ \mu_{x_{p+1}}^{p+1}((x_1, .., x_p, 0, .., 0), \nu)) \\
&\quad \text{(by the definition of } \Phi) \\
&= \frac{\partial}{\partial x_j} f_i(\mu_{x_j}^j \circ \mu_{x_n}^n \circ \cdots \circ \widehat{\mu_{x_j}^j} \circ \cdots \mu_{x_{p+1}}^{p+1}((x_1, .., x_p, 0, .., 0), \nu)) \\
&\quad \text{(since the flows commute; the term with the hat is omitted)} \\
&= (Z_j f_i)(\Phi(x,\nu)) \\
&= \left(\frac{\partial f_i}{\partial x_j} + \sum_{k \in K} \beta_{kj} \frac{\partial f_i}{\partial \omega_k} \right)(x,\nu).
\end{aligned}$$

If we evaluate both sides of (4.11) on $\Phi(x,\nu)$ and note that $(h_{j\ell} \circ \Phi)(x,\nu) = h_{j\ell}(x)$ by the form of Φ and the fact that the functions $h_{j\ell}$ depend only on x, then we get

$$\sum_{j=1}^n (f_j \circ \Phi) \frac{\partial h_{i\ell}}{\partial x_j} - \sum_{j=p+1}^n h_{j\ell} \frac{\partial}{\partial x_j}(f_i \circ \Phi) = \sum_{j=p+1}^n h_{ij}(\alpha_{j\ell} \circ \Phi)$$

for $p+1 \le i \le n, p+1 \le \ell \le n$. This is precisely the invariance condition (I2) for the family of vector fields $\tilde{f}^\nu$ of the feedback adjusted control system $\tilde{f} = f \circ \Phi$. By Prop. 4.1 S is invariant under the vector fields $\tilde{f}^\nu$, and the proof is complete.

REMARK 4.2. (i) It is not difficult to see that the invariance condition (INV) does not depend on the choice of the $n-p$ vector fields $X_{p+1}, \ldots, X_n$ which span the family of vector fields S in the sense that if (INV) holds for one spanning (over $C^\infty(M)$) set of $n-p$ vector fields, then it holds for any other spanning set of $n-p$ vector fields (but with possibly different functions $h_{j\ell}$).

(ii) The invariance condition (INV) reduces to the invariance condition given in Thm. 1.1 when the distribution dist[S] is regular (which is equivalent to the vector fields $X_{p+1}, \ldots, X_n$ being linearly independent throughout M).

(iii) If the vector fields Z_j are complete, then the feedback mapping α is defined throughout $M \times \Omega = \mathbb{R}^n \times \mathbb{R}^m$. This is always the case for systems affine in the control (see [9] for details).

EXAMPLE 4.2. Let $M = \mathbb{R}^3, \Omega = \mathbb{R}$ and let $\mathrm{S} = \mathrm{span}_{C^\infty(M)}\{X_2, X_3\}$ where

$$X_2 = x_3\frac{\partial}{\partial x_2} - x_2\frac{\partial}{\partial x_3}, \quad X_3 = x_1x_2\frac{\partial}{\partial x_2} + x_1x_3\frac{\partial}{\partial x_3}.$$

It is easy to see that S has singularities near the origin. However, S is involutive, since $[X_2, X_3] = 0$, and thus S is integrable because it is locally finitely generated. In fact the leaves of S are of three types: single points on the x_1-axis, circles centered at the origin in the x_2x_3 plane, and punctured planes parallel to the x_2x_3-plane of the form $P_a = \{(a, x_2, x_3) \mid (x_2, x_3) \in \mathbb{R}^2 \setminus (0,0)\}$ for $a \neq 0$. Consider the nonlinear control system

$$\begin{aligned}
\dot{x}_1 &= x_1p(x_1) - x_1x_2x_3 + x_1\omega \\
\dot{x}_2 &= -x_3 + 4x_2 - x_1x_2x_3^2 - x_1x_2^3x_3 + (x_1x_3 + x_1x_2^2)\omega \\
\dot{x}_3 &= x_2 + 4x_3 + x_1x_2^2x_3 - x_1x_2^2x_3^2 + (-x_1x_2 + x_1x_2x_3)\omega,
\end{aligned}$$

which we also write as $\dot{x} = f(x,\omega)$ with $x = (x_1, x_2, x_3)$; here $p(x_1)$ denotes any smooth function of one variable. If we let

$$Y_2 = x_3\frac{\partial f}{\partial \omega}, \quad Y_3 = x_2\frac{\partial f}{\partial \omega},$$

then obviously

$$Y_2, Y_3 \in \mathrm{span}_{C^\infty(M\times\Omega)}\left\{\frac{\partial f}{\partial \omega}\right\}$$

and a routine computation shows that for every $\omega \in \mathbb{R}$ we have

$$\begin{aligned}
[f^\omega, X_2] &= x_3(x_2x_3 - \omega)X_3 + x_3Y_2 - x_2Y_3, \\
[f^\omega, X_3] &= (p(x_1) - x_2x_3 + x_1x_2^2x_3 + \omega - x_1x_2\omega)X_3 + x_1x_2Y_2 + x_1x_3Y_3,
\end{aligned}$$

which is precisely the invariance condition (INV) for this system. If the procedure developed in the proof of Thm. 4.1 is applied to this system, then one obtains for the vector fields Z_j in equation (4.13)

$$Z_2 = \frac{\partial}{\partial x_2} + x_3\frac{\partial}{\partial \omega}, \quad Z_3 = \frac{\partial}{\partial x_3} + x_2\frac{\partial}{\partial \omega}.$$

The flows of these vector fields are easily computed and result in the feedback mapping $\alpha(x,\nu) = \nu + x_2x_3$, which in turn yields the feedback adjusted system $\dot{x} = \tilde{f}(x,\nu) = f(x, \alpha(x,\nu))$ given by

$$\dot{x}_1 = x_1(p(x_1) + \nu)$$

$$\dot{x}_2 = -x_3 + 4x_2 + (x_1x_3 + x_1x_2^2)\nu$$
$$\dot{x}_3 = x_2 + 4x_3 + (-x_1x_2 + x_1x_2x_3)\nu.$$

The family of vector fields S is evidently invariant under this system.

REFERENCES

[1] C. I. Byrnes. "Toward a global theory of (f,g)–invariant distributions with singularities," in *Mathematical Theory of Networks and Systems, Proceedings of the MTNS-83 International Symposium, Beer-Sheba, Israel, 1983* (P. A. Fuhrmann, Ed.), Springer-Verlag, New York, NY, 1984, pp. 149–165.

[2] C. I. Byrnes and A. J. Krener. "On the existence of globally (f,g)–invariant distributions," in *Differential Geometric Control Theory* (R. W. Brockett, R. S. Millman, H. J. Sussmann, Eds.), Birkhäuser, Boston, MA, 1983, pp. 209–225.

[3] C. Chevalley. *Theory of Lie Groups*, Princeton Univesity Press, Princeton, NJ, 1946.

[4] D. Cheng and T. J. Tarn. "New result on (f,g)–invariance," Systems Control Lett., **12**(1989), pp. 319–326.

[5] W. Dayawansa, D. Cheng, W. M. Boothby, and T. J. Tarn. "On the global (f,g)–invariance of a class of nonlinear systems" in *Proceedings of the 25th IEEE Conference on Decision and Control*, IEEE Control Systems Society, 1986, pp. 2065–2068.

[6] W. Dayawansa, D. Cheng, W. M. Boothby, and T. J. Tarn. "Global (f,g)–invariance of nonlinear systems," SIAM J. Control Optim., **26**(1988), pp. 1119–1132.

[7] K. A. Grasse. "A notion of global decomposition for systems of vector fields," in *Analysis and Control of Nonlinear Systems* (C. I. Byrnes, C. F. Martin, R. E. Saeks, Eds.), North Holland, New York, NY, 1988, pp. 323–330.

[8] K. A. Grasse. "Invariant distributions, global decompositions of nonlinear systems, and applications," in *Proceedings of the 28th IEEE Conference on Decision and Control*, IEEE Control Systems Society, Piscataway, NJ, 1989, pp. 1981–1986.

[9] K. A. Grasse. "On controlled invariance for fully nonlinear systems," Internat. J. Control, **56**(1992), pp. 1121–1137.

[10] R. Hermann. "The differential geometry of foliations II," J. Math. Mech., **11**(1962), pp. 303–315.

[11] R. M. Hirschorn. "Global controllability of nonlinear systems," SIAM J. Control Optim., **14**(1976), pp. 700–711.

[12] R. M. Hirschorn. "(A,B)–invariant distributions and disturbance decoupling of nonlinear systems," SIAM J. Control Optim., **19**(1981), pp. 1–19.

[13] A. Isidori. *Nonlinear Control Systems: An Introduction*, Springer-Verlag, New York, NY, 1985.

[14] A. Isidori, A. J. Krener, C. Gori-Giorgi, and S. Monaco. "Nonlinear decoupling via feedback: a differential geometric appraoch," IEEE Trans. Automat. Control, **AC-26**(1981), pp. 331–345.

[15] A. Isidori, A. J. Krener, C. Gori-Giorgi, and S. Monaco. "Locally (f,g)–invariant distributions," Systems Control Lett., **1**(1981), pp. 12–15.

[16] A. J. Krener. "$(\mathrm{Ad}_{f,g}), (\mathrm{ad}_{f,g})$ and locally $(\mathrm{ad}_{f,g})$ invariant and controllability distributions," SIAM J. Control Optim., **23**(1985), pp. 523–549.

[17] T. Nagano. "Linear differential systems with singularities and an application to transitive Lie algebras," J. Math. Soc. Japan, **18**(1966), pp. 398–404.

[18] H. Nijmeijer. "Controlled invariance for affine control systems," Internat. J. Control, **34**(1981), pp. 825–833.

[19] H. Nijmeijer. "The triangular decoupling problem for nonlinear control systems," Nonlinear Anal., **8**(1984), pp. 273–279.

[20] H. Nijmeijer and A. J. van der Schaft. "Controlled invariance for nonlinear systems," IEEE Trans. Automat. Control, **AC-27**(1982), pp. 904–914.

[21] H. Nijmeijer and A. J. van der Schaft. "The disturbance decoupling problem for nonlinear control systems," IEEE Trans. Automat. Control, **AC-28**(1983), pp. 621–623.

[22] H. Nijmeijer and A. J. Van der Schaft. *Nonlinear Dynamical Control Systems*, Springer-Verlag, New York, NY, 1990.

[23] V. Ramakrishna. "Controlled invariance for singular distributions," SIAM J. Control Optim., **32**, No. 3 (1994), pp. 790–807.

[24] V. Ramakrishna. "Local solvability of degenerate, overdetermined, first order partial differential equations—A control theoretic perspective," preprint.

[25] W. Respondek. "On decomposition of nonlinear control systems," Systems Control Lett., **1**(1982), pp. 301–308.

[26] P. Stefan. *Accessibility and Singular Foliations*, Ph.D. Thesis, University of Warwick, 1973.

[27] H. Sussmann. "Orbits of families of vector fields and integrability of distributions," Trans. Amer. Math. Soc., **180**(1973), pp. 171–188.

[28] F. W. Warner. *Foundations of Differentiable Manifolds and Lie Groups*, Scott, Foresman and Company, Glenview, IL, 1971.

[29] M. Wonham. *Linear Multivariable Control*, Springer-Verlag, New York, NY, 1974.

DYNAMIC FEEDBACK STABILIZATION*

HENRY HERMES†

1. Introduction. The affine control system

$$\dot{x}_1 = u, \quad \dot{x}_2 = x_1^3, \quad \dot{x}_3 = x_1^2 x_2 \tag{1.1}$$

does not admit a continuous, asymptotically stabilizing, state feedback control, i.e., there is no continuous function $u^*(x)$ which makes the rest solution $x = 0$ of (1.1) locally asymptotically stable. The main result, here, is the explicit construction of a smooth (actually real analytic) globally asymptotically stabilizing "dynamic" feedback control $u^*(\tau, x)$ for system (1.1) obtained from the dynamic extension which adjoins the single equation $\dot{\tau} = |x|^2$.

In more generality consider an n-dimensional, real analytic, affine system

$$\dot{x} = X(x) + uY(x), \qquad X(0) = 0. \tag{1.2}$$

For system (1.1), $X(x) = x_1^3\, \partial/\partial x_2 + x_1^2 x_2\, \partial/\partial x_3$, $Y = \partial/\partial x_1$. System (1.2) is small time locally controllable (STLC) at zero if for any $t_1 > 0$ all points in a neighborhood of zero can be reached in time t_1 by solutions of (1.2) which begin at zero at time zero and correspond to measurable open loop controls $t \to u(t)$. Let $[X, Y] = (\operatorname{ad} X, Y)$ denote the Lie bracket of the vector fields X, Y and inductively $(\operatorname{ad}^{k+1} X, Y) = [X, (\operatorname{ad}^k X, Y)]$. For system (1.1), the brackets which do not vanish at zero (modulo the Jacobi identity) are $Y(0)$, $(\operatorname{ad}^3 Y, X)(0)$ and $(\operatorname{ad}^2 Y, [X, (\operatorname{ad}^3 Y, X)])(0)$ giving what is called an "odd system" and hence we have STLC at zero. (See [He 1], [S]). When X and Y are real analytic, STLC also implies the origin can be reached, starting from all points in a neighborhood of the origin, by solutions corresponding to open loop controls. This is a first necessary condition for the existence of a continuous, asymptotically stabilizing (state) feedback control (ASFC) given by Brockett, [B]. Suppose $u^*(x)$ is such a control for (1.2). Then system (1.2), with u replaced by u^*, has $x = 0$ as an asymptotically stable solution hence it admits a (C^∞) Lyapunov function v. (See [Ku].) Let $W(x) = X(x) + u^*(x)Y(x)$; our stability assumption implies $x = 0$ is an isolated zero of W. Then the degree of the map $W/|W| : S^{n-1} \to S^{n-1}$ on a small $(n-1)$ sphere S^{n-1} about the origin (i.e., a level manifold of v which is a homotopy sphere) will be $(-1)^n$. This implies the values $W(x)$, for x in a neighborhood $\mathcal{N}$ of zero, must cover a neighborhood of zero giving a second necessary condition of Brockett,

* This research was supported by NSF grant DMS 93-01039.

† Department of Mathematics, Box 395, University of Colorado, Boulder, CO 80309-0395.

i.e., that $\{X(x)+uY(x) : x \in \mathcal{N},\ u \in \mathbb{R}^1\}$ must cover a neighborhood of zero. System (1.1) is easily seen to violate this necessary condition for the existence of a continuous ASFC.

In dimension $n = 2$, STLC at zero for system (1.2) does imply the existence of a continuous ASFC, [Ka], [He 2]. In [C 1], Coron showed that if (1.2) is STLC at zero, one can adjoin a k-dimensional system

$$\dot{y} = v, \quad v \text{ a } k\text{-dimensional control}, \quad y \in \mathbb{R}^k, \tag{1.3}$$

in such a way that there will exist a continuous, "dynamic," asymptotically stabilizing control $u^*(t,x,y)$ for the extended system (1.2), (1.3). Here the dimension k was related to the Lie bracket structure needed to invoke Sussmann's general theorem on STLC, [S]. In general, one seeks a dynamic feedback control which provides asymptotic stability only for the original system (1.2), i.e., drives the x-variables to zero. In a later paper, [C 2], Coron shows that if system (1.2) is STLC at zero, one can choose $k = 1$ in (1.3) and there exists a continuous, dynamic feedback control $u^*(t,x,y)$, periodic in t, which drives a full neighborhood of initial data points to zero in finite time. Of course, u^* is not C^1.

Our purpose, here, is to illustrate how the addition of a one-dimensional dynamic extension to the system (1.1) removes the topological obstruction to the existence of a continuous ASFC, and, furthermore, to give an explicit construction of the dynamic ASFC.

System (1.1) may be viewed as a homogeneous approximation to a more general system of the form (1.2). Indeed (see [He 3], [He 4]) let $L(X,Y)$ denote the Lie algebra generated by the vector fields X, Y of (1.2) and assign weights (integers m, $-\infty < m < \infty$) to these, denoted $\operatorname{wt} X, \operatorname{wt} Y$. Let the weight of a Lie product be the sum of the weights of its factors. If these weights can be assigned so that $W \in L(X,Y)$ with $\operatorname{wt} W \leq 0$ implies $W(0) = 0$, the weight assignment induces a filtration, at zero, of $L(X,Y)$, specifically a sequence $\{F_j : -\infty < j < \infty\}$ of subspaces of $L(X,Y)$ such that F_j consists of elements of weight $\leq j$, $F_j \subset F_{j+1}$, $\bigcup F_j = L(X,Y)$ and $[F_i, F_j] \subset F_{i+j}$. Let $F_j(0)$ denote the elements of F_j evaluated at zero. The filtration induces a dilation $\delta_\varepsilon^r : \mathbb{R}^n \to \mathbb{R}^n$, $\varepsilon > 0$, $\delta_\varepsilon^r x = (\varepsilon^{r_1}x_1, \ldots, \varepsilon^{r_n}x_n)$ as follows: $r_j = 1$ for $1 \leq j \leq n_1 = \dim \operatorname{span} F_1(0)$; $r_j = 2$ for $n_1 + 1 \leq j \leq n_2 = \dim \operatorname{span} F_2(0)$, etc. The filtration also produces "preferred" local coordinates relative to which we can write the vector fields as $\widehat{X}, \widehat{Y}$ plus "higher order remainders." If the homogeneous approximating system $\dot{x} = \widehat{X}(x) + u\widehat{Y}(x)$ admits an ASFC (and this will be global asymptotic stability by homogeneity), then this control is also a local ASFC for the original system. (See [He 4].) System (1.1) has the special feature that the weight assignment $\operatorname{wt} X = -2$, $\operatorname{wt} Y = 1$ induces the standard dilation $\delta_\varepsilon^r x = (\varepsilon x_1, \ldots, \varepsilon x_n)$, i.e., all $r_i = 1$ or equivalently $\dim \operatorname{span} F_1(0) = n$. This is very relevant in our construction! In general, homogeneity of the lowest order term relative to the standard dilation may be disguised if the local coordinates are not properly chosen, i.e., are not those induced by the filtration. Other examples of nonlinear systems (already exhibited in preferred local coordinates) which are homogeneous with respect to the standard dilation and STLC at zero include

$$\dot{x}_1 = u, \quad \dot{x}_2 = x_1^p, \ldots, \dot{x}_n = x_{n-1}^p, \quad p \geq 1 \text{ an odd integer.} \tag{1.4}$$

Here the weight assignment $\operatorname{wt} X = (1-p)$, $\operatorname{wt} Y = 1$ induces the standard dilation. System (1.4) is STLC at zero and does satisfy the second Brockett necessary condition mentioned above. In [DM], Dayawansa and Martin show there is an ASFC $u^*(x)$ for (1.4) of the form $u^*(x) = \left(\sum_{i=1}^{n} \alpha_i x_i\right)^p$, α_i constants. Note $p = 1 - \operatorname{wt} X$. Also, the Lyapunov function for the asymptotically controlled system can be taken as a quadratic form $v(x) = x^\top P x$, P positive definite, symmetric and constant. For our system (1.1), with its asymptotically stabilizing control $u^*(\tau, x)$, we will exhibit a Lyapunov function (actually a value function corresponding to an optimization problem) of the form $v(\tau, x) = x^\top P(\tau) x$, $P(\tau)$ positive definite, symmetric, with $\underline{c}|x|^2 \leq x^\top P(\tau) x \leq \bar{c}|x|^2$ for $\underline{c}$, $\bar{c} > 0$. For STLC systems of the form (1.2) with X homogeneous relative to the standard dilation, $Y = \partial/\partial x_1$, it is natural to expect a dynamic ASFC such that the controlled system admits a Lyapunov function of the above form. This is, in a sense, an extension of the (time varying) "linear quadratic regulator problem" to a class of nonlinear systems. Indeed, this underlies our approach [He 3]. As in the linear, quadratic regulator problem, or as in system (1.4), the stabilizing control will have the form $u^*(\tau, x) = \left(\sum_{i=1}^{n} \alpha_i(\tau) x_i\right)^p$, with $p = 1 - \operatorname{wt} X$.

2. The dynamic extension and stabilizing control. One explanation of the advantage of dynamic control over static control is that one can, for example, feed back both state and velocity information. This gives an advantage in the feedback decoupling problem [Hu], but we will next illustrate that this is not the case for feedback stabilization. Indeed, suppose we extend (prolong) system (1.1) by differentiation, i.e.,

$$\dot{x}_1 = u, \ \dot{x}_2 = x_4, \ \dot{x}_3 = x_5, \ \dot{x}_4 = \ddot{x}_2 = 3x_1^2 u, \ \dot{x}_5 = \ddot{x}_3 = x_1^5 + 2x_1 x_2 u \tag{2.1}$$

We could now consider a feedback control $u(x_1, \ldots, x_5)$ and the variables x_4, x_5 are essentially the velocities of the original states x_2, x_3 respectively. If we write (2.1) in the form of equation (1.2), the input vector field Y becomes $Y(x) = \partial/\partial x_1 + 3x_1^2\, \partial/\partial x_4 + 2x_1 x_2\, \partial/\partial x_5$. It is often useful to choose new local coordinates, here say $y = (y_1, \ldots, y_5)$, relative to which the input vector field is $\partial/\partial y_1$. A coordinate change which accomplishes this is:

$$y_1 = x_1, \ y_2 = x_2, \ y_3 = x_3, \ y_4 = x_4 - x_1^3, \ y_5 = x_5 - x_1^2 x_2.$$

In these new local coordinates, the system (2.1) becomes

$$\dot{y}_1 = u, \ \dot{y}_2 = y_1^3, \ \dot{y}_3 = y_1^2 y_2, \ \dot{y}_4 = 0, \ \dot{y}_5 = 0, \tag{2.2}$$

which shows that system (2.1) is essentially system (1.1) in disguise. Indeed, the solutions of (2.1) of interest correspond to initial data $x_1(0)$, $x_2(0)$, $x_3(0)$ arbitrary while $x_4(0) = x_1^3(0)$, $x_5(0) = x_1^2(0)x_2(0)$, hence lie on a three-manifold in $\mathbb{R}^5$. Our point is that differential prolongation is not the way to obtain a dynamic extension for the purpose of continuous, asymptotic stabilization. Instead, extra dimensions can be used to remove the topological obstruction to the existence of a continuous ASFC, referred to as Brockett's second necessary condition in the introduction.

Rescale time in system (1.1), which introduces a factor $\alpha > 0$ in the equations (since u is arbitrary we exclude α in the first equation) and introduce the dynamic extension $\dot{\tau}(t) = |x|^2$ giving the extended system

$$\dot{\tau}(t) = |x|^2, \quad \dot{x}_1 = u, \quad \dot{x}_2 = \alpha x_1^3, \quad \dot{x}_3 = \alpha x_1^2 x_2. \tag{2.3}$$

This extension, rather than just time via the equation $\dot{\tau} = 1$, is motivated by the desire to have a Hamilton–Jacobi–Bellman (HJB) equation (associated with an optimization problem for system (2.3)) which admits a dilation symmetry group. Indeed, consider the cost functional

$$C(u) = \int_0^\infty [e(u(\sigma))^{4/3} + h_4^+(x(\sigma))]d\sigma, \quad e > 0 \tag{2.4}$$

where h_4^+ is an arbitrary, positive definite function, homogeneous of degree 4 with respect to the standard dilation. Let $v(\tau, x)$ denote the value function for the optimization problem of minimizing the cost C along a solution of (2.3) starting at the arbitrary initial data (τ, x). The standard use of the Pontriagin maximum principle, e.g., see [He 3], shows the optimal control is

$$u^*(\tau, x) = -(3v_{x_1}(\tau, x)/4e)^3 \tag{2.5}$$

where $v(\tau, x)$ satisfies the HJB equation

$$|x|^2 v_\tau(\tau, x) - \gamma v_{x_1}^4(\tau, x) + \alpha x_1^3 v_{x_2}(\tau, x) + \alpha x_1^2 x_2 v_{x_3}(\tau, x) = -h_4^+(x), \tag{2.6a}$$

with data

$$v(\tau, 0) = 0, \qquad v(\tau, x) > 0 \text{ if } x \neq 0. \tag{2.6b}$$

Here

$$\gamma = (1/4)(3/4e)^3, \qquad \gamma \to \infty \text{ as } e \to 0^+. \tag{2.7}$$

If $v(\tau, x)$ is a solution of (2.6) for some positive definite function $h_4^+(x)$, it is a Lyapunov function for the system (2.3) with u replaced by u^* as given in (2.5). Note that we are only interested in asymptotic stability with respect to the x-variables, i.e., we do not expect, or require, the extended variable τ to approach zero as $t \to \infty$. Also, if $u^*(\tau, x)$, as given by (2.5), is such an ASFC for system (2.3), $(1/\alpha)u^*$ is an ASFC for (1.1).

Rescale the x-variables in (2.6) by $x = \varepsilon y$, $\varepsilon > 0$, and let $v(\tau, \varepsilon y) = \varepsilon^2 w(\tau, y)$. Then $v_\tau = \varepsilon^2 w_\tau$ while $\partial/\partial y\ v(\tau, \varepsilon y) = \varepsilon v_x(\tau, x) = \varepsilon^2 w_y(\tau, y)$ or $v_x(\tau, x) = \varepsilon w_y(\tau, y)$. Substitution in (2.6) shows that if $v(\tau, x)$ is a solution, then so is $(1/\varepsilon^2)v(\tau, \varepsilon x)$, i.e,. (2.6) admits a dilation group as a symmetry group (in the x-variables). Thus if the solution, v, of (2.6) is unique, $\varepsilon^2 v(\tau, x) = v(\tau, \varepsilon x)$, or v is homogeneous of degree 2 with respect to the standard dilation δ_ε^1 in the x-variables. This suggests we try, as a solution of (2.6),

$$v(\tau, x) = (1/2)x^\top P(\tau)x, \quad P(\tau) = (p_{ij}(\tau)) \text{ symmetric, positive definite.} \tag{2.8}$$

In particular, we require (from standard Lyapunov theory)

CONDITION I. $0 \leq \underline{c}|x|^2 \leq x^\top P(\tau)x \leq \bar{c}|x|^2$, $\underline{c}$, $\bar{c} > 0$ where the inequality $\underline{c}|x|^2 \leq x^\top P(\tau)x$ is usually stated as "$v(\tau, x)$ is positive definite" while $2v(\tau, x) = x^\top P(\tau)x \leq \bar{c}|x|^2$ is termed "$v(\tau, x)$ admits an infinitely small upper bound."

Since h_4^+ is not specified, $v(\tau, x)$ will be the desired Lyapunov function if it satisfies (2.6b) and the HJB inequality

$$|x|^2 v_\tau(\tau, x) - \gamma v_{x_1}^4(\tau, x) + \alpha x_1^3 v_{x_2}(\tau, x) + \alpha x_1^2 x_2 v_{x_3}(\tau, x) \leq -h_4^+(x) \tag{2.6a$'$}$$

for some positive definite function $h_4^+(x)$.

For additional results on the use of symmetry groups, with respect to arbitrary dilations, for solutions of Hamilton–Jacobi equations arising from systems which satisfy the Brockett necessary conditions for a continuous ASFC (see [He 3], [He 4], [He 5]).

Let $p^1(\tau)$ denote the first row of $P(\tau)$. Substitute $v(\tau, x)$ as given by (2.8) into the left side of (2.6a) and write the result as

$$(1/2)|x|^2 x^\top P'(\tau)x - \gamma(p^1(\tau)\cdot x)^4 + \alpha x_1^2 \left[(x_1, x_2) \begin{pmatrix} p_{21} & p_{22} & p_{23} \\ p_{31} & p_{32} & p_{33} \end{pmatrix} \begin{pmatrix} x_1 \\ x_2 \\ x_3 \end{pmatrix} \right]. \tag{2.9}$$

If $p^1(\tau) \cdot x \neq 0$, we can choose $e > 0$ sufficiently small to make $\gamma > 0$ large enough so the term $-\gamma(p^1(\tau)\cdot x)^4$ dominates (2.9). This will be made precise following the construction of $P(\tau)$. Hence our goal is to get (2.9) negative definite ($\leq -h_4^+(x)$ for some positive definite h_4^+) on the τ-varying subspace $p^1(\tau) \cdot x = 0$, or since we will require $p_{11}(\tau) > 0$, on the subspace

$$x_1 = \left(\frac{1}{p_{11}(\tau)}\right)(-p_{12}(\tau)x_2 - p_{13}(\tau)x_3). \tag{2.10}$$

On this subspace, we compute

$$\left(\frac{\alpha x_1^2}{p_{11}^2}\right)\left[(x_1, x_2)\begin{pmatrix} p_{21} & p_{22} & p_{23} \\ p_{31} & p_{32} & p_{33}\end{pmatrix}\begin{pmatrix} x_1 \\ x_2 \\ x_3\end{pmatrix}\right] \tag{2.11}$$

$$\begin{aligned} = \frac{\alpha x_1^2}{p_{11}^2}\{&(p_{12}^3 + p_{11}^2 p_{23} - p_{11}p_{12}p_{22} - p_{11}p_{12}p_{13})x_2^2 \\ &+(2p_{13}p_{12}^2 + p_{11}p_{33} - p_{11}p_{13}p_{22} - p_{11}p_{12}p_{23} - p_{11}p_{13}^2)x_2x_3 \\ &+ (p_{12}p_{13}^2 - p_{11}p_{13}p_{23})x_3^2\}\,. \end{aligned}$$

Let $K_1(x_2, x_3)$ denote the quadratic form in braces on the right side of (2.11).

CONDITION II. $K_1(x_2, x_3) \leq -k_1(x_2^2 + x_3^2)$ for some $k_1 > 0$.

The term $(\alpha x_1^2/p_{11}^2)K_1(x_2, x_3)$ contains the specific information about system (1.1), e.g., the fact that it is STLC at zero and an "odd" system. For an arbitrary system (1.2) one could not hope to make the term equivalent to (2.11) negative definite.

If condition II is satisfied and $x_1 \neq 0$, one can choose $\alpha > 0$ sufficiently large so that this term dominates $(|x|^2/2)x^\top P'(\tau)x$. Thus our final condition will be to have $(|x|^2/2)x^\top P'(\tau)x$ negative definite on the subspace determined by (2.10) and $x_1 = 0$. Assume $x_1 = 0$ and $p_{13} \neq 0$ so, in (2.10), $x_3 = -p_{12}x_2/p_{13}$. Subject to these conditions

$$\left(\frac{|x|^2}{2}\right)x^\top P(\tau)x = \left(\frac{p_{13}^2 + p_{12}^2}{2p_{13}^4}\right)x_2^4[p_{13}^2p'_{22} - 2p_{13}p_{12}p'_{23} + p_{12}^2p'_{33}].$$

CONDITION III. $(p_{13}^2p'_{22} - 2p_{13}p_{12}p'_{23} + p_{12}^2p'_{33} + p_{12}^2p'_{33}) \leq -k_2,\ k_2 > 0.$

Here the extra dimension obtained by the dynamic extension allows us to remove the topological obstruction to a continuous ASFC presented by Brockett's second necessary condition. Specifically, for any fixed value τ_1 a level surface of $v(\tau_1, x)$ will be a (homotopy) two sphere S^2. The vector field $W(\tau_1, x) = u^*(\tau_1, x)\,\partial/\partial x_1 + x_1^3\,\partial/\partial x_2 + x_1^2x_2\,\partial/\partial x_3$ cannot "point in" on this S^2 (i.e., have index -1). On the other hand, the level surface $\{(\tau, x) : v(\tau, x) = c > 0\}$ can be viewed as a (topological) cylinder. This will be constructed so that the vector field $|x|^2\,\partial/\partial\tau + u^*(\tau, x)\,\partial/\partial x_1 + x_1^3\,\partial/\partial x_2 + x_1^2x_2\,\partial/\partial x_3$ "points into the cylinder" even though its projection onto the plane $\tau = \tau_1$, i.e., $W(\tau_1, x)$ above does not point into $\{(\tau_1, x) : v(\tau_1, x) = c\} \cong S^2$.

We next proceed with an explicit choice of $P(\tau) = (p_{ij}(\tau))$ such that conditions I, II, III are satisfied. Condition III is the most interesting and the first to be considered. One may note that condition II does suggest $p_{23}(\tau) < 0$, $p_{13}(\tau) < 0$ to aid in making the quadratic form negative definite.

Choose:

$$(2.12)\quad \begin{cases} p_{23}(\tau) = (-c_{23} + \sin \pi m \tau), & c_{23} > 0,\ m \geq 1 \text{ integer} \\ p_{12}(\tau) = -c_{12} \cos \pi m \tau, & c_{12} > 0. \\ p_{13}(\tau) & = -c_{13},\ c_{13} > 0. \end{cases}$$

Then, in condition III, $-2p_{13}p_{12}p'_{23} = -2c_{12}c_{13}\pi m \cos^2 \pi m \tau$ which is negative except at the values $\tau = (2k+1)/2$, $k = 0, 1, \ldots$ at which points it is zero. To compensate for these values choose

$$\begin{aligned} p_{33}(\tau) &= c_{33} > 0 \\ p_{22}(\tau) &= c_{22} - \sin(2\pi m(\tau + 1/2m)),\ c_{22} > 1. \end{aligned}$$

This gives, for the expression on the left side of condition III,

$$(2.13)\qquad -2\pi m c_{13}^2 \cos(2\pi m(\tau + 1/2m)) - 2c_{12}c_{13}\pi m \cos^2 \pi m \tau$$

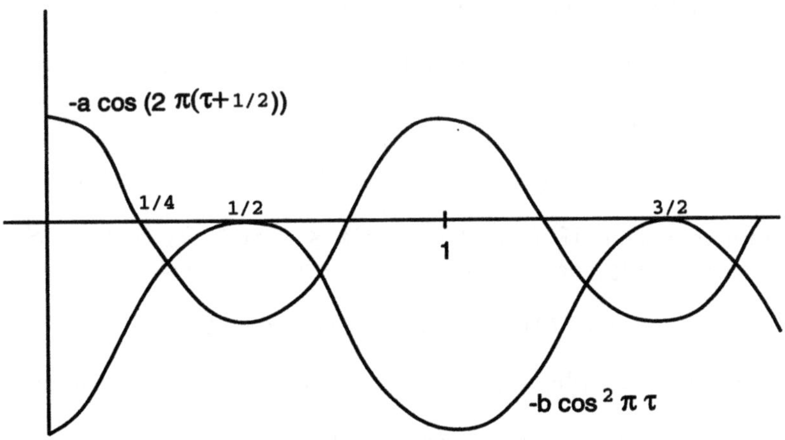

FIG. 2.1. $m = 1$

From FIG. 2.1, clearly if the "amplitude" $2\pi m c_{13}^2$ of $\cos(2\pi m(\tau + 1/2m))$ is smaller than the "amplitude" $2c_{12}c_{13}\pi m$ of $\cos^2 \pi m \tau$, expression (2.13) $\leq k_2$ for some $k_2 > 0$, as desired for condition III.

Next, in condition II, we want the negative of the quadratic form $K_1(x_2, x_3)$ to be positive definite, and we shall invoke Sylvester's condition to check this. Specifically, we will want $-p_{12}p_{13}^2 + p_{11}p_{13}p_{23} > 0$

and the negative of the matrix of this form to have positive determinant. If we choose $p_{33} \sim 1/\sqrt{p_{11}}$, $p_{11} > 0$, this determinant behaves like $-p_{13}p_{23}^2p_{11}^3 - p_{11}^3$ and hence will be positive if $c_{23} > 0$ is sufficiently large since, see (2.10), $p_{13} < 0$. We can then choose $c_{11} > 0$ sufficiently large so condition I holds. A choice which satisfies conditions I, II, III is

$$(2.14)\quad P(\tau) = \begin{pmatrix} 25 & -(1/2)\cos\pi\tau & -1/4 \\ -(1/2)\cos\pi\tau & 10 - \sin(2\pi(\tau + 1/2)) & -5 + \sin\pi\tau \\ -1/4 & -5 + \sin\pi\tau & 5 \end{pmatrix}$$

From (2.5), and including the factor $(1/\alpha)$ to deal with system (1.1), this gives

$$(2.15)\quad u^*(\tau,x) = -(1/\alpha)(3/4e)^3 v_{x_1}^3(\tau,x) = -c(25x_1 - (x_2/2)\cos\pi\tau - x_3/4)^3$$

where $c = (1/\alpha)(3/4e)^3 > 0$, $\alpha > 0$ is large, $e > 0$ is small. Numerical calculations with $c = 15$ and a variety of (small in absolute value) initial conditions for (1.1) show monotone convergence to zero for all three components x_1, x_2, x_3 of (1.1). The convergence is not, however, fast.

3. Verification that conditions I, II, III imply v satisfies (2.6a′) and (2.6b). Condition I shows v satisfies the initial conditions (2.6b) for the HJB equation (inequality).

Next, let $D = \{x \in \mathbb{R}^3 : |x| \leq 1\}$ and $T = [0,\infty) \times D$; picture T as a cylinder about the τ-axis to which we restrict our analysis. Define

$$S^1 = \{(\tau,x) \in T : p^1(\tau)\cdot x = 0\}, \quad S^2 = \{(\tau,x) \in S^1 : x_1 = 0\}$$

and for any $\varepsilon > 0$,

$$S^1_\varepsilon = \{(\tau,x) \in T : |p^1(\tau)\cdot x| < \varepsilon|x|\}, \quad S^2_\varepsilon = \{(\tau,x) \in S^1_\varepsilon : |x_1| < \varepsilon\}.$$

From condition III, $(|x|^2/2)x^\top P'(\tau)x \leq -k_3x_2^4$, $k_3 > 0$, on S^1 and hence by continuity, for $\varepsilon > 0$ small there is a $k_4 > 0$ such that

$$(3.1)\quad (|x|^2/2)x^\top P'(\tau)x \leq -k_4x_2^4 \quad \text{on} \quad S^2_\varepsilon.$$

Next, let $g(\tau,x) = \alpha x_1^3 v_{x_2}(\tau,x) + \alpha x_1^2 x_2 v_{x_3}(\tau,x)$. Then from condition II

$$g(\tau,x) = (\alpha x_1^2/c_{11}^2)K_1(x_2,x_3) \leq -(\alpha x_1^2/c_{11}^2)k_1(x_2^2+x_3^2), \quad k_1,\ \alpha > 0, \quad \text{on } S^1.$$

Choose $\alpha > 0$ sufficiently large that $(\alpha x_1^2/c_{11}^2)k_1(x_2^2 + x_3^2) > (|x|^2/2)x^\top P'(\tau)x$ on the complement, relative to T, of S^2_ε. Then we have, for some $k_5 > 0$,

$$(3.2)\quad (|x|^2/2)x^\top P'(\tau)x + \alpha x_1^3 v_{x_2}(\tau,x) + \alpha x_1^2 x_2 v_{x_3}(\tau,x) \leq \\ \leq \begin{cases} -k_4x_2^4 & \text{on } S^2_\varepsilon \\ -k_5x_1^2(x_2^2 + x_3^2) & \text{on } S^1_\varepsilon - S^2_\varepsilon. \end{cases}$$

Finally, on S^1_ε, $|p^1(\tau)\cdot x| < \varepsilon|x|$ hence on $T - S^1_\varepsilon$ we have $|p^1(\tau)\cdot x| \geq \varepsilon|x|$ and $-\gamma(p^1(\tau)\cdot x)^4 \leq -\gamma\varepsilon^4|x|^4$. Pick $\gamma > 0$ sufficiently large so that on $(T - S^1_\varepsilon)$ we have $\gamma(p^1(\tau)\cdot x)^4$ is larger than the absolute value of the left side of (3.2). Then there is a $k_6 > 0$ such that

$$\frac{|x|^2}{2}x^\top P'(\tau)x - \gamma(p^1(\tau)\cdot x)^4 + g(\tau,x) \leq \begin{cases} -k_6|x|^4 \text{ on } (T - S^1_\varepsilon) \\ -k_5x_1^2(x_2^2+x_3^2) \text{ on } (S^1_\varepsilon - S^2_\varepsilon) \\ -k_4x_2^4 \text{ on } S^2_\varepsilon \end{cases} \tag{3.3}$$

This gives the required inequality (2.6a′).

REFERENCES

[B] R. W. BROCKETT, *Asymptotic Stability and Feedback Stabilization*, in Differential Geometric Control Theory, (R.W. Brockett, R.S. Millman and H.J. Sussmann, eds.), **27**, Birkhauser, Boston, 1983, pp. 181–191.

[C1] J. M. CORON, *Links Between Local Controllability and Local Continuous Stabilization*, NOLCOS '92, Bordeaux, June 1992, pp. 477–482.

[C2] J. M. CORON, *On the Stabilization in Finite Time of Locally Controllable Systems by Means of Continuous Time-Varying Feedback Laws*, (preprint).

[DM] W. P. DAYAWANSA AND C. F. MARTIN, *Asymptotic Stabilization of Low-Dimensional Systems*, in Progress in Systems and Control Theory, (C. I. Byrnes and A. Kurzhansky, eds.), Birkhauser, Boston, 1991, pp. 53–67.

[He1] H. HERMES, *Control Systems which Generate Decomposable Lie Algebras*, J. Diff. Eqs. **44** (1982), pp. 166–187.

[He2] H. HERMES, *Homogeneous Coordinates and Continuous Asymptotically Stabilizing Feedback Controls*, in Differential Equations, Stability and Control, (S. Elaydi, ed.), Lecture Notes in Pure & Appl. Math. **127**, Marcel Dekker, New York, 1991, pp. 249–260.

[He3] H. HERMES, *Nilpotent and High-Order Approximations of Vector Field Systems*, SIAM Review **33** (1991), pp. 238–264.

[He4] H. HERMES, *Asymptotically Stabilizing Feedback Controls*, J. Diff. Eqs. **92** (1991), pp. 76–89.

[He5] H. HERMES, *Asymptotic Stabilization of Planar Systems*, Systems & Control Letters **17** (1991), pp. 437–444.

[Hu] H. HUIJBERTS, *Dynamic Feedback in Nonlinear Synthesis Problems*, Doctoral Dissertation, Universiteit Twente, 1991.

[Ka] M. KAWSKI, *Stabilization of Planar Systems in the Plane*, System & Control Letters **12** (1989), pp. 169–175.

[Ku] J. KURZWEIL, *The Converse Second Lyapunov's Theorem Concerning the Stability of Motion*, Czech Math. J. **6** (1956) (English summary), pp. 475–485.

[S] H. J. SUSSMANN, *A General Theorem on Local Controllability*, SIAM J. Control & Optimization **25** (1987), pp. 158–194.

INTRODUCTION TO A PAPER OF M.Z. SHAPIRO: HOMOTOPY THEORY IN CONTROL

RICHARD MONTGOMERY*

1. The results. A curve is called regular if its derivative is never zero. We associate to any regular curve $\mathbf{x}(t)$ on the unit two-dimensional sphere a moving frame $f(t) = [\mathbf{x}(t), \mathbf{T}(t), \mathbf{N}(t)]^t$ whose row vectors are $x(t)$, its unit tangent vector $\mathbf{T}(t) = \dot{\mathbf{x}}(t)/|\dot{\mathbf{x}}(t)|$ and its righthanded normal $\mathbf{N}(t) = \mathbf{x}(t) \times \mathbf{T}(t)$. Thus $f(t)$ is a curve in the three-dimensional rotation group $S0(3)$. It satisfies the Frenet–Serret equations

$$\frac{d}{dt} f = \begin{pmatrix} 0 & v & 0 \\ -v & 0 & k \\ 0 & -k & 0 \end{pmatrix} f \tag{1.1}$$

(These are the standard Frenet-Serret equations for the space curve $\int_0^t x(s)ds$.) Here v is the speed of the curve $x(t)$ and k/v is its curvature. (1) defines a right-invariant distribution of two-planes on $SO(3)$. (It is the distribution mentioned by John Baillieul at the beginning of his talk.) It defines a control system with controls v, k. We must impose the bound $v > 0$ since we are interested in regular curves.

Fix an initial frame and a final frame. Consider the space of all solutions $f(t)$ to the Frenet-Serret control system (1) which have these as their initial and final frames.

Question: How many connected components does this space have?

THEOREM 1.1 (Smale, 1958). *Two, as indicated by figure 1.1.*

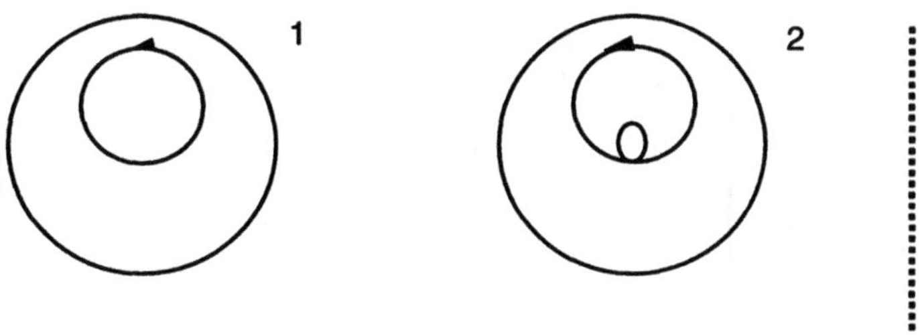

FIG. 1.1.

* Mathematics Department, University of California, Santa Cruz, CA 95064.

Now suppose we also impose the constraint $k > 0$. The resulting class of curves on the two-sphere are called right-handed nondegenerate. (Their curvature is always positive.) And let us ask the same question.

THEOREM 1.2 (J. Little, 1970). *There are* **three** *components of the space of solutions to the control system with the constraints $v > 0$, $k > 0$ imposed, provided the the inital and final frames are equal. The representatives of these components are indicated in figure 1.2.*

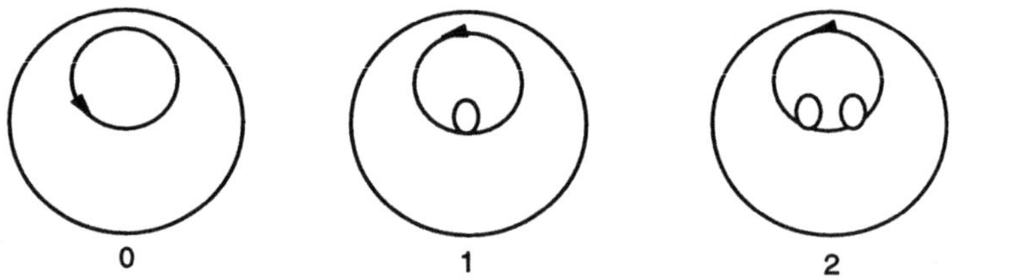

FIG. 1.2.

B.Z. and M.Z. Shapiro [BZMZ] have shown that the difference $1 = 3-2$ between Little's and Smale's theorem is a consequence of whether or not curves cross the boundary of the small-time accessible set. Our goal is to popularize their results and suggest that questions of homotopy theory may be important to control theory.

M.Z. Shapiro investigated the following generalization of Little's problem to n-dimensions:

$$f(t) \in SO(n),$$

the n-dimensional rotation group.

$$\frac{df}{dt} = \begin{pmatrix} 0 & k_1 & 0 & & & 0 \\ -k_1 & 0 & k_2 & & & \\ 0 & -k_2 & 0 & \ddots & & \\ & \ddots & \ddots & \ddots & & \\ & & \ddots & \ddots & \ddots & \\ & 0 & & \ddots & \ddots & k_{n-1} \\ & & & & -k_{n-1} & 0 \end{pmatrix} f$$

$$k_i > 0, i = 1, 2, \ldots, n-1$$

We call this the Frenet-Serret distribution, or the Cartan distribution. It is a right-invariant distribution of cones on $S0(n)$.

THEOREM 1.3 (M.Z. Shapiro, 1992). *The space of solution curves to the above control system on $SO(n)$ which connect a frame $f_0 \in SO(n)$ to itself has exactly two components if n is even and exactly three components when n is odd.*

2. Why should we care? 1) These theorems count the number of connected components of solutions to control problems with fixed endpoints. Two curves lie in the same component if and only if it is possible to find a one-parameter family of control strategies, $u_s(t)$, $0 \leq s, t \leq 1$ (and so a two-parameter family of controls) such that $u_0(t)$ leads to the first curve, $u_1(t)$ to the second, and all of the intermediate curves have the same endpoints. To put it more briefly, the first curve can be deformed into the second by a control-induced homotopy which fixes the endpoints.

For example, consider a man with two different control strategies which lead to the same position of his hand gripping a bar.

FIG. 2.1.

You cannot homotope from one to the other without breaking contact. Try it!

2) There are typically an uncountable number of solutions to the problem of finding controls connecting two given points. Counting the connected components of this solution space provides a meaningful way to count distinct solutions.

3) One of the main tools used in proving the theorems mentioned is the "covering homotopy property". This is really already part of a control theorist's toolbox. (See eg. Sussmann's talk in this proceedings.) A topological perspective should provide insight into the use and importance of this tool in control.

3. Why two? We begin by recalling some basic notions from homotopy theory. The set of path-components of a space X is denoted by $\pi_0(X)$. The space of closed continuous loops γ of a connected space Q based at $q_0 \in Q$ ($\gamma(0) = \gamma(1)) = q_0$) is denoted $\Omega(Q)$ or sometimes $\Omega(Q, q_0)$. The fundamental group of Q is

$$\pi_1(Q) = \pi_0(\Omega(Q)).$$

Its elements are called homotopy classes (of based loops) and it forms a group.

It is well known that $\pi_1(SO(3))$ is the two-element group so that

$$\#\pi_0(\Omega(SO(3)) = 2.$$

This is the "2" in the theorem of Smale.

Remark. The identity element $e \in \pi_1(SO(3))$ is represented by the constant path $f(t) \equiv f_0$. The nontrivial element $\sigma \in \pi_1(SO(3))$ is represented by rotation through 2π radians about any axis of space.

Let $\Omega = \Omega(SO(3))$, and let $\Omega_K \subset \Omega$ denote the loop space of Smale's theorem. The answer "2" is a corollary of a deeper result of Smale which states that the inclusion of Ω_K in Ω induces an isomorphism on π_0:

$$i_* : \pi_0(\Omega_K) \simeq \pi_0(\Omega)$$

Here i_* denotes the map which assigns to each connected component of Ω_K the corresponding connected component of Ω which contains it.

This is a surprising result, for given a pair of topological spaces $A \subset B$ there is no reason for the corresponding i_* to be 1-to-1 (see figure 1.4). (The black blobs represents $A = \Omega_K$ and the blobs encircling them $B = \Omega$.)

4. Covering homotopies. The notion of a covering homotopy is central to the proofs of the theorems above. It appears naturally in control theory.

Let $p : \mathcal{S} \to Q$ be a continuous map between connected spaces. We have in mind the endpoint map which assigns to each controlled path beginning at q_0 its endpoint. In other words, for each control strategy $u(\cdot)$, solve the control system $\dot{q} = f(q(t), u(t))$, with initial condition $q(0) = q_0$. Then $p(u(\cdot)) = q(1)$.

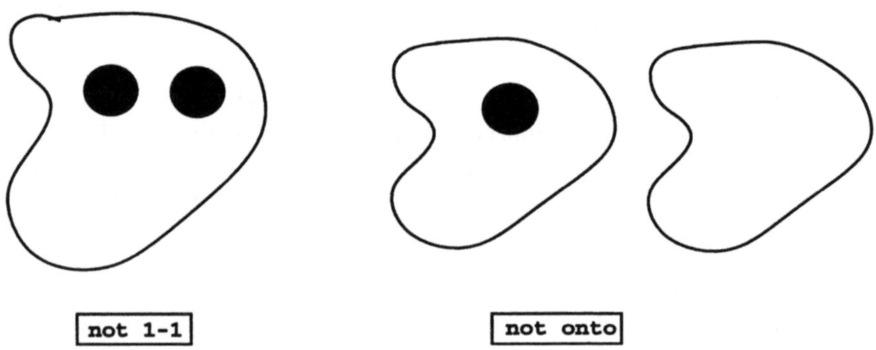

FIG. 4.1.

DEFINITION 4.1. *We say that p satisfies the 1-parameter covering homotopy property, or CHP for short, if for each path $q(s)$, $0 \leq s \leq 1$ in Q and any $\gamma_0 \in \mathcal{S}$ with $p(\gamma_0) = q(0)$ there exists a path $\gamma(s)$, $0 \leq s \leq 1$ covering $q(s)$: $p(\gamma(s)) = q(s)$.*

In other words, the 1-parameter CHP holds if we can follow any motion $q(s)$ of final states by an appropriate two-parameter families of controls $u(t, s)$.

The salient result from homotopy theory is that if $\mathcal{S}$ is contractible, and p satisfies the 1-parameter CHP then

$$\pi_0(p^{-1}(q_0)) = \pi_1(Q).$$

(This follows immediately from the exact homotopy sequence.) Since $\pi_0(Q) = \pi_0(\Omega(Q))$ this in turn implies that $\#(\pi_0(\Omega(Q)) = \#\pi_0(\Omega_K)$ as in Smale's theorem.

It follows from Little's theorem that the 1-parameter CHP must fail for his system. Let us see **how** it fails. Consider the following set-up for testing the CHP:

Here we are to swing the final frame f_0 of the initial nondegenerate curve $\gamma_0(t)$ across the equator defined by $\dot{\gamma}_0(0)$. This equator is indicated by the vertical dashed curve.

Now consider the central projection of this figure on to a tangent plane. (By a central projection we mean a stereographic projection with light source at the sphere's center.)

Central projection preserves nondegeneracy of curves. Now any planar curve with initial frame f_0, and final frame f_1, and *no self-intersections* must have *an inflection point*. See the following figure, 4.2.

(Cf. Arnol'd, [1].) At inflection points the curvature is zero and so the control bound $k > 0$ is violated. Thus any homotopy γ_s which follows the frames f_s must leave the space Ω_K of solution curves. This shows that the endpoint map for Little's distribution violates the 1-parameter CHP.

Remark The reason behind considering central projections comes from projective geometry. The sphere is the universal cover of the projective plane. The central projections then become the standard affine charts of projective geometry. The Frenet-Serret distribution is perhaps most properly thought of as having to do with projective geometry. In particular, as shown in the last section, it induces a distribution on projective frames and is invariant under projective transformations.

If instead, the initial choice γ_0 has a self-intersection then it becomes possible to cover the curve f_s with a homotopy γ_s of γ_0. This is indicated in figure 4.3.

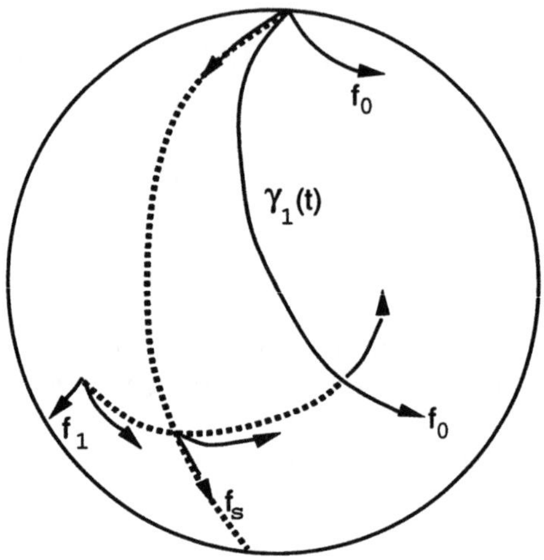

FIG. 4.2.

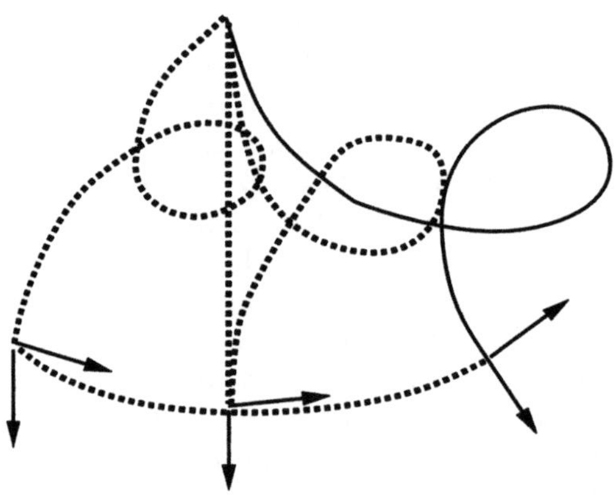

FIG. 4.3.

The central step in Shapiro's proof is isolating that subset of the space of nondegenerate curves for which the 1-parameter CHP fails. These are the disconjugate curves which we now describe.

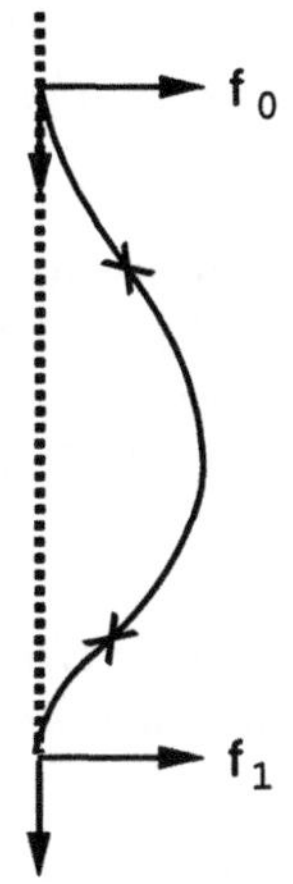

FIG. 4.4.

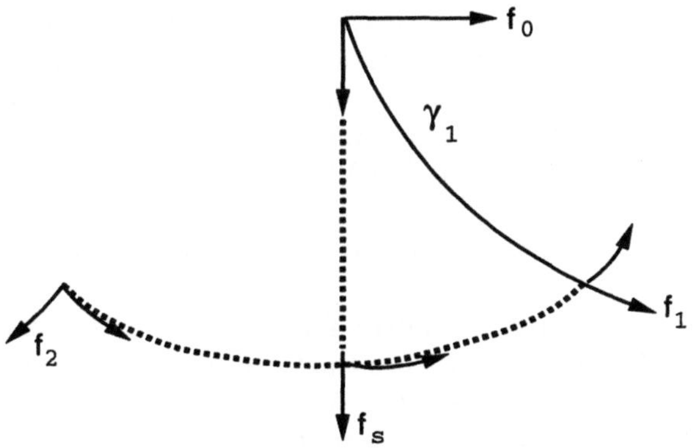

FIG. 4.5.
Illustration of the relation between inflection points and disconjugacy

5. Conjugate and disconjugate curves

DEFINITION 5.1. *A nondegenerate curve $x(t)$ in S^1 is called disconjugate if it intersects any great circle no more than 2 times. It is called strictly conjugate if it intersects some great circle three times transversely.*

For the reason behind this terminology see the remark in the next subsection.

The theorem of Shapiro-Little follows directly from the following:

THEOREM 5.1. *The set of disconjugate loops and the set of conjugate loops are disconnected within the space of all nondegenerate loops. The disconjugate loops form a contractible set within the set of all nondegenerate loops. The 1-parameter CHP holds for the space of nondegenerate curves minus the disconjugate curves.*

6. Higher-dimensional homotopies and other generalities. The 1-parameter CHP requires us to follow a 1-parameter family of targets by an appropriate family of control strategies. Suppose instead that we want to follow k-parameter target sets. Let Σ^j denote the j-dimensional cube and I the unit interval.

DEFINITION 6.1. *The k-parameter CHP holds for $p : \mathcal{P} \to Q$ provided whenever $f : \Sigma^{k-1} \times I \to Q$ is a continuous map and $\gamma : \sum^{k-1} \to \mathcal{P}$ is another continuous map such that $\pi(\gamma(\sigma)) = f(\sigma, 0)$ then there is a map $\Gamma : \sum^{k-1} \times I \to \mathcal{P}$ with $\Gamma(\cdot, 0) = \gamma(\cdot)$ and $p \circ \Gamma = f$.*

If the k-parameter CHP holds for all k we will say, that p satisfies the CHP.

Let us return to Smale's paper. Let Q be a connected Riemannian manifold. Fix a point q_0 in Q and a unit vector v_0 attached there. Recall a curve is called regular if its derivative is nowhere zero. Let $\mathcal{P}^{\text{reg}}$ denote the space of all continuously differentiable regular paths in Q beginning at a point q_0 and with initial direction v_0. Let STQ denote the space of all unit tangent vectors. Consider the map $p : \mathcal{P}^{\text{reg}} \to STQ$ which assigns to each path the value $(\gamma(0), \dot\gamma(0)\ /||\dot\gamma(0)||)$ of its final tangent direction.

THEOREM 6.1. [Smale] *This map p satisfies the CHP.*

Now it is a general fact (again following from the exact homotopy sequence) that if $p : \mathcal{P} \to Y$ satisfies the CHP, if $\mathcal{P}$ is contractible, and if Y is connected then the 'fibers' $p^{-1}(y)$ and the space of loops $\Omega(Y)$ on Y are *weakly homotopy equivalent*. To say that two spaces are weakly homotopy equivalent means that all their homotopy groups agree:

$$\pi_k(p^{-1}(y)) \simeq \pi_k(\Omega(Y)) = \pi_{k+1}(Y).$$

(If Ω is a connected topological space, then $\pi_k(\Omega)$ is the space of path-connected components of the space of maps of a k-sphere into Ω.)

Smale showed that $\mathcal{P}^{\text{reg}}$ was contractible. Since $STS^2 = S0(3)$ his theorem stated at the beginning follows immediately from this one.

Consider the particular case when Q is a two-dimensional surface. Let $e_0(q)$, $e_1(q)$ be a local orthonormal frame on Q so that any unit vector u can be written $\vec{u} = \cos\varphi e_0 + \sin\varphi e_1$. Then $\dot q = v\vec{u}, v > 0$ is our control law. It can be rewritten as the Pfaffian system

$$\omega \equiv -\sin\varphi\theta_0 + \cos\varphi\theta_1 = 0$$

where θ_0, θ_1 are the dual basis to e_0, e_1.

This system is of contact type: $w \wedge dw \neq 0$. As a baby version of his main theorem, Smale proves the following

THEOREM 6.2. *Let D be a contact distribution on a connected 3-manifold. Let Ω_D denote the set of all absolutely continuous Legendrian ($\dot\gamma \in D$) loops through a fixed point. Then the inclusion of this space into the space of all loops is a weak homotopy equivalence.*

In this same vein, Ge Zhong and independently Sarychev have proved the following.

THEOREM 6.3. *Let D be a bracket generating distribution on a connected manifold Q. Then the inclusion $\Omega_D \hookrightarrow \Omega$ of the horizontal ($\dot\gamma \in D$) absolutely continuous loops through a fixed point into the space of all loops through that point induces a weak homotopy equivalence.*

Their proofs follow the main lines of Smale's. All additional difficulties are taken care of by invoking Chow's theorem, as the reader may have guessed.

As we can see from the results of Little and Shapiro, the situation becomes much more interesting when we impose inequality constraints on the controls. In fact, the situation become more interesting if we simply impose more smoothness on our controls. This is evidenced by the existence of C^0-rigid curves as defined by Bryant-Hsu. (The simplest example of such a curve is any segment of the x-axis for the control system $dz - y^2dx = 0$ on $\mathbf{R}^3$.)

7. Problems. In order to organize our thoughts we will now state some general problems.

Suppose we are given a distribution $K \subset D \subset TQ$ of cones where Q is a smooth connected manifold, D a bracket generating distribution and $K_q, q \in Q$ a family of cones varying smoothly with q. Fix two points q_0, q_1 and let $\Omega_K^r = \Omega_K^r(q_0, q_1)$ denote the set of all r-times continuously differentiable paths γ joining q_0 to q_1 and satisfying the control system $\dot\gamma \in K_\gamma$. When $r = 0$ take the paths to be absolutely continuous and write $\Omega_K^0 = \Omega_K$, $\Omega_{TQ}^r = \Omega^r$.

Problem 1: How many path-connected components does Ω_K^r have?

Problem 2: Let Ω denote the space of *all* paths joining q_0 to q_0 (no conditions on controls or smoothness). Is $\Omega_K^r \hookrightarrow \Omega$ a weak homotopy equivalence?

Problem 3: Does the answer to problem 1 depend on the degree of smoothness r?

Problem 4: How does the answer to problem 1 vary as we vary the end points?

Problem 5: How does the answer to Problem 1 vary as we vary the opening angle of the cone? If the original cone is open in D, can the answer change if we take its closure?

Regarding problems 1,3,4. In the case of the stable abnormal simple curve the in $\mathbb{R}^3$ mentioned above, when $q_0 \neq q_1$ are two points on this curve we have

$$\#\pi_0(\Omega^0_D) = 0$$
$$\#\pi_0(\Omega^0_D) = 1$$

But if q_0 is not on the curve then

$$\#\pi_0(\Omega^r_0) = 0, \quad r = 0, 1, 2, 3, \ldots$$

Such phenomena are impossible when there are no controls ($D = TQ$):

$$\pi_0(\Omega^r) = \pi_0(\Omega).$$

Regarding problem 5. The results of Smale-Little-Shapiro show clearly that indeed the answer can depend on the cone's opening angle.

Regarding problem 2. *Relaxing controls and the h-principle of Gromov.* To say that the answer to this problem is 'yes' means that by completely relaxing the controls, we get the "right answer" for the topology. In this situation we say that Gromov's weak h-principle applies. His regular *h-principle* is that

$$i_* : \pi_0(\Omega_k) \to \pi_0(\Omega)$$

is onto. This means every homotopy class of path is represented by a control path. His *1-parameter h-principle* states that i_* is 1-to-1. This means that if two control paths (with fixed end points) are homotopic, disregarding the controls, then that homotopy can be realized through a one-parameter family of control strategies (all having the same end points).

(He also has C^r-versions of his principle.)

Thus we can summarize: In Smale's case the weak h-principle applies. The 1-parameter h-principle fails for the examples of Shapiro and Little.

Warning: To say that the weak h-principle holds is stronger than saying that the regular or 1-parameter h-principle holds. This is an accident of historical nomenclature we are stuck with.

8. Disconjugacy in higher dimensions. It is not immediately clear how to generalize the Little result to higher dimensions. To do this the viewpoint of projective geometry appears essential.

Call a curve $\gamma(t)$ in $\mathbb{R}^n$ vector nondegenerate, or VN for short, if $\gamma(t)$, $\dot{\gamma}(t)$, $\ddot{\gamma}(t), \ldots, \gamma^{(n-0)}(t)$ are linearly independent vectors in $\mathbb{R}^n$ for each t. Call it right-handed or (RHVN for short) if in addition this basis is positively oriented.

The Graham-Schmidt procedure allows us to pass from RHVN curves to a curves in $S0(n)$ satisfying the Frenet-Serret equations $(*)$. We can also associate to a RHVN curve a moving family of subspaces of $\mathbb{R}^n$. Namely, set $f_0(t) = \text{span } \{\dot{\gamma}(t)\}$ and let $f_j(t)$ denote the linear span of the first j derivatives of γ. By assumption, $dim(f_j(t)) = j$, and

$$0 \subset f_0(t) \subset f_1(t) \subset \ldots \subset f_j(t) \subset \ldots f_{n-1}(t) \subset \mathbb{R}^n.$$

A collection of such subspaces is called a (complete) flag. (See figure 8.1).

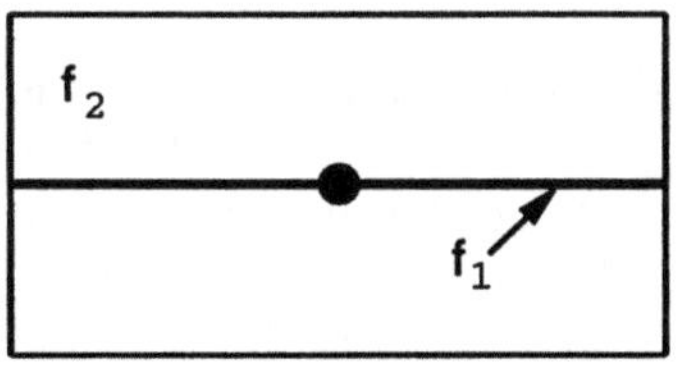

FIG. 8.1.

Each subspace f_j is *oriented* by the ordering of the derivative vectors $\gamma^{(j)}(t)$. Denote the set of all oriented flags in $\mathbb{R}^n$ by SF_n. $S0(n)$ acts freely and transitely on SF_n, and so this action defines a diffeomorphism $S0(n) \simeq SF(n)$. The Frenet-Serret control system on $SO(n)$ mentioned in the beginning has a beautiful description on SF_n. Let e_0 denote the vector field on SF_n defined by rotating the line f_0 of any given flag within the plane f_1 according to the positive sense of rotation defined by the orientations. And in general let e_i, $i = 0, 1, \ldots, n-0$ be the vector field defined by rotating the i-dimensional subspace f_i about the $(i-1)$st keeping it within the $(i+1)$st, and so that the rotation is in a positive sense. Then our control system is

$$\dot{f} = \sum_{i=0}^{n-1} k_i e_i(f) \quad ; k_i > 0.$$

We will call this description the "projective description" of our control system.

From the projective description it becomes obvious that the full linear group, $Gl(n)$, is the symmetry group of our control system. From our original description we could only see that the smaller group $SO(n)$ was a symmetry group. These additional symmetries allow Shapiro to construct explicit covering homotopies. (Any positive scalar multiple of the identity I acts trivially on SF_n, so that the action of $Gl(n)$ actually factors through an action of $GL(n)$ modulo this one-dimensional subgroup. This quotient group is the disconnected double cover of the projective linear group $\mathbb{P}GL(n) = GL(n)/\mathbb{R}I$ which is the group of projective transformations.) More importantly, the projective viewpoint allows Shapiro and Shapiro to pinpoints the higher dimensional disconjugate curves, that is, the set of curves on which the 1-parameter CHP fails.

Remark

The motivation for Shapiro to study this system came from a certain Poisson structure called the Gelfand-Dikii structure which arises in the study of completely integrable PDE such as KdV. The underlying manifold for this structure is the affine space of all linear n-th order differential operators:

$$L(y) = y^{(n)} + u_{n-1}(t)y^{(n-1)} + \ldots u_0(t)y(t).$$

Let $y_0, \ldots y_n$ be any basis for the space of solutions to the equation $L(y) = 0$. Then $(y(t), \ldots, y_n(t))$ is a VN curve in $\mathbb{R}^n$. We think of the coefficients u_i as the controls. Let $Y(t)$ be the fundamental matrix solution to such a DE. Thus the ij entry of Y can be taken to be the jth derivation of the ith solution y_i. We will say that $Y(0)Y(0)^{-0}$ is the monodromy of the nth order differential operator. We will say that two operators are isomonodromic if there monodromy operators are constant. The symplectic leaves of the Gelfand-Dickii Poisson structure consists of the connected components of the isomonodromy classes. Shapiro and Khesin show how to reduce the problem of seperating the connected components to the problem we have been discussing on $SO(n)$ or SF_n.

Exercise. Relate the controls u_i to the Frenet-Serret controls k.

Exercise. Relate this to standard linear control theory, cf. Sontag, p. 133.

Now suppose that $\gamma : I = [0,1] \to \mathbb{R}^n$ is a VN curve, let H be any hyperplane and ℓ a linear function defining H; $H = \{\ell = 0\}$. Define the multiplicity of H at t_0 to be the order of vanishing of $\ell(\gamma(t))$ at t_0. In particular the multiplicity is zero if $\gamma(t_0) \notin H$. The multiplicity is a nonnegative integer less than or equal to $n-1$. It is equal to $n-1$ if and only if H is the osculating plane of γ at t_0. To see this, observe that relative to the basis defined by its derivatives at t_0 we have

$$\gamma(t) = \left(1, (t-t_0), \frac{1}{1}(t-t_0)^1, \ldots, \frac{1}{(n-1)!}(t-t_0)^{n-1}\right) + 0(t-t_0)^n).$$

If we perturb H a bit in the correct direction then it will intersect γ transversely in $n-1$ points near p_0 and these points limit to p_0 as H approaches the osculating plane.

Let the multiplicity of γ relative to H be the sum of all nonzero multiplicities.

DEFINITION 8.1. *γ is a conjugate curve if for all hyperplanes the total multiplicity is at most $n-1$. Otherwise it is disconjugate.*

With this definition in place M.Z. Shapiro proves that the analog of the theorem of Little-Shapiro holds for higher dimensions. A key ingredient in the proof is the notion of the *train* of an initial flag f_0.

DEFINITION 8.2. *Two flags $f = f_0 \subset f_1 \subset \ldots$ and $e = e_0 \subset e_1 \subset \ldots$ in $\mathbb{R}^n$ are said to be transverse, or in general position, if the intersections of all their subspaces is as transverse as possible. In other words, they are transverse if for each i, j the dimension of $f_i \cap e_j$ is the minimum possible for such an intersection of subspaces, namely* $\max(i + j - n, 0)$. *The train of the flag f is the set of all flags e which are* **not** *transverse to it.*

THEOREM 8.1. *If the VN curve γ is not conjugate then its associated flag curve $f(t)$ must intersect the train of its initial flag $f(0)$.*

Remark If we put bounds, eg $\Sigma k_i < 0$, on the controls then near an initial flag f_0 its train is precisely the the boundary of its small-time accessible set.

Remark For those familiar with some elements of Lie group theory, the train is the union of all the lower-dimensional Schubert cells in the cell decomposition of $SO(n)$.

Finally, I should explain to the reader what the difference is between even and odd dimensions. Why do we get 2 when n is even and 3 when n is odd? Because when n is even there are no conjugate loops! The reason for this is simple. The curve must return to its starting place, and hence intersect any hyperplane an even number of times. But the multiplicity of the loop's initial osculating plane, and slight perturbations of it, is at least $n-1$ which is odd. Hence the multiplicity of the curve with respect to such planes is at least n and the loop is conjugate.

I hope I have given the reader enough background and motivation to read the papers of Shapiro et al. Bon voyage.

REFERENCES

[MZ] M.Z. Shapiro, ("Moscow Independent University") *Topology of the space of nondegenerate closed curves*, preprint 1993. A shortened version appeared in the journal Funktsional'nyi Analiz; Ego Prilozheniya, as translated by Plenum (same title). v. 26, no. 3, pp. 93–96 July-Sept. 1992.
An earlier version of the above appeared in the Bulletin of the A.M.S., 25, No. 1, 1991.

[BZMZ] B.Z. Shapiro and M.Z. Shapiro, *On the number of connected components in the space of closed nondegenerate curves on S^n*, Bulletin of the A.M.S. (1991), pp. 75–79.

The following references are in historical order:

[SS] S. Smale, *Regular curves on Riemannian Manifolds*, Trans. A.M.S. 87, (1958), pp. 492–512.

[JL] J. Little, *Nondegenerate Homotopies of Curves on the Unit Two-sphere*, J. Diff. Geom, 4 (1970), pp. 339–348.

[MG] M. Gromov, *Partial Differential Relations*, Springer-Verlag, New York (1986).

[HHL] J. Hilgert, K.H. Hofman, and J.D. Lawson, *Lie Groups Convex Cones and Semi-Groups*, Oxford Math. Monographs, series, Oxford Univ. Press, [1989].

[1] V. I. Arnol'd *A Branched Covering of $CP^2 \to S^4$, Hyperbolicity, and Projective Topology*, Siberian Math. J., 29, no. 2 (1988), pp. 36–47.

[KS] B.A. Khesin and B.Z. Shapiro *Homotopy Classification of Nondegenerate Quasiperiodic curves on 2-sphere*, preprint, [1993].

For background information on homotopy theory, I recommend D.B. Fuks and A.T. Fomenko, *Lecture in Homotopical Topology* Akademia Kiado, Budapest, Hungary, (1986) and M. Greenberg, *Introduction to Algebraic Topology*, 1st edition (not later), Benjamin, Reading, MA, (1973).

OPTIMIZATION AND FINITE DIFFERENCE APPROXIMATIONS OF NONCONVEX DIFFERENTIAL INCLUSIONS WITH FREE TIME*

BORIS S. MORDUKHOVICH†

Abstract. This paper is concerned with a free-time optimal control problem for nonconvex-valued differential inclusions with a nonsmooth cost functional in the form of Bolza and general endpoint constraints involving free time. We develop a finite difference method for studying this problem and focus on two major topics: 1) constructions of well-posed discrete approximations ensuring a strong convergence of optimal solutions, and 2) necessary optimality conditions for free-time differential inclusions obtaining by the limiting process from discrete approximations. As a result, we construct a sequence of discrete approximations with the strong convergence of optimal solutions in the $W^{1,2}$-norm. Then using the convergence result and appropriate tools of nonsmooth analysis, we prove necessary optimality conditions for differential inclusions in the refined Euler-Lagrange form with a new relation for an optimal free time.

1. Introduction. In this paper we study problem (P) of minimizing the real-valued Bolza functional

$$J[x,T] := \varphi(x(0), x(T), T) + \int_0^T f(t, x(t), \dot{x}(t))dt \tag{1.1}$$

on trajectories for the compact-valued and Lipschitz continuous differential inclusion

$$\dot{x}(t) \in F(t, x(t)) \tag{1.2}$$

over a varying time interval subject to general endpoint constraints on $(x(0), x(T), T)$. For brevity, we refer to this problem as to a problem of Bolza with *free time*. If $f = 0$, then (P) is said to be a Mayer free-time problem.

Such differential inclusion problems are natural generalizations of free-time problems in both the calculus of variations and optimal control. The latter case corresponds to the representation

$$Q(t,x) = \{g(t,x,U) |\ u \in U\} \tag{1.3}$$

with some vector function g and set U which may depend on the time variable. Moreover, the differential inclusion model (1.2) allows to consider closed-loop control systems with time-depended control regions $U = U(x)$ in (1.3).

---

* This research was supported in part by grant DMS–9206989 and DMS-9404128 from the National Science Foundation and the Career Development Chairs Award at Wayne State University and also by the Institute for Mathematics and its Applications with funds provided by the National Science Foundation.

† Department of Mathematics, Wayne State University, Detroit, Michigan 48202. E-mail: boris@math.wayne.edu

It is known even in classical settings that problems with free time have certain qualitative and technical distinctions from corresponding fixed-time problems. Of course, fixed-time problems can be considered as special cases of problems with free time. But usually results for fixed-time problems can be easier obtained under more general assumptions about the dependence on time variables.

On the other hand, in many situations necessary optimality conditions for free-time problems may be derived from corresponding results for fixed-time problems under additional regularity assumptions on the time-dependence of the data. Such transformation techniques are widely used in the classical variational and control problems; see, e.g., [26].

To our knowledge, the first set of necessary optimality conditions for differential inclusions with free time was published in Clarke's book [11, Section 3.6]. Considering the autonomous case for *convex* (this means convex-valued) differential inclusions, he obtained necessary conditions for a free-time Mayer problem by reducing it to a fixed-time one and using the so-called (true) *Hamiltonian*

$$\mathcal{H}(x,p) := \max\{\langle p, v\rangle \mid v \in F(x)\} \tag{1.4}$$

which is a Lipschitz continuous function in the state variable x and the adjoint variable p.

Clarke established the Hamiltonian necessary conditions for an optimal solution (strong minimum) $\{\bar{x}(t),\ 0 \le t \le \bar{T}\}$ in terms of his generalized gradients ∂_C of Lipschitz continuous functions. These conditions involve the *Hamiltonian inclusion*

$$(-\dot{p}(t), \dot{x}(t)) \in \partial_C \mathcal{H}(\bar{x}(t), p(t)) \text{ a.e. } t \in [0, \bar{T}] \tag{1.5}$$

with additional relations for $(p(0), p(\bar{T}), \bar{T})$ expressed in terms of the generalized gradients of φ and endpoint constraint functions. If, in particular, φ and endpoint constraints do not depend on free time T, then Clarke's conditions imply that

$$\mathcal{H}(\bar{x}(t), p(t)) \equiv 0 \text{ on } [0, \bar{T}]. \tag{1.6}$$

The latter constancy relation is well known for the classical problems of optimal control where it actually follows from the maximum condition in the Pontryagin maximum principle; see [26, 46, 57].

The nonautonomous case can be easily reduced to the autonomous one if F is Lipschitzian with respect to both variables (t, x). The case of merely continuous dynamics is more complicated even if a standard differential equation formulation is adopted; see Berkovitz [3] for smooth problems and Mordukhovich [34, 35, 37] for nonsmooth problems of optimal control with free time. We refer the reader to the papers of Clarke and Vinter [13, 14], Clarke, Loewen, and Vinter [15], and Rowland and Vinter [54] for studying as well as applications of free-time and related optimization problems for convex differential inclusions with discontinuous (in general measurable) time dependence. The mentioned papers contain necessary optimality conditions of the Hamiltonian type generalized the results of [11].

Another version of necessary optimality conditions for convex differential inclusions was developed by Mordukhovich first for fixed-time [33, 35] and then for free-time [39] Mayer problems. In this version, the main differential relation is obtained in the following form (all results in the rest of this section are formulated only for autonomous systems):

$$(1.7)\qquad (\dot{p}(t), \dot{\bar{x}}(t)) \in \text{co}\{(u,v) \mid (u, p(t)) \in N((\bar{x}(t), v); \text{gph } F),$$

$$v \in M(\bar{x}(t), p(t))\} \ \text{ a.e. } t \in [0, \bar{T}]$$

where "co" stands for the convex hull and

$$(1.8)\qquad M(x,p) := \{v \in F(x) \mid \langle p, x\rangle = \mathcal{H}(x,p)\}.$$

In (1.7), N is not Clarke's normal cone but its *nonconvex* counterpart which was first used in Mordukhovich [31] for obtaining transversality conditions in nonsmooth optimal control problems. Now it is clear that this normal cone and the corresponding nonconvex subdifferential ∂ are, probably, the most convenient tools for describing transversality and related conditions for dynamic optimization problems as well as necessary conditions in finite dimensional nonsmooth optimization; see, e.g., [12, 22, 28, 29, 33–39, 52–54] and references therein. We consider some properties of these objects in Section 4.

It is rather surprising that the dynamic relationships (1.5) and (1.7) turn out to be *equivalent* in the general framework of necessary conditions for convex differential inclusions. This has been recently proved by Rockafellar [51] in the direction (1.5) $\Longrightarrow$ (1.7) and Ioffe (personal communication; see also [24, Section 3.5]) in the opposite direction.

Let us consider condition (1.7) and its further improvements in more details. First observe that (1.7) implies for a.e. $t \in [0, \bar{T}]$ both the *maximum condition*

$$(1.9)\qquad \langle p(t), \dot{\bar{x}}(t)\rangle = \mathcal{H}(\bar{x}(t), p(t))$$

and an analogue of the *Euler-Lagrange inclusion* in the form

$$(1.10)\ \dot{p}(t) \in \text{co}\{u \mid (u, p(t)) \in N((\bar{x}(t), v); \text{gph } F),\ v \in M(\bar{x}(t), p(t))\}.$$

Note that version (1.10) is, in general, independent of Clarke's version of the Euler-Lagrange inclusion [10]

$$(1.11)\qquad (\dot{p}(t), p(t)) \in \text{cl co } N((\bar{x}(t), \dot{\bar{x}}(t)); \text{gph } F);$$

see examples in [29]. On the other hand, if the maximum set $M(\bar{x}(t), p(t))$ in (1.8) is a *singleton* for a.e. $t \in [0, \bar{T}]$ (in particular, when the sets $F(x)$ are *strictly convex* along $\bar{x}(\cdot)$), then (1.10) is reduced to the following *refined* form

$$\dot{p}(t) \in \mathrm{co}\{u \mid (u, p(t)) \in N((\bar{x}(t), \dot{\bar{x}}(t)); \mathrm{gph}\, F)\} \tag{1.12}$$

which obviously implies (1.11). Observe that the refined condition (1.12) requires less convexification: only to the components involving derivatives of the adjoint function instead of to all components at once. This makes (1.12) essentially stronger than (1.11) in certain situations; see Remark 4.5 in Section 4.

It has been recently proved by Loewen and Rockafellar [29] that the general case of fixed-time Mayer problems for convex differential inclusions can be actually reduced to the case when the sets $F(x)$ are strictly convex along the optimal trajectory under consideration. In this way, using the Hamiltonian analysis, they establish the refined condition (1.12) with no single-valuedness assumption on the maximum set (1.8).

Note that in the convex-valued setting for F, the refined Euler-Lagrange inclusion (1.12) automatically implies the maximum condition (1.9); see Proposition 4.6 stated below. Therefore, it also implies the "fuzzy" inclusion (1.7) as well as the Hamiltonian inclusion (1.5), and thus it appears to be the strongest result in this direction for convex problems.

The principal necessary conditions for differential inclusions considered above are obtained under the convexity assumption on $F(x)$ which is essentially used in their proofs. What happens if the sets $F(x)$ are no longer assumed to be convex? First note that in this setting neither Clarke's form (1.11) of the Euler-Lagrange inclusion nor the refined form (1.12) implies the maximum condition (1.9). Could one ensure that these inclusions themselves are necessary for optimality in *nonconvex* problems?

The positive answer for the case of (1.11) can be found in Clarke's paper [10] under the *calmness* hypothesis imposed on a Mayer fixed-time problem. The latter hypothesis is a kind of regularity (constraint qualification) assumption for problems with endpoint constraints which ensures *normality* in transversality conditions; see Section 2 for more information. Recently Kaskosz and Lojasiewicz [27] have released the calmness assumption proving that Clarke's form of the Euler-Lagrange inclusion holds for any boundary trajectory in nonconvex differential inclusions.

We have considered fixed-time optimization problems for nonconvex differential inclusions in the recent paper [43]. In that paper, we prove that the refined form (1.12) of the Euler-Lagrange inclusion is a necessary condition for optimality in Mayer problems as well as for boundary trajectories. (Actually this result provides a necessary condition for the *weak* local minimum; see Remark 7.3 stated below.) In [43] we also prove that the following analogue

$$(1.13)\quad \dot{p}(t) \in \text{co}\{u|\,(u,p(t)) \in \lambda\partial f(\bar{x}(t),\dot{\bar{x}}(t)) + N((\bar{x}(t),\dot{\bar{x}}(t));\text{gph } F)\}$$

of the refined Euler-Lagrange inclusion (1.12) with a multiplier $\lambda \geq 0$ appears to be a necessary optimality condition for fixed-time Bolza problems involving nonconvex differential inclusions. The latter result (in contrast to those for Mayer problems and boundary trajectories) is proved under a certain *relaxation stability* assumption; cf. Section 2. The core of our approach in [43] consists in using a method of *discrete (finite difference) approximations* together with appropriate tools of *nonsmooth analysis*.

The primal objective of the present paper is to extend the main constructions and results of [43] to *free-time Bolza problems for nonconvex differential inclusions*. To the best of our knowledge, such problems have never been considered in the literature and results for them cannot be directly derived from those for fixed-time problems.

Here we pursue a twofold goal. First, to develop a discrete approximation approach for free-time Bolza problems involving nonconvex differential inclusions. And second, to obtain necessary optimality conditions in the refined Euler-Lagrange form (1.13) with an additional characterization of the optimal time interval $[0,\bar{T}]$.

Discrete approximation techniques have been long time recognized as a powerful tool for studying and solving infinite dimensional variational problems. This approach goes back to Euler (1744) who used finite differences (broken lines) to prove the classical Euler-Lagrange equation in the calculus of variations as well as for proving the existence of solutions to differential equations.

In further applications, Euler's finite difference method and its modifications are mostly employed for computational purposes. There is a number of works devoted to numerical aspects of discrete approximations for optimal control and differential inclusion problems. We refer the reader to the recent papers of Dontchev and Lempio [19] and Polak [45] containing surveys and new developments in this area; see also Dontchev's contribution to this volume [17].

Besides a numerical analysis of variational problems, consistent discrete approximations provide a possibility to obtain qualitative (theoretical) results for infinite dimensional problems by passing to the limit in corresponding results for their finite dimensional approximations. For instance, in this way one can derive necessary optimality conditions for variational problems using those in finite dimensional optimization. Such an approach may be successful if it is possible:

1) to construct "right" discrete approximations with appropriate convergence properties;

2) to derive "robust" necessary conditions in the finite dimensional optimization problems obtained; and then

3) to justify the convergence procedure in those optimality conditions.

Some implementations of this approach applied to optimal control and differential inclusion problems can be found in Mordukhovich [33, 35, 36, 39, 43], Pshenichnyi [47], and Smirnov [56]. Observe that finite dimensional approximations for such problems always contain many equality and/or geometric constraints arising from finite difference replacements of differential relations.

Due to the natural presence of many geometric constraints, the finite dimensional problems obtained in this way turn out to be objects of nonsmooth analysis and optimization even for the case of smooth functional data in the original models. Moreover, to achieve the purposes 2) and 3) stated above, they require to use only generalized differential constructions with special properties (or postulate such properties as in [47]). One can see that our constructions in Section 4 just fit all the requirements. At the same time, other widely spread constructions in nonsmooth analysis cannot be employed without some restrictive assumptions; see, in particular, discussions in Remark 4.9.

Previous efforts in using discrete approximations to obtain necessary optimality conditions for differential inclusions were mostly concerned with fixed-time Mayer problems under the convexity assumption on $F(x)$. In [39], we consider a free-time problem of Mayer with convexity and formulate necessary optimality conditions grouped around the "fuzzy" Euler-Lagrange inclusion (1.7). The recent paper [43] employs discrete approximations to prove the refined Euler-Lagrange inclusion (1.13) for fixed-time Bolza problems in the nonconvex setting.

In the present paper we develop a discrete approximation procedure for free-time Bolza problems involving nonconvex differential inclusions. This procedure and the results obtained have some essential distinctions from the previous considerations.

To perform finite difference approximations for differential inclusion problems with free time, we use the simplest uniform Euler scheme but on a *varying time interval.* This implies that for each step of approximation, a *discrete grid* (stepsize) is variable and becomes a *subject to optimization.* The latter requires *more regularity* with respect to the time variable in order to prove necessary conditions involving optimal time.

In this way, we construct a sequence of discrete approximation problems (P_K) with a varying grid whose optimal solutions *strongly* converge in $W^{1,2}$-norm to a reference optimal solution $\{\bar{x}(\cdot), \bar{T}\}$ for the original Bolza problem (P) with general endpoint constraints. Problems (P_K) with discrete-time dynamics can be reduced to special static problems of nonsmooth optimization in finite dimensions with many equality, inequality, and geometric type constraints. Necessary optimality conditions for (P_K) are directly derived from a *generalized Lagrange multiplier rule* in nondifferentiable programming. Passing to the limit as $K \to \infty$ and using convergence results together with appropriate tools of nonsmooth analysis, we get a set of necessary optimality conditions for the original nonconvex problem (P) with free time.

In these necessary conditions, the differential relation is expressed in the form of the *refined Euler-Lagrange inclusion* (1.13) for a.e. $t \in [0, \bar{T}]$ with additional relations for $(p(0), p(\bar{T}), \bar{T})$ in terms of (nonconvex) subdifferentials of the cost function φ in (1.1) and endpoint constraints. In the case when φ and endpoint constraints do not depend on T, the relations obtained imply that

$$\int_0^{\bar{T}} H(\bar{x}(t), \dot{\bar{x}}(t), p(t), \lambda) dt = 0 \tag{1.14}$$

in terms of the so-called *pseudo-Hamiltonian*

$$H(x, v, p, \lambda) := \langle p, v \rangle - \lambda f(x, v) \tag{1.15}$$

of the Bolza problem (P) calculated on the optimal solution $\{\bar{x}(\cdot), \bar{T}\}$ and the corresponding adjoint pair $\{p(\cdot), \lambda\}$.

Note that in general nonconvex setting under consideration, the refined Euler-Lagrange inclusion (1.13) does not imply the maximum condition

$$H(\bar{x}(t), \dot{\bar{x}}(t), p(t), \lambda) = \mathcal{H}(\bar{x}(t), p(t), \lambda) := \max\{\langle p, v \rangle - \lambda f(x, v) | \ v \in F(x)\}$$

for a.e. $t \in [0, \bar{T}]$ as well as the constancy condition

$$\mathcal{H}(\bar{x}(t), p(t), \lambda) \equiv const \ \text{ on } \ [0, \bar{T}].$$

If the latter conditions hold, then (1.14) is reduced to (1.6) for the Hamiltonian $\mathcal{H}$ of the Bolza problem. In general, (1.14) appears to be an independent integral condition in terms of the pseudo-Hamiltonian (1.15) over the optimal time interval $[0, \bar{T}]$.

The remainder of the paper is organized as follows. In Section 2 we formulate the problem and consider the property of relaxation stability employed for obtaining the principal results. Section 3 is devoted to constructing correct discrete approximations of the original problem and proving the strong convergence of optimal solutions. In Section 4 we review some concepts and results in nonsmooth analysis used in the paper. Section 5 is concerned with necessary optimality conditions for discrete approximation problems. In Section 6 we prove the main theorem about necessary optimality conditions for the differential inclusion problem under consideration. In the concluding Section 7 we discuss some open questions and further generalizations of the results obtained.

In this paper we basically use standard notation. Some special symbols are introduced and explained in Section 4. Throughout the paper, the set B stands for the unit closed ball of the space in question; the adjoint (transposed) matrix to A is denoted by $A^{\star}$. Note that we consider all finite dimensional vectors to be vector-columns although they may be written as vector-rows for the purpose of convenience.

2. Problem formulation and relaxation. Let us consider the following problem (P) of dynamic optimization with a varying time interval:

$$\text{(2.1)}\quad \text{minimize } J[x,T] := \varphi_0(x(0),x(T),T) + \int_0^T f(t,x(t),\dot{x}(t))dt$$

over all arcs $x(\cdot) \in W^{1,\infty}[0,T]$ satisfying the differential inclusion

$$\text{(2.2)}\qquad \dot{x}(t) \in F(t,x(t)) \text{ a.e. } t \in [0,T]$$

and the general endpoint constraints

$$\text{(2.3)}\qquad \varphi_i(x(0),x(T),T) \leq 0 \text{ for } i = 1,\ 2,\dots,q;$$

$$\text{(2.4)}\qquad \varphi_i(x(0),x(T),T) = 0 \text{ for } i = q+1,\ q+2,\dots,q+r;$$

$$\text{(2.5)}\qquad (x(0),x(T),T) \in \Omega \subset \mathbf{R}^{2n+1}.$$

Here F is a set-valued mapping (multifunction) from $\mathbf{R}^{n+1}$ into $\mathbf{R}^n$ and f, φ_i are real-valued functions defined on $\mathbf{R}^{2n+1}$. We call problem (P) the *Bolza problem for differential inclusion with free time.* If $f \equiv 0$ in (2.1), then (P) is reduced to the corresponding *Mayer problem* for differential inclusions.

Any solution $x(\cdot) \in W^{1,\infty}$ to (2.2) is called an *(original) trajectory* for the differential inclusion, and any trajectory for (2.2) satisfying constraints (2.3)–(2.5) is called a *feasible solution* to problem (P). A feasible solution $\bar{x}(t)$, $0 \leq t \leq \bar{T}$, minimizing the cost functional (2.1) over all other feasible solutions $x(t)$, $0 \leq t \leq T$, (with different T) is called an *(original) optimal solution* to (P).

Along with problem (P), we consider its *relaxation* (variational extension) which is defined by the following way going back to the classical works of Bogoljubov and Young in the 30s; cf. [4, 7, 21, 25, 57, 59]. Let

$$\text{(2.6)}\qquad f_F(t,x,v) := f(t,x,v) + \delta(v,F(t,x))$$

where $\delta(v,\Lambda) = 0$ if $v \in \Lambda$ and $\delta(v,\Lambda) = \infty$ if $v \notin \Lambda$ (the indicator function). Denote by $\hat{f}_F(t,x,v)$ the *convexification* (the biconjugate function) for f_F in the v variable, i.e., the largest convex function majorized by $f_F(t,x,\cdot)$ for each x and t. The *relaxed problem* (R) is defined as follows:

$$\text{(2.7)}\quad \text{minimize } \hat{J}[x,T] := \varphi_0(x(0),x(T),T) + \int_0^T \hat{f}_F(t,x(t),\dot{x}(t))dt$$

over all arcs $x(\cdot) \in W^{1,\infty}[0,T]$ subject to constraints (2.3)–(2.5). Observe that if $\hat{J}[x,T] < \infty$, then $x(\cdot)$ satisfies the convexified differential inclusion

$$\text{(2.8)}\qquad \dot{x}(t) \in \text{co } F(t,x(t)) \text{ a.e. } t \in [0,T].$$

Any trajectory for (2.8) is called a *relaxed trajectory* for (2.2) in contrast to original trajectories satisfying (2.2). It is well known (cf., e.g., [1, 7, 21, 25]) that under natural assumptions involving the Lipschitz continuity of F in x, the following *approximation property* holds:

Every relaxed trajectory $x(t)$, $0 \le t \le T$, can be uniformly in $[0,T]$ approximated by original trajectories $x_k(t)$, $0 \le t \le T$, starting with the same initial point (but may be not satisfy endpoint constraints) such that

$$\liminf_{k\to\infty} \int_0^T f(t, x_k(t), \dot{x}_k(t))dt \le \int_0^T \hat{f}_F(t, x(t), \dot{x}(t))dt. \tag{2.9}$$

Probably the first result in this vein has been obtained by Bogoljubov [4] who proved such a property for the classical problem of the calculus of variations (no differential inclusions). The approximation property is related to the so-called "hidden convexity" of continuous-time systems under consideration which is reflected (from an abstract viewpoint) by the celebrated Lyapounov theorem about the range convexity of nonatomic vector measures [30] (the same as the convexity of Aumann's multivalued integral; see, e.g., [11, 26]).

The discussions above make natural the following definition.

2.1. Definition. Denote by $\inf(P)$ and $\inf(R)$ the infima of the cost functionals in problems (P) and (R) respectively. Then one says that (P) possesses the property of *relaxation stability* (or (P) is *stable with respect to relaxation*) if

$$\inf(P) = \inf(R). \tag{2.10}$$

Note that by virtue of the convexity with respect to velocities in the relaxed problem, the infimum in (R) is attained under well known growth conditions of the Tonelli type; see, e.g., [12, 26]. This means the existence of optimal solutions to the relaxed problem which is not the case for the original problem (P). On the other hand, the approximation property allows to obtain a minimizing sequence of original trajectories which, however, may mot exactly satisfy the boundary conditions.

Obviously, we always have $\inf(R) \le \inf(P)$. To establish the relaxation stability, one should prove the opposite inequality which is somehow connected with the approximation property. Let us present a result in this direction going back to the classical Bogoljubov theorem.

2.2. Proposition. *Let $F \equiv 0$ in (P), and let the integrand f be continuous. Then (P) is stable with respect to relaxation and*

$$f(t, \bar{x}(t), \dot{\bar{x}}(t)) = \hat{f}(t, \bar{x}(t), \dot{\bar{x}}(t)) \text{ a.e. } t \in [0, \bar{T}] \tag{2.11}$$

for any optimal solution $\bar{x}(t)$, $0 \le t \le \bar{T}$, to (P).

Proof. Assume that $\inf(R) < \inf(P)$. Then there exist $T > 0$ and a function $x(\cdot) \in W^{1,\infty}[0,T]$ which satisfy all constraints (2.3)–(2.5) and

$$\varphi_0(x(0), x(T), T) + \int_0^T \hat{f}(t, x(t), \dot{x}(t))dt < \inf(P). \tag{2.12}$$

According to the version of Bogoljubov's theorem in [26, Section 9.2.4], for this $x(\cdot)$ we can find a sequence of $x_k(\cdot) \in W^{1,\infty}[0,T]$ such that $x_k(0) = x(0)$, $x_k(T) = x(T)$, $x_k(\cdot)$ converges to $x(\cdot)$ uniformly in $[0,T]$, and (2.9) holds with $f_F = f$. So, all $x_k(t)$, $0 \le t \le T$, turns out to be feasible solutions to problem (P) in (2.1), (2.3)–(2.5) with

$$\liminf_{k\to\infty} I[x_k, T] < \inf(P)$$

due to (2.9) and (2.12). This contradiction proves the relaxation stability of (P).

If $\bar{x}(t)$, $0 \le t \le \bar{T}$, is an optimal solution to (P), then by virtue of (2.10) it solves the relaxed problem (R) as well. This ensures

$$\int_0^{\bar{T}} [f(t, \bar{x}(t), \dot{\bar{x}}(t)) - \hat{f}(t, \bar{x}(t), \dot{\bar{x}}(t))]dt = 0. \tag{2.13}$$

From the definition one always has

$$f(t, \bar{x}(t), \dot{\bar{x}}(t)) - \hat{f}(t, \bar{x}(t), \dot{\bar{x}}(t)) \ge 0 \text{ a.e. } t \in [0, \bar{T}].$$

Therefore, (2.13) is equivalent to (2.11). □

2.3. *Remark.* Proposition 2.2 implies the corresponding result of Clarke [7, Theorem 1] where f is assumed to be Lipschitz continuous in (x, v). As it was observed in [7], (2.11) is the essence of the necessary condition of Weierstrass in the calculus of variations.

It follows directly from the approximation property stated above that the relaxation stability is inherent in any problem (P) involving (Lipschitz) differential inclusions with endpoint constraints at either $t = 0$ or $t = T$ (one of the ends is *free*). In general, the relaxation stability is clearly related to a kind of the value stability (regularity) of (P) with respect to perturbations of endpoint constraints. Such a regularity condition has been developed by Clarke and Rockafellar under the name of *calmness*; see [6–11, 50] and references therein. As it has been proved in Clarke [7, 9], the calmness property implies the relaxation stability and actually allows to reduce problems involving differential inclusions to the classical ones as in Proposition 2.2.

Moreover, the calmness property is fulfilled for *most* endpoint constraints (at least of inequality type; cf. [6, 8]) and shows that the relaxation stability may fail only for ill-posed problems where small perturbations of endpoint constraints produce proportionally unbounded variations of the minimum (value function).

Note also that according to Clarke [9, 10], the calmness hypothesis implies that corresponding necessary optimality conditions can be taken *normal.* A general result that *normality implies relaxation stability* for optimal control systems has been obtained by Warga [57, 58].

For special classes of problems (P) with arbitrary endpoint constraints, the relaxation stability holds with *no calmness or normality* assumptions. In particular, let differential inclusion (2.2) be represented in the *linear form:*

$$\dot{x}(t) \in F_1(t)x(t) + F_2(t)$$

where the multifunctions F_1 is convex-valued while F_2 is not. If, in addition, the function f in (2.1) is convex in v, then any of such problems possesses the property of relaxation stability. This can be proved by using the Lyapounov-Aumann theorem about the convexity of set-valued integrals; cf. the arguments in Mordukhovich [35, Theorem 19.7]. Similarly, the relaxation stability holds for general problems (P) involving *one-dimensional* differential inclusions; see Remark 19.2 in [35].

In the subsequent sections of the paper, we use the property of relaxation stability to establish convergence results for discrete approximations and to obtain necessary optimality conditions for differential inclusions.

3. Finite difference approximations. This section is devoted to constructing finite difference (discrete) approximations of the original problem of Bolza for differential inclusions with free time and to establishing the principal theorem about the strong convergence of optimal solutions to discrete approximations in the $W^{1,2}$-norm.

In what follows we use the simplest uniform *Euler scheme*

$$\dot{x}(t) \approx \frac{x(t+h) - x(t)}{h}$$

for the replacement of the derivative in (2.2). Note that most of the qualitative results obtained below hold for many other first-order and higher-order finite difference approximations.

For any positive integer $K = 1\ 2, \ldots$, we consider a real number T_K approximating T and the uniform grid

$$t_0 = 0,\ \ t_{j+1} = t_j + h_K \ \text{ for } \ j = 0,\ 1, \ldots, K-1$$

with the stepsize $h_K = T_K/K$ (so $t_K = T_K$). We define a *discrete approximation inclusion* as follows

$$\text{(3.1)} \qquad x_K(t_{j+1}) \in x_K(t_j) + h_K F(t_j, x_K(t_j)) \ \text{ for } \ j = 0,\ 1, \ldots, K-1$$

with some initial state $x_K(0) = x_{0K}$. A collection of vectors $\{x_{jK} := x_K(t_j) |\ j = 0, \ldots, K\}$ satisfying (3.1) is called a *discrete trajectory* for

(3.1). The corresponding collection $\{v_{jK} := (x_K(t_{j+1}) - x_K(t_j))/h_K |$ $j = 0, \ldots, K-1\}$ is called a *discrete velocity.*

We also consider piecewise-constant extensions of discrete velocities

$$(3.2)\quad v_K(t) := \frac{x_K(t_{j+1}) - x_K(t_j)}{h_K} \text{ for } t \in [t_j, t_{j+1}) \ \& \ j = 0, \ldots, K-1$$

and the corresponding piecewise-linear extensions of discrete trajectories

$$(3.3)\qquad x_K(t) := x_{0K} + \int_0^t v_K(s)ds \text{ for } t \in [0, T_K]$$

to the continuous time interval $[0, T_K]$. We call (3.2) and (3.3), respectively, *extended discrete velocities* and *extended discrete trajectories* for (3.1). It follows from (3.3) that

$$\dot{x}_K(t) = v_K(t) \text{ a.e. } t \in [0, T_K]$$

for extended discrete trajectories and velocities. So, given an extended discrete trajectory $x_K(\cdot)$, one can always identify the corresponding discrete velocity with $\dot{x}_K(\cdot)$.

Now let us fix arbitrary original trajectory $\bar{x}(t)$, $0 \le t \le \bar{T}$, for the differential inclusion (2.2) and formulate an important result about its strong approximation by discrete trajectories. For this purpose, we assume that the multifunction F is bounded and locally Lipschitzian in x around $\bar{x}(\cdot)$ and it is Hausdorff continuous in t a.e. on $[0, \bar{T}]$. In the sequel we actually need these assumptions and the result only for the case when $\bar{x}(t)$, $0 \le t \le \bar{T}$, is an optimal solution to the initial problem (P).

More precisely, we impose the following hypotheses:

(H1) There are an open set $U \subset \mathbf{R}^n$ and positive numbers m_F, l_F such that $\bar{x}(t) \in U$ for any $t \in [0, \bar{T}]$, the sets $F(t, x)$ are closed for all $(t, x) \in [0, \bar{T}] \times U$, and

$$(3.4)\qquad F(t, x) \subset m_F B \ \ \forall (t, x) \in [0, \bar{T}] \times U,$$

$$(3.5)\quad F(t, x_1) \subset F(t, x_2) + l_F|x_1 - x_2|B \ \ \forall x_1, x_2 \in U, \ \ t \in [0, \bar{T}].$$

(H2) The multifunction $F(\cdot, x)$ is Hausdorff continuous for a.e. $t \in [0, \bar{T}]$ uniformly in $x \in U$.

Following Dontchev and Farkhi [18], we consider the so-called *averaged modulus of continuity* for the multifunction $F(t, x)$ in $t \in [0, \bar{T}]$ when $x \in U$. This modulus $\tau(F; h)$ depending on the parameter $h > 0$ is defined as

$$(3.6)\qquad \tau(F; h) := \int_0^{\bar{T}} \sigma(F; t, h)dt$$

where $\sigma(F;t,h) := \sup\{\omega(F;t,x,h) | \, x \in U\}$,

$\omega(F;t,x,h) := \sup\{\text{haus}(F(t',x),F(t'',x)) | \, t',t'' \in [t-h/2,t+h/2] \cap [0,\bar{T}]\}$,

haus$(\cdot,\cdot)$ is the Hausdorff distance between compact sets.

It is proved in [18] that *if* $F(\cdot,x)$ *is Hausdorff continuous for a.e.* $t \in [0,\bar{T}]$ *uniformly in* $x \in U$, *then* $\tau(F;h) \to 0$ *as* $h \to 0$. Moreover, $\tau(F;h) = O(h)$ if $F(\cdot,x)$ has a bounded variation [18] uniformly in $x \in U$ (in particular, if F is Lipschitz continuous in t with a uniform Lipschitz constant).

Note that in the case of single-valued bounded functions $f(t)$ not depending on x, the construction (3.6) has been originally developed in Sendov and Popov [55] under the name of "averaged modulus of smoothness". It has been proved in [55] that $\tau(f;h) \to 0$ as $h \to 0$ if and only if f is Riemann integrable on $[0,\bar{T}]$, i.e., f is continuous for a.e. $t \in [0,\bar{T}]$. If f is of bounded variation on $[0,\bar{T}]$, then $\tau(f;h) = O(h)$. In this paper we use the name "averaged modulus of continuity" for both single-valued and multi-valued cases.

Now we formulate an auxiliary approximation result which is of independent interest for qualitative and numerical aspects of discrete approximations.

3.1. LEMMA. *Let* $\bar{x}(t)$, $0 \le t \le \bar{T}$, *be an original trajectory for the differential inclusion* (2.2) *under hypotheses* (H1) *and* (H2). *Then there exists a sequence of solutions* $\{z_N(t_j) | \, j = 0,\dots,K\}$ *to discrete inclusions* (3.1) *with* $T_K \equiv \bar{T}$ *such that* $z_K(0) = \bar{x}(0)$ *for any* $K = 1, 2, \dots$, *and the extended discrete trajectories* $z_K(t)$, $0 \le t \le \bar{T}$, *converge to* $\bar{x}(\cdot)$ *as* $K \to \infty$ *in the norm topology of* $W^{1,2}[0,\bar{T}]$.

Proof. The complete proof of this result can be found in Mordukhovich [43]. The main idea is related to the so-called *proximal algorithm* to construct discrete trajectories for (3.1) by using projections of the derivative $\dot{\bar{x}}(t)$ on the admissible velocity sets $F(t_j, z_K(t_j))$; cf. [35, 36, 56]. In this way, we establish the *strong* $L^2[0,\bar{T}]$-convergence of the extended discrete velocities $\dot{z}_K(\cdot)$ to $\dot{\bar{x}}(\cdot)$ with effective *error estimates.* The latter estimates involve the boundedness and Lipschitz constants m_F and l_F in (3.4), (3.5) as well as the averaged modulus of continuity (3.6). □

Now let $\bar{x}(t)$, $0 \le t \le \bar{T}$, be a given optimal solution to the original problem (P) for the differential inclusion (3.1) satisfying (H1) and (H2) around $\bar{x}(\cdot)$. Because of U in (H1) is an open neighborhood of $\bar{x}(t)$ for all $t \in [0,\bar{T}]$, one can find a number $\epsilon > 0$ such that

$$B_\epsilon(\bar{x}(t)) \subset U \quad \forall t \in [0,\bar{T}]. \tag{3.7}$$

Using Lemma 3.1, we construct a sequence of discrete approximation problems for finite difference inclusions whose optimal solutions strongly $W^{1,2}$-converge to the given trajectory $\bar{x}(\cdot)$.

For any $K = 1, 2, \ldots$, we consider the numbers

$$\alpha_{iK} := \varphi_i(\bar{x}(0), z_K(\bar{T}), \bar{T}) - \varphi_i(\bar{x}(0), \bar{x}(\bar{T}), \bar{T}) \text{ for } i = 1, \ldots, q; \tag{3.8}$$

$$\beta_{iK} := |\varphi_i(\bar{x}(0), z_K(\bar{T}), \bar{T})| \text{ for } i = q+1, \ldots, q+r; \tag{3.9}$$

$$\gamma_K := |\bar{x}(\bar{T}) - z_K(\bar{T})| \tag{3.10}$$

where $z_K(t)$, $0 \le t \le \bar{T}$, is the extended discrete trajectory from Lemma 3.1.

According to this lemma, $\gamma_K \to 0$ as $K \to \infty$, and also α_{iK}, $\beta_{iK} \to 0$ as $K \to \infty$ when the functions φ_i are continuous in the second variable at the point $(\bar{x}(0), \bar{x}(\bar{T}), \bar{T})$. It follows directly from (3.8)–(3.10) that

$$|\alpha_{iK}| \le l_i \gamma_K \text{ for } i = 1, \ldots, q \ \& \ \beta_{iK} \le l_i \gamma_K \text{ for } i = q+1, \ldots, q+r$$

if the corresponding functions $\varphi_i(\bar{x}(0), \cdot, \bar{T})$ are Lipschitz continuous around $\bar{x}(\bar{T})$ with constants l_i. Note that the number γ_K in (3.10) can be effectively estimated in terms of the initial data of the problem; see [43].

Now for each $K = 1, 2, \ldots$, we define the *discrete approximation problem* (P_K) as follows:

$$\text{minimize } J_K[x_K, T_K] := \varphi_0(x_K(0), x_K(T_K), T_K) + |x_K(0) - \bar{x}(0)|^2 + (T_K - \bar{T})^2 + h_K \sum_{j=0}^{K-1} f\left(t_j, x_K(t_j), \frac{x_K(t_{j+1}) - x_K(t_j)}{h_K}\right) + \sum_{j=0}^{K-1} \int_{t_j}^{t_{j+1}} \left| \frac{x_K(t_{j+1}) - x_K(t_j)}{h_K} - \dot{\bar{x}}(t) \right|^2 dt \tag{3.11}$$

over trajectories $\{x_K(t_j) | \ j = 0, \ldots, K\}$ for the finite difference inclusion (3.1) subject to the constraints

$$\varphi_i(x_K(0), x_K(T_K), T_K) \le \alpha_{iK} \text{ for } i = 1, \ldots, q; \tag{3.12}$$

$$-\beta_{iK} \le \varphi_i(x_K(0), x_K(T_K), T_K) \le \beta_{iK} \text{ for } i = q+1, \ldots, q+r; \tag{3.13}$$

$$(x_K(0), x_K(T_K), T_K) \in \Omega_K := \Omega + \gamma_K B; \tag{3.14}$$

$$x_K(t_j) \in B_\epsilon(\bar{x}(t_j)) \ \forall j = 0, \ldots, K; \ T_K \le \bar{T} + \varepsilon \tag{3.15}$$

where ϵ satisfies (3.7) and ε is a given (arbitrarily small) positive number.

Let us emphasize that in each problem (P_K) the final time T_K and the discretization step h_K are *variable* for any fixed $K = 1, 2, \ldots$.

We assume that the multifunction F satisfies hypotheses (H1) and (H2) along the given optimal solution $\bar{x}(\cdot)$ for $t \in [0, \bar{T}+\varepsilon]$. In addition to this, we impose the following hypotheses on f, φ_i, and Ω:

(H3) $f(\cdot, x, v)$ is continuous for a.e. $t \in [0, \bar{T}+\varepsilon]$ and bounded uniformly in $(x, v) \in U \times (m_F B)$.

(H4) There exists $\nu > 0$ such that the function $f(t, \cdot, \cdot)$ is continuous on the set

$$A_\nu(t) := \{(x, v) \in U \times (m_F + \nu)B \mid v \in F(t', x) \text{ for some } t' \in (t-\nu, t]\}$$

uniformly in $t \in [0, \bar{T}+\varepsilon]$.

(H5) The functions $\varphi_0, \varphi_1, \ldots, \varphi_q$ are lower semicontinuous on $U \times U \times [0, \bar{T}+\varepsilon]$ being continuous in the second variable at $(\bar{x}(0), \bar{x}(\bar{T}), \bar{T})$.

(H6) The functions $\varphi_{q+1}, \ldots, \varphi_{q+r}$ are continuous on $U \times U \times [0, \bar{T}+\varepsilon]$.

(H7) The set Ω is closed around $(\bar{x}(0), \bar{x}(\bar{T}), \bar{T})$.

3.2. THEOREM. *Let $\bar{x}(t)$, $0 \le t \le \bar{T}$, be an original optimal solution to problem (P) which possesses the property of relaxation stability. Assume that hypotheses* (H1)–(H7) *hold. Then:*

(i) *The discrete approximation problems (P_K) admit optimal solutions $\{\bar{x}_K(t_j) \mid j = 0, \ldots, K\}$ for all K large enough.*

(ii) *For any sequence of extended optimal trajectories $\{\bar{x}_K(t)$, $0 \le t \le \bar{T}_K\}$ in (P_K) one has*

$$\bar{T}_K \to \bar{T} \text{ as } K \to \infty, \tag{3.16}$$

$$\max\{|\bar{x}_K(t) - \bar{x}(t)| : t \in \bar{T}_K\} \to 0 \text{ as } K \to \infty, \tag{3.17}$$

$$\int_0^{\bar{T}_K} |\dot{\bar{x}}_K(t) - \dot{\bar{x}}(t)|^2 dt \to 0 \text{ as } K \to \infty. \tag{3.18}$$

Proof. First let us show that the discrete approximation problem (P_K) has a feasible solution for any K large enough. In fact, we check that the trajectory $\{z_K(t_j) \mid j = 0, \ldots, K\}$ constructed in Lemma 3.1 satisfies all constraints (3.12)–(3.15). For the case of (3.12)–(3.14), it immediately follows from the definitions of the perturbations α_{iK}, β_{iK}, and γ_K in (3.8)–(3.10).

According to Lemma 3.1, the extended trajectories $z_K(t)$, $0 \le t \le \bar{T}$, converge to $\bar{x}(t)$ uniformly in $[0, \bar{T}]$. Now taking any ϵ satisfying (3.7), one can find a natural number $\bar{K}$ such that

$$|z_K(t) - \bar{x}(t)| \le \epsilon \;\; \forall K \ge \bar{K}.$$

Therefore, we ensure (3.15) for $z_K(\cdot)$ when $K \ge \bar{K}$. Now the existence of optimal solutions to (P_K) for $K \ge \bar{K}$ follows directly from the classical

Weierstrass theorem due to the compactness and continuity assumptions made. This proves assertion (i).

To prove (ii), let us start with establishing the inequality

$$\limsup_{K\to\infty} J_K[\bar{x}_K,\bar{T}_K] \le J[\bar{x},\bar{T}] \tag{3.19}$$

for any sequence of optimal solutions to (P_K). If (3.19) is not true, then there is a sequence $\mathcal{N}$ of natural numbers $K \to \infty$ such that

$$\varphi_0(\bar{x}(0),\bar{x}(\bar{T}),\bar{T}) + \int_0^{\bar{T}} f(t,\bar{x}(t),\dot{\bar{x}}(t))dt < J_K[\bar{x}_K,\bar{T}_K] \ \forall K \in \mathcal{N}. \tag{3.20}$$

To get a contradiction from (3.20), let us again use Lemma 3.1 and consider an approximating sequence of the discrete trajectories $\{z_K(t_j)|\ j = 0,\ldots,K;\ T_K = \bar{T}\}$ which are proved to be feasible for (P_K). By virtue of $z_K(0) = \bar{x}(0)$ and the continuity of φ_0 in the second variable, one has

$$\varphi_0(z_K(0),z_K(T_K),T_K) \to \varphi_0(\bar{x}(0),\bar{x}(\bar{T}),\bar{T}) \text{ as } K \to \infty.$$

Observe that in the cost functional expression (3.11) for $z_K(\cdot)$, the second and third terms vanish. For the last term therein we get

$$\sum_{j=0}^{K-1}\int_{t_j}^{t_{j+1}} \left|\frac{z_K(t_{j+1})-z_K(t_j)}{h_K} - \dot{\bar{x}}(t)\right|^2 dt = \int_0^{\bar{T}} |\dot{z}_K(t)-\dot{\bar{x}}(t)|^2 dt \to 0 \text{ as} K \to \infty$$

by virtue of the extension rules (3.2), (3.3) and the convergence result in Lemma 3.1.

It remains to estimate the fourth term in (3.11). Let us prove that

$$\eta_K := h_K \sum_{j=0}^{K-1} f(t_j, z_K(t_j), \frac{z_K(t_{j+1})-z_K(t_j)}{h_K}) \to \int_0^{\bar{T}} f(t,\bar{x}(t),\dot{\bar{x}}(t))dt \text{ as } K\to\infty$$

under assumptions (H1)–(H4). Note that (H3) implies that $\tau(f;h_K) \to 0$ as $K \to \infty$ for the averaged modulus of continuity (3.6).

In what follows we use the sign "$\sim$"for expressions which are equivalent as $K \to \infty$. Due to (3.2), (3.3), (3.6), and Lemma 3.1, one gets

$$\eta_K = \sum_{j=0}^{K-1}\int_{t_j}^{t_{j+1}} f(t_j,z_K(t_j),\dot{z}_K(t))dt \sim \sum_{j=0}^{K-1}\int_{t_j}^{t_{j+1}} f(t,z_K(t_j),\dot{z}_K(t))dt + \tau(f;h_K) \sim$$

$$\sum_{j=0}^{K-1}\int_{t_j}^{t_{j+1}} f(t,\bar{x}(t),\dot{z}_K(t))dt = \int_0^{\bar{T}} f(t,\bar{x}(t),\dot{z}_K(t))dt \sim \int_0^{\bar{T}} f(t,\bar{x}(t),\dot{\bar{x}}(t))dt.$$

To justify the last conclusion, we observe that the strong $L^2[0,\bar{T}]$- convergence of $\dot{z}_K(\cdot) \to \dot{\bar{x}}(\cdot)$ in Lemma 3.1 implies the existence of a subsequence

$\{\dot{z}_K(t)\}$ converging to $\dot{\bar{x}}(t)$ for a.e. $t \in [0, \bar{T}]$. Now one can employ the classical Lebesgue theorem about the limit under the integral sign.

Thus we have proved that

$$J_K[z_K, \bar{T}] \to J[\bar{x}, \bar{T}] \text{ as } K \to \infty.$$

By virtue of (3.20) and the feasibility of $z_K(\cdot)$ for (P_K), this contradicts the optimality of $\bar{x}_K(\cdot)$ for the discrete approximations under consideration. Therefore, we ensure (3.19).

In the discussions above we have heavily used Lemma 3.1 but have not yet used the property of relaxation stability for the original problem (P). Now let us prove that the relaxation stability property together with result (3.19) imply the desirable convergences (3.16)–(3.18). We are going to show that

$$(3.21)\ \lim_{K\to\infty} [\delta_K := |\bar{x}_K(0) - \bar{x}(0)|^2 + |\bar{T}_K - \bar{T}|^2 + \int_0^{\bar{T}_K} |\dot{\bar{x}}_K(t) - \dot{\bar{x}}(t)|^2 dt] = 0.$$

Suppose that (3.21) doesn't hold and consider any limiting point $\delta > 0$ of the sequence $\{\delta_K\}$ in (3.21). For the purpose of simplicity, we assume that $\delta = \lim \delta_K$ for all $K \to \infty$.

Using the boundedness of $\bar{T}_K$ by virtue of (3.7), one can find a real number $\tilde{T} \le \bar{T} + \varepsilon$ such that $\bar{T}_K \to \tilde{T}$ as $K \to \infty$ (here and on we take all natural K without loss of generality). Let us consider the extended discrete trajectories $\bar{x}_K(t)$ on the time interval $[0, \tilde{T}]$ defining $\bar{x}_K(t) \equiv \bar{x}_K(T_K)$ for $t \in (T_K, \tilde{T}]$ when $T_K < \tilde{T}$.

Taking into account the boundedness conditions (3.7) and (3.4) for $t \in [0, \tilde{T}]$ and employing the classical compactness results, we claim the existence of an absolutely continuous function $\tilde{x}(t)$ in $[0, \tilde{T}]$ such that $\bar{x}_K(\cdot) \to \tilde{x}(t)$ uniformly in $[0, \tilde{T}]$ and $\dot{\bar{x}}_K(\cdot) \to \dot{\tilde{x}}(\cdot)$ weakly in $L^2[0, \tilde{T}]$ as $K \to \infty$. Due to assumptions (H5)–(H7), the limiting function $\tilde{x}(t)$, $0 \le t \le \tilde{T}$, satisfies endpoint constraints (2.3)–(2.5). Now we study the limit of the cost functional $J_K[\bar{x}_K, \bar{T}_K]$ in (3.11) as $K \to \infty$.

According to the well-known Mazur theorem, there is a sequence of convex combinations of $\dot{\bar{x}}_K(\cdot)$ which converges to $\dot{\tilde{x}}(\cdot)$ in the norm topology of $L^2[0, \tilde{T}]$. Hence it contains a subsequence converging to $\dot{\tilde{x}}(\cdot)$ for a.e. $t \in [0, \tilde{T}]$.

From here one can easily conclude that $\tilde{x}(\cdot)$ satisfies the convexified differential inclusion (2.8). Moreover, taking into account that

$$h_K \sum_{j=0}^{K-1} f(t_j, \bar{x}_K(t_j), \frac{\bar{x}_K(t_{j+1}) - \bar{x}_K(t_j)}{h_K}) \sim \int_0^{\tilde{T}} f(t, \bar{x}_K(t), \dot{\bar{x}}_K(t)) dt \text{ as } K \to \infty$$

and also the definition of the convexified function $\hat{f}_F$ for (2.6), we get

$$\int_0^{\tilde{T}} \hat{f}_F(t, \tilde{x}(t), \dot{\tilde{x}}(t))dt \leq \liminf_{K\to\infty} h_K \sum_{j=0}^{K-1} f(t_j, \bar{x}_K(t_j), \frac{\bar{x}_K(t_{j+1}) - \bar{x}_K(t_j)}{h_K}).$$

(3.22)

Let us observe that

$$\sum_{j=0}^{K-1} \int_{t_j}^{t_{j+1}} |\frac{\bar{x}_K(t_{j+1})-\bar{x}_K(t_j)}{h_K} - \dot{\tilde{x}}(t)|^2 dt = \int_0^{\bar{T}_K} |\dot{\bar{x}}_K(t) - \dot{\tilde{x}}(t)|^2 dt \sim \int_0^{\tilde{T}} |\dot{\bar{x}}_K(t) - \dot{\tilde{x}}(t)|^2 dt \text{ as } K \to \infty.$$

By virtue of the convexity in v of the function $g(v,t) := |v - \dot{\tilde{x}}(t)|^2$, the integral functional

$$I[v] := \int_0^{\tilde{T}} |v(t) - \dot{\tilde{x}}(t)|^2 dt$$

is lower semicontinuous in the weak topology of $L^2[0,\tilde{T}]$. Therefore.

$$(3.23) \quad \int_0^{\tilde{T}} |\dot{\tilde{x}}(t) - \dot{\tilde{x}}(t)|^2 dt \leq \liminf_{K\to\infty} \sum_{j=0}^{K-1} |\frac{\bar{x}_K(t_{j+1}) - \bar{x}_K(t_j)}{h_K} - \dot{\tilde{x}}(t)|^2 dt.$$

Now passing to the limit in (3.11) for $J_K[\bar{x}_K, \bar{T}_K]$ and taking into account (3.22), (3.23), and the lower semicontinuity of φ_0, one has

$$(3.24) \varphi_0(\tilde{x}(0), \tilde{x}(\tilde{T}), \tilde{T}) + \int_0^{\tilde{T}} \hat{f}_F(t, \tilde{x}(t), \dot{\tilde{x}}(t))dt + \delta \leq \liminf_{K\to\infty} J_K[\bar{x}_K, \bar{T}_K].$$

By virtue of (3.19) and our assumption about $\delta > 0$, (3.24) implies that

$$\hat{J}[\tilde{x}, \tilde{T}] < J[\bar{x}, \bar{T}].$$

But the latter is impossible because

$$J[\bar{x}, \bar{T}] = \inf(P) = \inf(R)$$

due to the relaxation stability property for (P). Therefore, $\delta = 0$ and we get relationship (3.21).

The relationship obtained directly implies (3.16), (3.18), and the convergence $\bar{x}_K(0) \to \bar{x}(0)$ as $K \to \infty$. The latter and (3.18) provide (3.17). This ends the proof of the theorem. □

3.3. *Remark.* Denote by $\bar{J}_K$ the optimal value of the cost functional in the discrete approximation problem (P_K) for each $K = 1, 2, \ldots$. We have actually proved in Theorem 3.2 that

$$(3.25) \qquad \inf(R) \leq \liminf_{K\to\infty} \bar{J}_K \leq \limsup_{K\to\infty} \bar{J}_K \leq \inf(P)$$

with no assumption about the relaxation stability. Moreover, (3.25) implies that *the relaxation stability property for* (P) *is in fact equivalent to the value convergence* $\bar{J}_K \to \inf(P)$ *of the discrete approximations under appropriate perturbations of constraints.* We refer to Mordukhovich [32, 35, 36] and the recent book of Dontchev and Zolezzi [20] for related properties and discussions in other problems of optimal control.

3.4. *Remark.* In the convergence results stated above, one may avoid the continuity hypothesis (H3) on f in t by changing the approximation

$$h_K \sum_{j=0}^{K-1} f(t_j, x_K(t_j), \frac{x_K(t_{j+1})-x_K(t_j)}{h_K}) \text{ for}$$
$$\sum_{j=0}^{K-1} \int_{t_j}^{t_{j+1}} f(t, x_K(t_j), \frac{x_K(t_{j+1})-x_K(t_j)}{h_K}).$$

Indeed, we can handle the latter approximation in the same way as the last term in (3.11) under the measurability (summability) assumption on f in t. On the other hand, one can change

$$\sum_{j=0}^{K-1} \int_{t_j}^{t_{j+1}} |\frac{x_K(t_{j+1})-x_K(t_j)}{h_K} - \dot{\bar{x}}(t)|^2 dt \text{ for}$$
$$h_K \sum_{j=0}^{K-1} |\frac{x_K(t_{j+1})-x_K(t_j)}{h_K} - \dot{\bar{x}}(t_j)|^2$$

assuming that $\dot{\bar{x}}(\cdot)$ is continuous for a.e. $t \in [0, \bar{T}]$. These two kinds of approximations are treated by different techniques from the viewpoint of necessary optimality conditions; cf. [43] and Section 5 below.

3.5. *Remark.* The convergence results obtained in this section allow to make a bridge between variational problems for differential inclusions with free time and dynamic optimization problems in finite dimensions. The latter can be reduced to special finite dimensional problems of mathematical programming; see Section 5. In this paper we develop this procedure to obtain necessary optimality conditions for the original variational problem (P) by passing to the limit in corresponding necessary conditions for finite dimensional optimization problems (P_K).

Note that the convergence results in Theorem 3.2 ensure that optimal solutions to problems (P_K) belong to the *interiority* of the constraints in (3.15). Therefore, from the viewpoint of necessary conditions, constraints (3.15) can be omitted without any loss of generality. *In what follows we shall always consider problems* (P_K) *with no constraints* (3.15).

3.6. *Remark.* Problems (P_K) and equivalent problems of mathematical programming always have *many geometric constraints* arising from the approximation of differential inclusions. Such problems turn out to be objects of *nonsmooth analysis and optimization.*

For the variational analysis of these problems and then for passing to the limit in necessary optimality conditions as $K \to \infty$, we need to use generalized differential constructions with special properties which are the subject of the next section.

4. Generalized differentiation. In this section we briefly review some constructions and results on the generalized differentiation of nonsmooth and set-valued mappings which are widely employed in the paper. Most of these results with detailed proofs and discussions can be found in Mordukhovich [35, 40, 41]. We also refer the reader to Clarke [12], Ioffe [22–24], Loewen [28], Rockafellar [48, 52], and Rockafellar and Wets [53] for related and additional material.

Developing a geometric approach to the generalized differentiation, we begin with the definition of a normal cone to arbitrary sets in finite dimensions.

Let Ω be a nonempty set in $\mathbf{R}^n$ and let

$$\Pi(x,\Omega) := \{\omega \in \text{cl}\,\Omega \mid \; |x-\omega| = \text{dist}(x,\Omega)\}$$

be the (multi-valued) *Euclidean projector* of x on the set cl Ω. In the following definition, "cone" stands for the conic hull of a set and "Limsup" denotes the well-known *Kuratowski-Painlevé upper limit* for multifunctions, i.e., the collection of all limiting points in their values (see, e.g., [2, p. 41]).

4.1. DEFINITION. Given $\bar{x} \in \text{cl}\,\Omega$, the closed cone

$$N(\bar{x};\Omega) := \text{Limsup}_{x\to\bar{x}}[\text{cone}(x - \Pi(x,\Omega))] \tag{4.1}$$

is called the *normal cone* to the set Ω at the point $\bar{x}$. If $\bar{x} \notin \text{cl}\,\Omega$, we put $N(\bar{x};\Omega) = \emptyset$.

This concept of (nonconvex) normal cone was first adopted and utilized in Mordukhovich [31] for the purpose of obtaining necessary optimality conditions in nonsmooth finite dimensional problems of mathematical programming and optimal control. Actually this construction turns out to be a by-product of a special metric approximation method for proving necessary conditions in nonsmooth optimization. Then this method was employed to obtain comprehensive calculus rules for the normal cone (4.1) and related differential constructions. We refer the reader to the book [35] and recent papers [38–43] for more details and comments.

Note that such a construction (sometimes called the cone of limiting proximal normals or the prenormal cone) has been used in Clarke's nonsmooth analysis for an intermediate device and applications of his well-known normal cone $N_C(\bar{x};\Omega)$ coinciding with the convex closure of (4.1):

$$N_C(\bar{x};\Omega) = \text{cl co}\, N(\bar{x};\Omega); \tag{4.2}$$

see [11, 12, 28, 48] and references therein. Note also that in the finite dimensional setting the normal cone (4.1) coincides with the "approximate normal cone" defined by Ioffe [22, 23].

The normal cone (4.1) always admits the following useful representation

$$N(\bar{x};\Omega) = \text{Limsup}_{x\to\bar{x}}\hat{N}(x;\Omega)$$

where the convex cone

$$\text{(4.3)}\qquad \hat{N}(x;\Omega) := \{x^\star \in \mathbf{R}^n \mid \limsup_{x'(\in\Omega)\to x} \frac{\langle x^\star, x'-x\rangle}{|x'-x|} \le 0\} \text{ for } x \in \operatorname{cl}\Omega$$

turns out to be dual (in finite dimensions) to the well-known Bouligand *contingent* (tangent) cone.

Observe that our basic normal cone in (4.1) is *not dual* to any tangent cone because it is not convex, except for situations where all three normal cones (4.1)– (4.3) coincide at $\bar{x}$. (The latter fails in many important settings, e.g., for sets Ω which can be locally represented as graphs of nonsmooth Lipschitz continuous functions; see Remark 4.5.)

Despite its nonconvexity, the normal cone (4.1) possesses many nice properties essential for applications. First, it is always *robust* with respect to perturbations of $\bar{x}$, i.e., the multifunction $N(\cdot,\Omega)$ has closed graph. What is really surprising a priori, that this nonconvex normal cone and related differential objects enjoy *rich calculi* which is even better than ones for convex-valued counterparts; see below. The progress in this direction has been achieved by using a *variational approach* instead of convex analysis.

Now we consider generalized differentiation constructions for multifunctions and nonsmooth mappings induced by the normal cone (4.1) to their *graphs*. The following notion was first introduced in [33].

4.2. Definition. Let F be a multifunction from $\mathbf{R}^n$ into $\mathbf{R}^m$ whose graph

$$\operatorname{gph} F := \{(x,y) \in \mathbf{R}^n \times \mathbf{R}^m \mid y \in F(x)\}$$

is nonempty, and let $(\bar{x},\bar{y}) \in \operatorname{cl}(\operatorname{gph} F)$. The multifunction $D^\star F(\bar{x},\bar{y})$ from $\mathbf{R}^m$ into $\mathbf{R}^n$ defined by

$$\text{(4.4)}\quad D^\star F(\bar{x},\bar{y})(y^\star) := \{x^\star \in \mathbf{R}^n \mid (x^\star, -y^\star) \in N((\bar{x},\bar{y}); \operatorname{gph} F)\}$$

is called the *coderivative* of F at $(\bar{x},\bar{y})$. We put $D^\star F(\bar{x},\bar{y})(y^\star) = \emptyset$ if $(\bar{x},\bar{y}) \notin \operatorname{cl}(\operatorname{gph} F)$. The symbol $D^\star F(\bar{x})$ is used in (4.4) when F is single-valued at $\bar{x}$ and $\bar{y} = F(\bar{x})$.

One can see that the coderivative $D^\star F(\bar{x},\bar{y})(\cdot)$ is a positive homogeneous multifunction with closed values. These values may be not convex by virtue of the nonconvexity of (4.1). Therefore, the coderivative (4.4) is *not dual* to any tangentially generated *derivative* of multifunctions (see, e.g., [2, Section 5]).

Let us present two useful representations of the coderivative (4.4) in the following classical settings.

4.3. Proposition. *Let F be single-valued around $\bar{x}$ and strictly differentiable at $\bar{x}$ with the Jacobian $\nabla F(\bar{x}) \in \mathbf{R}^{m\times n}$, i.e.,*

$$\lim_{x,x'\to\bar{x}} \frac{F(x)-F(x')-\nabla F(\bar{x})(x-x')}{|x-x'|} = 0.$$

Then one has

$$D^\star F(\bar{x})(y^\star) = \{(\nabla F(\bar{x}))^\star y^\star\} \ \ \forall y^\star \in \mathbf{R}^m.$$

4.4. PROPOSITION. *Let F be a multifunction of convex graph. Then for any point $(\bar{x},\bar{y}) \in \mathrm{cl}(\mathrm{gph}\, F)$ and for any $y^\star \in \mathbf{R}^m$ one has*

$$D^\star F(\bar{x},\bar{y})(y^\star) = \{x^\star \in \mathbf{R}^n |\ \langle x^\star, \bar{x}\rangle - \langle y^\star, \bar{y}\rangle = \sup_{(x,y)\in gph\ F} [\langle x^\star, x\rangle - \langle y^\star, y\rangle]\}.$$

4.5. *Remark.* Note that in both cases considered above the coderivative (4.4) coincides with the *Clarke coderivative* $D^\star_C F(\bar{x},\bar{y})(\cdot)$ which is obtained in the scheme (4.4) by replacing the normal cone (4.1) with its Clarke's counterpart (4.2). However, this is no longer true for a broad class of multifunctions whose graphs are nonsmooth *Lipschitzian manifolds* in the sense of Rockafellar [50], i.e., they are locally homeomorphic around $(\bar{x},\bar{y})$ to graphs of nonsmooth Lipschitz continuous functions. Besides locally Lipschitzian vector functions, this class includes maximal monotone relations, in particular, subdifferential operators for convex, concave, and saddle functions; see [50].

Indeed, for such multifunctions the coderivative (4.4) is always *strictly smaller* than its Clarke's counterpart; moreover, they may be distinguished *in dimensions.* (We refer the reader to [42, Section 3] for more details and related discussions.) That is is why the refined form of the Euler-Lagrange inclusion in (1.12) involving the coderivative (4.4) is essentially stronger than Clarke's one (1.11) involving $D^\star_C F$.

Let us consider some general properties of the coderivative (4.4) which are useful in the framework of this paper. First we note that it is a *robust* construction with respect to perturbations of the data $(\bar{x},\bar{y},y^\star)$. The next result [35, Theorem 3.1] shows that the considered Euler-Lagrange conditions for differential and discrete inclusions automatically imply the maximum (minimum) conditions in problems with convex velocities. In what follows, we use a conventional concept of lower (inner) semicontinuity for multifunctions; see, e.g., [2, p. 39].

4.6. PROPOSITION. *Let F be convex-valued around $\bar{x}$ and lower semicontinuous at $\bar{x}$. Then one has*

$$[D^\star F(\bar{x},\bar{y})(y^\star) \neq \emptyset] \Longrightarrow [\langle y^\star, \bar{y}\rangle = \min\{\langle y^\star, y\rangle |\ y \in F(\bar{x})\}].$$

One of the most important advantages of the coderivative (4.4) in the general setting consists of effective using this construction for *complete dual characterizations of Lipschitzian properties* of multifunctions and nonsmooth mappings. These results play a *crucial role* to justify the convergence of adjoint functions in necessary optimality conditions for discrete approximations; see the proof of Theorem 6.1 in Section 6.

Recall that the multifunction F is said to be *pseudo-Lipschitzian* around $(\bar{x}, \bar{y}) \in \operatorname{gph} F$ if there is a neighborhood U of $\bar{x}$, a neighborhood V of $\bar{y}$, and a constant $l \geq 0$ such that

$$F(x') \cap V \subset F(x) + l|x' - x|B \quad \forall x, x' \in U. \tag{4.5}$$

This definition goes back to Aubin who imposed the additional condition $F(x) \cap V \neq \emptyset$ for all $x \in U$; see [2, Definition 1.4.5]. Rockafellar [49] obtained some characterizations of the pseudo-Lipschitzian property and established its interrelations with other Lipschitzian properties of multifunctions. Note that the pseudo-Lipschitzian property of F turns out to be equivalent to such fundamental properties as *metric regularity* and *openness at linear rate* for the *inverse mapping* F^{-1}; see Borwein and Zhuang [5], Penot [44], and Mordukhovich [40] for various modifications.

The next dual characterizations were obtained in Mordukhovich [38, Theorem 5.1] and [40, Theorem 5.7]. Moreover, therein one can find formulae for calculating the *exact bound* of Lipschitz moduli l in (4.5).

4.7. PROPOSITION. *Let F be a multifunction from $\mathbf{R}^n$ into $\mathbf{R}^m$ whose graph is closed around $(\bar{x}, \bar{y})$. Then the following are equivalent:*

(a) *F is pseudo-Lipschitzian around $(\bar{x}, \bar{y})$;*

(b) *there exist a neighborhood U of $\bar{x}$, a neighborhood V of $\bar{y}$, and a constant $l \geq 0$ such that*

$$\sup\{|x^*| : x^* \in D^*F(x,y)(y^*)\} \leq l|y^*| \tag{4.6}$$

for any $x \in U$, $y \in F(x) \cap V$, and $y^* \in \mathbf{R}^m$;

(c) $D^*F(\bar{x}, \bar{y})(0) = \{0\}$.

If F is locally bounded around $\bar{x}$, then its classical (Hausdorff) Lipschitzian behavior ($V = \mathbf{R}^m$ in (4.5)) is equivalent to F being pseudo-Lipschitzian around $(\bar{x}, \bar{y})$ for *every* $\bar{y} \in F(\bar{x})$; see [49]. Therefore, we get dual criteria for the classical local Lipschitz continuity of multifunctions.

4.8. COROLLARY. *Let a closed-graph multifunction F be locally bounded around the point $\bar{x}$ where $F(\bar{x}) \neq \emptyset$. Then the following are equivalent:*

(a) *F is locally Lipschitzian around $\bar{x}$;*

(b) *there are a neighborhood U of $\bar{x}$ and a number $l \geq 0$ such that estimate* (4.6) *holds for any $x \in U$, $y \in F(x)$, and $y^* \in \mathbf{R}^m$;*

(c) $D^*F(\bar{x}, \bar{y})(0) = \{0\} \;\; \forall \bar{y} \in F(\bar{x})$.

4.9. *Remark.* If one replaces the coderivative (4.4) by its Clarke's counterpart in the "null-criteria" (c) of Propositions 4.7 and 4.8, then the conditions obtained are far removed from the necessity for F to possess the Lipschitzian properties. Indeed, they are *never fulfilled* even in the case of single-valued Lipschitz continuous functions which do not happen to be strictly differentiable at $\bar{x}$ (and also in more general settings with nonsmooth Lipschitzian manifolds in Remark 4.5). In fact, the condition $D^\star_C F(\bar{x}, \bar{y})(0) = \{0\}$ implies a "strictly smooth" property of multifunctions whose graphs are Lipschitzian manifolds; see [42, 50]. So, one cannot ensure an analogue of the basic estimate (4.6) in terms of $D^\star_C$ for such nonsmooth multifunctions.

Now let us consider some *calculus rules* for the coderivative (4.4) and related differential constructions which are valid under natural assumptions and turns out to be of great importance for applications. The proofs of the following and other calculus results based on an *extremal principle* for systems of sets can be found in Mordukhovich [41].

4.10. PROPOSITION. *Let F_1 and F_2 be closed-graph multifunctions from $\mathbf{R}^n$ into $\mathbf{R}^m$, and let $\bar{y} \in F_1(\bar{x})+F_2(\bar{x})$. Assume that the multifunction S from $\mathbf{R}^{n+m}$ into $\mathbf{R}^{2m}$ defined by*

$$S(x, y) := \{(y_1, y_2) \in \mathbf{R}^{2m} |\, y_1 \in F_1(x),\ y_2 \in F_2(x),\ y_1 + y_2 = y\}$$

is locally bounded around $(\bar{x}, \bar{y})$ and the qualification condition

$$D^\star F_1(\bar{x}, y_1)(0) \cap (-D^\star F_2(\bar{x}, y_2)(0)) = \{0\} \ \ \forall (y_1, y_2) \in S(\bar{x}, \bar{y}) \tag{4.7}$$

is fulfilled. Then for any $y^\star \in \mathbf{R}^m$ one has

$$D^\star(F_1 + F_2)(\bar{x}, \bar{y})(y^\star) \subset \bigcup_{(y_1,y_2)\in S(\bar{x},\bar{y})} [D^\star F_1(\bar{x}, y_1)(y^\star) + D^\star F_2(\bar{x}, y_2)(y^\star)]$$

where equality holds if either F_1 or F_2 is strictly differentiable at $\bar{x}$.

Observe that according to Proposition 4.7, the qualification condition (4.7) is automatically fulfilled if for each $(y_1, y_2) \in S(\bar{x}, \bar{y})$ *either F_1 is pseudo-Lipschitzian around $(\bar{x}, y_1)$ or F_2 is pseudo-Lipschitzian around $(\bar{x}, y_2)$.*

The next calculus result is concerned with general *chain rules* for the composition

$$(G \circ F)(x) = G(F(x)) := \bigcup_{y\in F(x)} G(y)$$

of multifunctions F from $\mathbf{R}^n$ into $\mathbf{R}^m$ and G from $\mathbf{R}^m$ into $\mathbf{R}^q$.

4.11. PROPOSITION. *Let F and G be of closed graphs and let $\bar{z} \in (G \circ F)(\bar{x})$. Assume that the multifunction*

$$(x, z) \to F(x) \cap G^{-1}(z) = \{y \in F(x) |\, z \in G(y)\}$$

is locally bounded around $(\bar{x}, \bar{z})$ *and the qualification condition*

$$(4.8)\quad D^*G(y,\bar{z})(0) \cap \ker D^*F(\bar{x},y) = \{0\} \ \ \forall y \in F(\bar{x}) \cap G^{-1}(\bar{z})$$

is fulfilled. Then one has

$$D^*(G \circ F)(\bar{x},\bar{z}) \subset \bigcup_{y \in F(\bar{x}) \cap G^{-1}(\bar{z})} [D^*F(\bar{x},y) \circ D^*G(y,\bar{z})]$$

where equality holds in each of the following cases:

(i) *F is strictly differentiable at $\bar{x}$ with the quadratic ($n = m$) and nonsingular Jacobian matrix $\nabla F(\bar{x})$;*

(ii) *F is single-valued and Lipschitz continuous around $\bar{x}$ while G is strictly differentiable at $\bar{y} = F(\bar{x})$.*

Note that according to criterion (c) in Proposition 4.7 and the inverse criterion for the (local) metric regularity in [40, Corollary 4.3], the qualification condition (4.8) is automatically fulfilled if for each $y \in F(\bar{x}) \cap G^{-1}(\bar{z})$ *either G is pseudo-Lipschitzian around $(y, \bar{z})$ or F is metrically regular around $(\bar{x}, y)$.*

These major calculus results implies some other calculus rules for the coderivatives of multifunctions and associated constructions of the first and second order subdifferentials for *extended-real-valued functions* $f : \mathbf{R}^n \to \bar{\mathbf{R}} = [-\infty, \infty]$; see [35, 41]. Now we consider some results for the (first order) subdifferentials of such functions important in what follows.

4.12. Definition. Let $|f(\bar{x})| < \infty$, and let

$$E_f(x) := \{\mu \in \mathbf{R} |\ \mu \geq f(x)\}$$

be the epigraphical multifunction (gph E_f = epi f) associated with f. The set

$$(4.9)\quad \begin{aligned} &\partial f(\bar{x}) := D^*E_f(\bar{x}, f(\bar{x}))(1) = \\ &\{x^* \in \mathbf{R}^n |\ (x^*, -1) \in N((\bar{x}, f(\bar{x})); \text{epi}\, f)\} \end{aligned}$$

is called the *subdifferential* of f at $\bar{x}$. We put $\partial f(\bar{x}) = \emptyset$ if $|f(\bar{x})| = \infty$.

The subdifferential (4.9) has been studied in details in the book of Mordukhovich [35] and previous publications starting with [31]. There are various analytic representations of (4.9) in terms of other subdifferential mappings named Dini, Frechét, viscosity, proximal subdifferentials, etc. We refer the reader to [12, 16, 22, 28, 35, 52, 53] for more information about these and related questions. Therein construction (4.9) is often used in some equivalent forms under different names (e.g., the approximate subdifferential, the presubdifferential, the set of limiting proximal subgradients or basic subgradients).

Note that for functions $f : \mathbf{R}^n \to \bar{\mathbf{R}}$ *continuous* around $\bar{x}$, the subdifferential (4.9) can be expressed directly in terms of the coderivative (4.4) for f:

$$\partial f(\bar{x}) = D^\star f(\bar{x})(1),$$

i.e., (epi f) in (4.9) can be replaced by (gph f); cf. [35, Proposition 2.1].

On the other hand, if $f : \mathbf{R}^n \to \mathbf{R}^m$ is a *vector function Lipschitz continuous* around $\bar{x}$, then the coderivarive of f can be expressed in terms of the subdifferential (4.9) for its *Lagrange scalarization*:

$$D^\star f(\bar{x})(y^\star) = \partial \langle y^\star, f\rangle(\bar{x}) \;\; \forall y^\star \in \mathbf{R}^m$$

where $\langle y^\star, f\rangle(x) := \langle y^\star, f(x)\rangle$; see the proofs and comments in [22, Section 5] and [35, Section 3].

For the case of real-valued Lipschitz continuous functions we have the following well-known results (see, e.g., [12, Propositions 1.1 and 1.2] and [35, Theorem 2.1]).

4.13. Proposition. *Let a real-valued function f be locally Lipschitzian around $\bar{x}$ with modulus l. Then:*

(i) $\partial f(\bar{x}) \neq \emptyset$ & $|x^\star| \leq l \;\; \forall x^\star \in \partial f(\bar{x})$;

(ii) *one has*

$$\partial_C f(\bar{x}) = \text{co}\, \partial f(\bar{x}) = \text{co}\{\lim_{\nu\to\infty} \nabla f(x_\nu)|\, f \text{ is differentiable at } x_\nu \to \bar{x},\ x_\nu \notin S\}$$

for the generalized gradient of Clarke, where S is arbitrary set of measure zero.

It is easy to see that

$$N(\bar{x};\Omega) = \partial\delta(\bar{x},\Omega) \;\text{ if }\; \bar{x} \in \Omega$$

for the *indicator function* of the set Ω. We also use another representation of the normal cone (4.1) in terms of the subdifferential (4.9) for the Lipschitz continuous *distance function* $\text{dist}(\cdot,\Omega)$. The following result is proved in [35, Proposition 2.7].

4.14. Proposition. *For any nonempty set Ω, one has*

$$N(\bar{x};\Omega) = \text{cone}[\partial \text{dist}(\bar{x},\Omega)] \;\text{ if }\; \bar{x} \in \text{cl}\,\Omega.$$

Various calculus results for the subdifferentials (4.9) and the normal cones (4.1) can be easily deduced from the coderivative results stated above; cf. [41]. A number of calculus rules for (4.1) and (4.9) have been proved in [12, 22, 28, 35, 38, 53] (see also references therein) by using different but somewhat close variational approximations. In this paper we employ the following simple corollaries of Proposition 4.10.

4.15. Proposition. *Let f_1 and f_2 be extended-real-valued and lower semicontinuous functions one of which is Lipschitz continuous around $\bar{x}$. Then one has*

$$\partial(f_1 + f_2)(\bar{x}) \subset \partial f_1(\bar{x}) + \partial f_2(\bar{x}). \tag{4.10}$$

Moreover, (4.10) *becomes an equality if either* f_1 *or* f_2 *is strictly differentiable at* $\bar{x}$.

4.16. PROPOSITION. *Let* Ω_1 *and* Ω_2 *be closed sets satisfying the qualification condition*

$$N(\bar{x};\Omega_1) \cap (-N(\bar{x};\Omega_2)) = \{0\}$$

at $\bar{x} \in \Omega_1 \cap \Omega_2$. *Then one has*

$$N(\bar{x};\Omega_1 \cap \Omega_2) \subset N(\bar{x};\Omega_1) + N(\bar{x};\Omega_2).$$

5. Nonsmooth optimization in finite dimensions. In this section we study nonsmooth optimization problems in finite dimensions and obtain necessary optimality conditions by using generalized differentiation constructions in Section 4. First we consider a broad class of mathematical programming problems with many inequality, equality, and geometric type constraints. We provide necessary conditions for such problems in the form of a *generalized Lagrange multiplier rule.* Then we employ this result for studying the finite difference problems (P_K) with a *variable stepsize* (varying discrete time interval) approximating the differential inclusion problem (P) with free time. We derive necessary conditions for the discrete approximation problems in a *refined Euler-Lagrange form* with *no convexity* in the original and/or adjoint inclusions.

For our purpose of applications to dynamic optimization problems like (P_K), we need to consider mathematical programming problems with *many geometric* constraints, *nonsmooth inequality* constraints, and *smooth equality* constraints.

Given real-valued functions ϕ_i, n-vector-valued functions g_j, and sets Δ_j in the space $\mathbf{R}^d$, we formulate the *mathematical programming* problem (MP) as follows:

$$\text{minimize } \phi_0(z) \text{ for } z \in \mathbf{R}^d \text{ subject to} \tag{5.1}$$

$$\phi_i(z) \le 0 \text{ for } i = 1, \ldots, s; \tag{5.2}$$

$$g_j(z) = 0 \in \mathbf{R}^n \text{ for } j = 0, \ldots, m; \tag{5.3}$$

$$z \in \Delta_j \text{ for } j = 0, \ldots, l. \tag{5.4}$$

5.1. PROPOSITION. *Let* $\bar{z}$ *be an optimal solution to problem (MP). Assume that the functions* ϕ_i *are Lipschitz continuous, the functions* g_j *are smooth, and the sets* Δ_j *are closed around* $\bar{z}$. *Then there exist real numbers* $\{\mu_i |\ i = 0, \ldots, s\}$ *as well as vectors* $\{\psi_j \in \mathbf{R}^n |\ j = 0, \ldots, m\}$ *and* $\{z_j^* \in \mathbf{R}^d |\ j = 0, \ldots, l\}$, *not all zero, such that*

$$z_j^* \in N(\bar{z};\Delta_j) \ \textit{for} \ j = 0, \ldots, l; \tag{5.5}$$

(5.6) $$\mu_i \geq 0 \ \ for \ \ i = 0, \ldots, s;$$

(5.7) $$\mu_i \phi_i(\bar{z}) = 0 \ \ for \ \ i = 1, \ldots, s;$$

(5.8) $$-z_0^\star - \cdots - z_l^\star \in \partial(\sum_{i=0}^{s} \mu_i \phi_i)(\bar{z}) + \sum_{j=0}^{m} (\nabla g_j(\bar{z}))^\star \psi_j.$$

Proof. This proposition follows from general necessary conditions in nondifferentiable programming proved in Mordukhovich [33, Theorem 1] and [35, Corollary 7.5.1] on the basis of the so-called *metric approximation method.* This method allows to approximate the nonsmooth constrained problem (MP) by a special parametric family of smooth unconstrained minimization problems and then to get necessary conditions (5.5)–(5.8) by passing to the limit in the classical Fermat stationary rule. In such a way, we obtain more general results in the presence of nonsmooth equality constraints, vector objective functions, non-Lipschitzian data, etc.; see [35, Section 2].

□

Now let us reduce the discrete approximation problems (P_K) in (3.11)–(3.14) to the (MP) form (5.1)–(5.4). For any fixed $K = 1, 2, \ldots$, we consider the following minimization problem with respect to variables $\theta > 0$ and $(x_0, x_1, \ldots, x_K) \in \mathbf{R}^{(K+1)n}$:

(5.9) $$\text{minimize } \varphi_0(x_0, x_K, \theta) + |x_0 - \bar{x}(0)|^2 + (\theta - \bar{T})^2 +$$
$$(\theta/K) \textstyle\sum_{j=0}^{K-1} f(j\theta/K, x_j, (K/\theta)(x_{j+1} - x_j)) +$$
$$\textstyle\sum_{j=0}^{K-1} \int_{j\theta/K}^{(j+1)\theta/K} |(K/\theta)(x_{j+1} - x_j) - \dot{\bar{x}}(t)|^2 dt$$

subject to the constraints

(5.10) $$x_{j+1} \in x_j + (\theta/K) F(j\theta/K, x_j) \ \text{ for } \ j = 0, \ldots, K-1;$$

(5.11) $$\varphi_i(x_0, x_K, \theta) \leq \alpha_{iK} \ \text{ for } \ i = 1, \ldots, q;$$

(5.12) $$-\beta_{iK} \leq \varphi_i(x_0, x_K, \theta) \leq \beta_{iK} \ \text{ for } \ i = q+1, \ldots, q+r;$$

(5.13) $$(x_0, x_K, \theta) \in \Omega_K := \Omega + \gamma_K B$$

where $\bar{x}(t)$, $0 \leq t \leq \bar{T}$, is the considered optimal solution to the original problem (P) and $(\alpha_K, \beta_K, \gamma_K)$ are the defined constraint perturbations in (P_K).

One can easily see that for each $K = 1, 2, \ldots$ problems (P_K) and (5.9)–(5.13) are equivalent by virtue of the following correspondence between decision variables:

$$\{x_K(t_j), T_K\} \longleftrightarrow \{x_j, \theta\} \text{ for } j = 0, \ldots, K. \tag{5.14}$$

For convenience, let us introduce the auxiliary variables

$$y_j := (K/\theta)(x_{j+1} - x_j) \text{ for } j = 0, \ldots, K-1.$$

Now problem (5.9)–(5.13) (and, therefore, (P_K)) can be rewritten in the equivalent form (5.1)–(5.4) with respect to variables $z = (x_0, \ldots, x_K, y_0, \ldots, y_{K-1}, \theta) \in \mathbf{R}^{(2K+1)n+1}$ as follows:

$$\text{minimize } \phi_0(z) := \varphi_0(x_0, x_K, \theta) + |x_0 - \bar{x}(0)|^2 + (\theta - \bar{T})^2 + \tag{5.15}$$

$$(\theta/K)(\sum_{j=0}^{K-1} f(j\theta/K, x_j, y_j) + \sum_{j=0}^{K-1} \int_{j\theta/K}^{(j+1)\theta/K} |y_j - \dot{\bar{x}}(t)|^2 dt$$

subject to

$$\phi_i(z) := \varphi_i(x_0, x_K, \theta) - \alpha_{iK} \leq 0 \text{ for } i = 1, \ldots, q; \tag{5.16}$$

$$\phi_i(z) := \varphi_i(x_0, x_K, \theta) - \beta_{iK} \leq 0 \text{ for } i = q+1, \ldots, q+r; \tag{5.17}$$

$$\phi_{r+i}(z) := -\varphi_i(x_0, x_K, \theta) - \beta_{iK} \leq 0 \text{ for } i = q+1, \ldots, q+r; \tag{5.18}$$

$$g_j(z) := x_{j+1} - x_j - (\theta/K)y_j = 0 \text{ for } j = 0, \ldots, K-1; \tag{5.19}$$

$$z \in \Delta_j := \{(x_0, \ldots, y_{K-1}, \theta) \in \mathbf{R}^{(2K+1)n+1} |\ y_j \in F(j\theta/K, x_j)\} \text{ for } j = 0, \ldots, K-1; \tag{5.20}$$

$$z \in \Delta_K := \{(x_0, \ldots, y_{K-1}, \theta) \in \mathbf{R}^{(2K+1)n+1} |\ (x_0, x_K, \theta) \in \Omega_K\}. \tag{5.21}$$

Obviously, (5.15)–(5.21) is a problem of mathematical programming in the form (5.1)–(5.4) where $d = (2K+1)n + 1$, $s = q + 2r$, $m = K - 1$, and $l = K$. Moreover, the real-valued functions ϕ_i, the smooth vector functions g_j, and the sets Δ_j have special structures. Now one can employ Proposition 5.1 to obtain necessary optimality conditions for the dynamic optimization problem (P_K) taking into account the special nature of (5.15)–(5.21) and calculus rules for the generalized differentiation constructions in Section 4.

For simplicity, in the following theorem we deal with autonomous systems where F and f do not depend on the time variable. Moreover, to avoid technical complications and taking into account applications in Section 6, we focus on the case when the derivative $\dot{\bar{x}}(\cdot)$ of the reference optimal solution to (P) is Riemann integrable on $[0,\bar{T}]$, i.e., continuous for a.e. $t \in [0,\bar{T}]$. The general case of $\bar{x}(\cdot) \in W^{1,\infty}[0,\bar{T}]$ will be discussed in Remark 5.5.

5.2. THEOREM. *Let $\{\bar{x}_K(\cdot),\bar{T}_K\}$ be an optimal solution to problem (P_K) of autonomous dynamics where $\dot{\bar{x}}(\cdot)$ is Riemann integrable on $[0,\bar{T}_K]$. Assume that the functions φ_i and f are Lipschitz continuous and the sets Ω and* gph F *are closed around $\{\bar{x}_K(\cdot),\bar{T}_K\}$. Then there exist real numbers $\{\lambda_{iK}|\ i=0,\dots,q+r\}$ and a discrete n-vector function $\{p_K(t_j)|\ j=0,\dots,K\}$, not all zero, such that*

$$\lambda_{iK} \geq 0 \quad for \quad i=0,\dots,q; \tag{5.22}$$

$$\lambda_{iK}(\varphi_i(\bar{x}_K(0),\bar{x}_K(\bar{T}_K),\bar{T}_K)-\alpha_{iK})=0 \quad for \quad i=1,\dots,q; \tag{5.23}$$

$$\begin{aligned}&\Big(\frac{p_K(t_{j+1})-p_K(t_j)}{h_K},\ p_K(t_{j+1})+\frac{\lambda_{0K}}{h_K}\rho_{jK}\Big) \in \lambda_{0K}\partial f\Big(\bar{x}_K(t_j),\\ &\qquad\qquad \frac{\bar{x}_K(t_{j+1})-\bar{x}_K(t_j)}{h_K}\Big)+\\ &N\Big(\Big(\bar{x}_K(t_j),\frac{\bar{x}_K(t_{j+1})-\bar{x}_K(t_j)}{h_K}\Big);\text{gph } F\Big) \quad for \quad j=0,\dots,K-1;\end{aligned} \tag{5.24}$$

$$\begin{aligned}&\big(p_K(0)+2\lambda_{0K}(\bar{x}(0)-\bar{x}_K(0)),\ -p_K(\bar{T}_K),\ \bar{H}_K+2\lambda_{0K}(\bar{T}-\bar{T}_K)\\ &\qquad +\lambda_{0K}\varrho_K\big) \in\\ &\partial\Big(\sum_{i=0}^{q+r}\lambda_{iK}\varphi_i\Big)(\bar{x}_K(0),\bar{x}_K(\bar{T}_K),\bar{T}_K)+N((\bar{x}_K(0),\bar{x}_K(\bar{T}_K),\bar{T}_K);\Omega_K)\end{aligned} \tag{5.25}$$

where

$$h_K := \bar{T}_K/K;\quad t_j := jh_K \quad for \quad j=0,\dots,K \quad (t_K=\bar{T}_K); \tag{5.26}$$

$$\begin{aligned}\bar{H}_K := \frac{1}{K}\sum_{j=0}^{K-1}\Big[\Big\langle p_K(t_{j+1}),\frac{\bar{x}_K(t_{j+1})-\bar{x}_K(t_j)}{h_K}\Big\rangle - \lambda_{0K} f\Big(\bar{x}_K(t_j),\\ \frac{\bar{x}_K(t_{j+1})-\bar{x}_K(t_j)}{h_K}\Big)\Big];\end{aligned} \tag{5.27}$$

$$\rho_{jK} := 2\int_{t_j}^{t_{j+1}}\Big(\dot{\bar{x}}(t)-\frac{\bar{x}_K(t_{j+1})-\bar{x}_K(t_j)}{h_K}\Big)dt \quad for \ j=0,\dots,K-1; \tag{5.28}$$

$$
\begin{aligned}
\varrho_K := \textstyle\sum_{j=0}^{K-1}[\frac{j}{K}|\frac{\bar{x}_K(t_{j+1})-\bar{x}_K(t_j)}{h_K} - \dot{\bar{x}}(t_j)|^2 \\
-\frac{(j+1)}{K}|\frac{\bar{x}_K(t_{j+1})-\bar{x}_K(t_j)}{h_K} - \dot{\bar{x}}(t_{j+1})|^2].
\end{aligned}
\tag{5.29}
$$

Proof. Let $\bar{z} = (\bar{x}_0, \ldots, \bar{x}_K, \bar{y}_0, \ldots, \bar{y}_{K-1}, \bar{\theta})$ be an optimal solution to the mathematical programming problem (5.15)–(5.21). According to Proposition 5.1, there exist real numbers $\{\mu_i | \, i = 0, \ldots, q + 2r\}$ as well as vectors $\psi_j \in \mathbf{R}^n$ $(j = 0, \ldots, K-1)$ and $z_j^* := (x_{j0}^*, \ldots, x_{jK}^*, \bar{y}_{j0}, \ldots, \bar{y}_{jK-1}, \theta_j^*) \in \mathbf{R}^{(2K+1)n+1}$ $(j = 0, \ldots, K)$, not all zero, such that conditions (5.5)–(5.8) are fulfilled for the initial data in (5.15)–(5.21). Note that these μ_i, ψ_j, and z_j^* as well as the solutions $\bar{z}$ depend on K but here we omit the index "K" for simplicity.

Without loss of generality one can always suppose that $\beta_{iK} > 0$ for $i = q + 1, \ldots, q + r$ and all $K = 1, 2, \ldots$. Then the complementary slackness conditions (5.17) and (5.18) obviously imply that

$$
\mu_i \cdot \mu_{r+i} = 0 \;\; \forall i = q + 1, \ldots, q + r.
$$

Denoting

$$
\lambda_i := \mu_i \text{ for } i = 0, \ldots, q \;\; \& \;\; \lambda_i := \mu_i - \mu_{r+i} \text{ for } i = q + 1, \ldots, q + r,
$$

we get the conditions

$$
\lambda_i \geq 0 \text{ for } i = 0, \ldots, q \text{ and} \tag{5.30}
$$

$$
\lambda_i(\varphi_i(\bar{x}_0, \bar{x}_K, \bar{\theta}) - \alpha_{iK}) = 0 \text{ for } i = 1, \ldots, q \tag{5.31}
$$

ensuring that $\{\lambda_i | \, i = 0, \ldots, q + r\}$ are not equal to zero simultaneously with $\{\psi_j | \, j = 0, \ldots, K - 1\}$ and $\{z_j^* | \, j = 0, \ldots, K\}$.

Now let us express condition (5.8) in terms of the initial data in problem (5.15)–(5.21). From the structure of (5.15)–(5.18) and the calculus rule in Proposition 4.15 one has

$$
\begin{aligned}
\partial(\sum_{i=0}^{q+2r} \mu_i \phi_i)(\bar{z}) \subset \partial(\sum_{i=0}^{q+r} \lambda_i \varphi_i)(\bar{z}) + \lambda_0 \partial\zeta(\bar{z}) + \lambda_0 \partial\xi(\bar{z}) + \\
2(\lambda_0(\bar{x}_0 - \bar{x}(0)), 0, \ldots, 0, \lambda_0(\bar{\theta} - \bar{T}))
\end{aligned}
\tag{5.32}
$$

where

$$
\zeta(z) = \zeta(x_0, \ldots, x_K, y_0, \ldots, y_{K-1}, \theta) := \frac{\theta}{K} \sum_{j=0}^{K-1} f(x_j, y_j), \tag{5.33}
$$

$$
\xi(z) = \xi(x_0, \ldots, x_K, y_0, \ldots, y_{K-1}, \theta) := \sum_{j=0}^{K-1} \int_{j\theta/K}^{(j+1)\theta/K} |y_j - \dot{\bar{x}}(t)|^2 dt. \tag{5.34}
$$

It is easy to see that

$$\partial(\textstyle\sum_{i=0}^{q+r}\lambda_i\varphi_i)(\bar z) = \{u_0, 0, \ldots, u_K, 0, \vartheta)|\,(u_0, u_K, \vartheta) \in \partial(\textstyle\sum_{i=0}^{q+r}\lambda_i\varphi_i)(\bar x_0, \bar x_K, \bar\theta)\}, \tag{5.35}$$

$$\partial\xi(\bar z) = \{(v_0, \ldots, v_{K-1}, w_0, \ldots, w_{K-1}, \tfrac{1}{K}|\textstyle\sum_{j=0}^{K-1} f(\bar x_j, \bar y_j))|\,(v_j, w_j) \in \partial f(\bar x_j, \bar y_j)\}. \tag{5.36}$$

Next let us consider the function ξ in (5.34). This function is obviously differentiable in the (x, y) variables. Moreover, under the Riemann integrability assumption on $\dot{\bar x}(\cdot)$, this function is also differentiable in θ. The latter follows from a variant of the fundamental theorem of the classical calculus where an integrand may not be continuous but admits a primitive and takes all the intermediate values of its range.

Now using the classical formulae for the differentiation under the integral sign and with respect to varying integral limits, one gets

$$\nabla_{x_j}\xi(\bar z) = 0 \ \text{ for } \ j = 0, \ldots, K; \tag{5.37}$$

$$\nabla_{y_j}\xi(\bar z) = \xi_j := 2\int_{j\bar\theta/K}^{(j+1)\bar\theta/K} (\bar y_j - \dot{\bar x}(t))dt \ \text{ for } \ j = 0, \ldots, K-1; \tag{5.38}$$

$$\nabla_\theta\xi(\bar z) = \xi_\theta := \sum_{j=0}^{K-1}[\frac{j+1}{K}|\bar y_j - \dot{\bar x}((j+1)\bar\theta/K)|^2 - \frac{j}{K}|\bar y_j - \dot{\bar x}(j\bar\theta/K)|^2]. \tag{5.39}$$

Note that to justify (5.39) one needs to assume the one-sided continuity of $\dot{\bar x}(\cdot)$ at the points $j\bar\theta/K$ for $j = 0, \ldots, K-1$ (otherwise, we should take $\dot{\bar x}(\cdot)$ at points nearby). But this does not restrict generality as $K \to \infty$ due to the procedure in (6.18).

Next let us compute the vectors $(\nabla g_j(\bar z))^*\psi_j$ for g_j in (5.19). For each $j = 0, \ldots, K-1$, we have the following expressions:

$$(\nabla_{x_j} g_j(\bar z))^*\psi_j = -\psi_j, \ (\nabla_{x_{j+1}} g_j(\bar z))^*\psi_j = \psi_j, \ (\nabla_{x_k} g_j(\bar z))^*\psi_j = 0 \ \text{ if } \ k \neq j; \tag{5.40}$$

$$(\nabla_{y_j} g_j(\bar z))^*\psi_j = -(\bar\theta/K)\psi_j, \ (\nabla_{y_k} g_j(\bar z))^*\psi_j = 0 \ \text{ if } \ k \neq j; \tag{5.41}$$

$$(\nabla_\theta g_j(\bar z))^*\psi_j = -\frac{1}{K}\sum_{j=0}^{K-1}\langle\psi_j, \bar y_j\rangle. \tag{5.42}$$

Thus we have calculated the left-hand side of (5.8) for the data in (5.15)–(5.19). To represent the right-hand side of (5.8), we observe that conditions (5.5) for the sets Δ_j in (5.20) and (5.21) are equivalent to

$$(5.43)\qquad (x^\star_{jj}, y^\star_{jj}) \in N((\bar{x}_j, \bar{y}_j); \text{gph } F) \ \ \&$$

$$x^\star_{jk} = y^\star_{jk} = 0 \ \text{ if } \ k \neq j, \ \ \forall j = 0, \ldots, K-1;$$

$$(5.44)\ (x^\star_{K0}, x^\star_{KK}) \in N((\bar{x}_0, \bar{x}_K); \Omega_K) \ \ \& \ \ x^\star_{Kj} = y^\star_{Kj} = 0 \ \text{ otherwise.}$$

Now putting all calculations (5.32)–(5.44) together in Proposition 5.1, one gets the equalities:

$$-x^\star_{00} - x^\star_{K0} = u_0 + 2\lambda_0(\bar{x}_K - \bar{x}(0)) + \frac{\lambda_0 \bar{\theta}}{K} v_0 - \psi_0;$$

$$-x^\star_{jj} = \frac{\lambda_0 \bar{\theta}}{K} v_j + \psi_{j-1} - \psi_j \ \text{ for } \ j = 1, \ldots, K-1;$$

$$-x^\star_{KK} = u_K + \psi_{K-1};$$

$$-y^\star_{jj} = \frac{\lambda_0 \bar{\theta}}{K} w_j - \frac{\theta}{K} \psi_j + \lambda_0 \xi_j \ \text{ for } \ j = 0, \ldots, K-1;$$

$$0 = \vartheta + \frac{\lambda_0}{K} \sum_{j=0}^{K-1} f(\bar{x}_j, \bar{y}_j) - \frac{1}{K} \sum_{j=0}^{K-1} \langle \psi_j, \bar{y}_j \rangle + 2\lambda_0(\bar{\theta} - \bar{T}) + \lambda_0 \xi_\theta$$

where ξ_j and ξ_θ are defined in (5.38) and (5.39),

$$(u_0, u_K, \vartheta) \in \partial\Big(\sum_{i=0}^{q+r} \lambda_0 \varphi_i\Big)(\bar{x}_0, \bar{x}_K, \bar{\theta}), \ \text{ and}$$

$$(v_j, w_j) \in \partial f(\bar{x}_j, \bar{y}_j) \ \text{ for } \ j = 0, \ldots, K-1.$$

Thus denoting

$$p_0 := x^\star_{K0} + u_0 + 2\lambda_0(\bar{x}_0 - \bar{x}(0)) \ \ \& \ \ p_j := \psi_{j-1} \ \text{ for } \ j = 1, \ldots, K,$$

we obtain the following necessary conditions for the solution $(\bar{x}_0, \ldots, \bar{x}_K, \bar{y}_0, \ldots, \bar{y}_{K-1}, \bar{\theta})$ to problem (5.15)–(5.21): there are real numbers $\{\lambda_i |\ i =$

$0,\ldots,q+r\}$ and n-vectors $\{p_j|\ j=0,\ldots,K\}$, not all zero, such that one has (5.30), (5.31), and

$$(5.45)\quad \begin{gathered}((K/\bar\theta)(p_{j+1}-p_j),\ p_{j+1}-(\lambda_0K/\bar\theta)\xi_j)\in\lambda_0\partial f(\bar x_j,(K/\bar\theta)(\bar x_{j+1}-\bar x_j))+\\ N((\bar x_j,(K/\bar\theta)(\bar x_{j+1}-\bar x_j));\text{gph }F)\ \text{ for }\ j=0,\ldots,K-1;\end{gathered}$$

$$(5.46)\quad \begin{gathered}(p_0+2\lambda_0(\bar x(0)-\bar x_0),\ -p_K,\ 2\lambda_0(\bar T-\bar\theta)-\xi_\theta)+\\ \tfrac{1}{K}\textstyle\sum_j^{K-1}[\langle p_{j+1},(K/\bar\theta)(\bar x_{j+1}-\ \bar x_j)\rangle_{\bar 0}-\lambda_0 f(\bar x_j,(K/\bar\theta)(\bar x_{j+1}-\bar x_j))]\big)\in\\ \partial(\textstyle\sum_{j=0}^{q+r}\lambda_i\varphi_i)(\bar x_0,\bar x_K,\bar\theta).\end{gathered}$$

Now let $\{\bar x_K(\cdot),\bar T_K\}$ be an optimal solution to the discrete approximation problem (P_K). Using correspondence (5.14), one gets an optimal solution $(\bar x_0,\ldots,\bar x_K,\bar\theta)$ to problem (5.15)–(5.21) for which the necessary optimality conditions (5.30), (5.31), (5.45), and (5.46) hold. For any fixed K in these optimality conditions, we mark the dependence of the multipliers λ_i on K and set

$$p_K(t_j):=p_j\ \text{ for }\ j=0,\ldots,K$$

taking (5.26) into account. Then conditions (5.30), (5.31), (5.45), and (5.46) turn into the necessary optimality conditions (5.22)–(5.25) for the problem (P_K) under consideration. This ends the proof of the theorem. □

5.3. Corollary. *In addition to the assumptions of Theorem* 5.2, *let us suppose that the multifunction F is pseudo-Lipschitzian around* $(\bar x_K(t_j),(\bar x_K(t_{j+1})-\bar x_K(t_j))/h_K)$ *for each* $j=0,\ldots,K-1$. *Then conditions* (5.22)–(5.25) *hold with* $(\lambda_{0K},\ldots,\lambda_{q+rK},p_K(\bar T_K))\neq 0$. *Therefore, one can set*

$$(5.47)\quad |p_K(\bar T_K)|+\sum_{i=0}^{q}\lambda_{iK}+\sum_{i=q+1}^{q+r}|\lambda_{iK}|=1\ \ \forall K=1,\ 2,\ldots$$

Proof. If $\lambda_{0K}=0$, then (5.24) is represented as

$$\frac{p_K(t_{j+1})-p_K(t_j)}{h_K}\in D^*F(\bar x_K(t_j),\frac{\bar x_K(t_{j+1})-\bar x_K(t_j)}{h_K})(-p_K(t_{j+1}))\quad\text{for }\ j=0,\ldots,K-1$$

in terms of the coderivative (4.4). Now using criterion (c) in Proposition 4.7, one gets that $p_K(\bar T_K)=0$ implies $p_K(t_j)=0$ for all $j=0,\ldots,K-1$. This proves the corollary. □

5.4. *Remark.* From Theorem 5.2, one can easily obtain necessary optimality conditions for nonautonomous problems (P_K) using a standard reduction to the autonomous case with respect to a new state variable $z = (t, x)$. The extended finite difference inclusion for z is written in the form

$$z_K(t_{j+1}) \in z_K(t_j) + h_K \tilde{F}(z_K(t_j)) \text{ for } j = 0, \ldots, K-1$$

where $\tilde{F}(z) := (1, F(z))$.

5.5. *Remark.* If the time T_K is fixed in problem (P_K), then θ is fixed in the equivalent problem (5.9)–(5.13). In this case we do not need the assumption about the Riemann integrability of $\dot{\bar{x}}(\cdot)$ in the proof of Theorem 5.2 and its applications in Section 6 which work for any $\bar{x}(\cdot) \in W^{1,\infty}$; cf. [43]. The only place where we use the Riemann integrability of $\dot{\bar{x}}(\cdot)$ in the latter proof is the differentiability of function (5.34) in the θ variable.

In the general setting when $\bar{x}(\cdot) \in W^{1,\infty}$, the function ξ in (5.34) is Lipschitz continuous in its variables jointly being differentiable at almost all points. Moreover, one can compute the gradients of ξ at the points of differentiability according to the classical results. This allows us to use effectively the construction of Clarke's generalized gradient and its connection with the subdifferential (4.9) for estimating the impact of the term (5.34) to the conditions of Theorem 5.2.

It follows from (5.34) and Proposition 4.15 that

$$\partial\xi(\bar{z}) \subset \partial\xi_0(\bar{z}) + \cdots + \partial\xi_{K-1}(\bar{z}) \tag{5.48}$$

where

$$\xi_j(z) = \xi_j(x_0, \ldots, x_K, y_0, \ldots, y_{K-1}, \theta) := \int_{j\theta/K}^{(j+1)\theta/K} |y_j - \dot{\bar{x}}(t)|^2 dt \tag{5.49}$$

for $j = 0, \ldots, K-1$. By virtue of Proposition 4.13(ii) one has

$$\partial\xi_j(\bar{z}) \subset \partial_C \xi_j(\bar{z}) = \text{co}\{\lim_{\nu\to\infty} \nabla\xi_j(z_\nu) |\ \xi_j \text{ is differentiable at } z_\nu \to \bar{z}\}. \tag{5.50}$$

where the last expression is invariant to "excluding sets of measure zero."

Now let us calculate the partial derivatives of ξ_j at $z_\nu = (x_{0\nu}, \ldots, x_{K\nu}, y_{0\nu}, \ldots, y_{K-1\nu}, \theta_\nu)$ for any $j = 0, \ldots, K-1$. From (5.49) one obviously gets

$$\nabla_{x_s}\xi_j(z_\nu) = 0 (s = 0, \ldots, K) \& \nabla_{y_s}\xi_j(z_\nu) = 0\ (s = 0, \ldots, K-1; s \neq j); \tag{5.51}$$

$$\nabla_{y_j}\xi_j(z_\nu) = 2\int_{j\theta_\nu/K}^{(j+1)\theta_\nu/K} (y_{j\nu} - \dot{\bar{x}}(t))dt. \tag{5.52}$$

Moreover, we can compute

$$\nabla_\theta \xi_j(z_\nu) = \frac{j+1}{K}|y_{j\nu} - \dot{\bar{x}}((j+1)\theta_\nu/K)|^2 - \frac{j}{K}|y_{j\nu} - \dot{\bar{x}}(j\theta_\nu/K)|^2. \tag{5.53}$$

at almost all z_ν where ξ_j is differentiable. Thus using (5.48)–(5.53), we obtain

$$\partial\xi(\bar z)\subset(\nabla_{x_0}\xi(\bar z),\ldots,\nabla_{x_K}(\bar z),\nabla_{y_0}\xi(\bar z),\ldots,\nabla_{y_{K-1}}\xi(\bar z),C_\theta(\bar z)) \tag{5.54}$$

where $\nabla_{x_j}\xi(\bar z)$ and $\nabla_{y_j}\xi(\bar z)$ are expressed in (5.37) and (5.38) while

$$\begin{aligned}C_\theta(\bar z) := \tfrac{1}{K}\textstyle\sum_{j=0}^{K-1}\operatorname{co}\{\lim_{\theta_\nu\to\bar\theta}[(j+1)|\bar y_j-\dot{\bar x}((j+1)\theta_\nu/K)|^2-\\ j|\bar y_j-\dot{\bar x}(j\theta_\nu/K)|^2]\}.\end{aligned} \tag{5.55}$$

(Note that similarly to Clarke [11, Example 2.2.5], one can express the convex hulls of the limiting points in (5.55) in terms of the essential supremum and essential infimum of the functions

$$\eta_{jK}(\theta):=(j+1)|\bar y_j-\dot{\bar x}((j+1)\theta/K)|^2-j|\bar y_j-\dot{\bar x}(j\theta/K)|^2 \quad\text{for } j=0,\ldots,K-1$$

at $\theta=\bar\theta$).

Now coming back to the notation of problem (P_K), we provide necessary optimality conditions in form (5.22)–(5.28) where instead of equality (5.29) for ϱ_K one has the inclusion

$$\begin{aligned}\varrho_K\in\tfrac{1}{K}\textstyle\sum_{j=0}^{K-1}\operatorname{co}\{\lim_{\theta\to\bar T_K}[j|\frac{\bar x_K(t_{j+1}^\theta)-\bar x_K(t_j^\theta)}{h_K}-\dot{\bar x}_K(t_j^\theta)|^2\\ -(j+1)|\frac{\bar x_K(t_{j+1}^\theta)-\bar x_K(t_j^\theta)}{h_K}-\dot{\bar x}_K(t_{j+1}^\theta)|^2]\}\end{aligned} \tag{5.56}$$

$$\text{with } t_j^\theta:=j\theta/K \text{ for } j=0,\ldots,K-1.$$

From (5.56) one can conclude that $\varrho_K\to 0$ as $K\to\infty$ when $\dot{\bar x}(\cdot)$ is Riemann integrable on $[0,\bar T]$; cf. the arguments in (6.18). The question is about such a conclusion in more general settings for problems with non-fixed time.

6. Necessary conditions for differential inclusions. In this section we prove necessary optimality conditions for the original Bolza problem (P) with free time by passing to the limit in the necessary conditions obtained above for its discrete approximations. The following two circumstances are most important in this procedure:

1) the strong $W^{1,2}$-convergence of discrete optimal solutions proved in Section 2; and

2) robustness of the generalized differential constructions and the dual differential characterization of the Lipschitzian behavior of multifunctions in Section 4 ensuring the convergence of the adjoint arcs in optimality conditions.

Note that the collection of generalized differentiation properties we need is inherent in objects of Section 4 but not in other known generalized differential constructions.

Now we obtain necessary optimality conditions for problem (P) in the refined Euler-Lagrange form. Using Theorem 5.1, we assume in what follows that the derivative $\dot{\bar{x}}(\cdot)$ of the optimal trajectory under consideration is *Riemann integrable* on $[0,\bar{T}]$. First we deal with the autonomous case when F and f do not depend on t.

6.1. THEOREM. *Let $\bar{x}(t)$, $0 \le t \le \bar{T}$, be an optimal solution to the autonomous problem (P) which possesses the property of relaxation stability. In addition to* (H1) *and* (H7), *we assume that the functions φ_i are Lipschitz continuous around $(\bar{x}(0), \bar{x}(\bar{T}), \bar{T})$ and the function f is Lipschitz continuous with modulus l_f around any point of the set*

$$A := \{(x,v) \in \mathbf{R}^{2n} \mid x \in U,\ v \in F(x)\}. \tag{6.1}$$

Then there exist real numbers $\lambda_0, \ldots, \lambda_{q+r}$ and an absolutely continuous function $p : [0,\bar{T}] \to \mathbf{R}^n$, not all zero, such that

$$\lambda_i \ge 0 \ \ for \ \ i = 0, \ldots, q; \tag{6.2}$$

$$\lambda_i \varphi_i(\bar{x}(0), \bar{x}(\bar{T}), \bar{T}) = 0 \ \ for \ \ i = 1, \ldots, q; \tag{6.3}$$

$$\begin{aligned} \dot{p}(t) \in \operatorname{co}\{u \mid (u, p(t)) \in \lambda_0 \partial f(\bar{x}(t), \dot{\bar{x}}(t)) + N((\bar{x}(t), \dot{\bar{x}}(t));& \\ \operatorname{gph} F) \ \ a.e. \ \ t \in [0,\bar{T}];& \end{aligned} \tag{6.4}$$

$$(p(0), -p(\bar{T}), \bar{H}) \in \partial\Big(\sum_{i=0}^{q+r} \lambda_i \varphi_i\Big)(\bar{x}(0), \bar{x}(\bar{T}), \bar{T}) + N((\bar{x}(0), \bar{x}(\bar{T}), \bar{T}); \Omega) \tag{6.5}$$

where

$$\bar{H} := \frac{1}{\bar{T}} \int_0^{\bar{T}} [\langle p(t), \dot{\bar{x}}(t)\rangle - \lambda_0 f(\bar{x}(t), \dot{\bar{x}}(t))]dt. \tag{6.6}$$

Proof. Let us consider a sequence of discrete approximations (P_K) in (3.1), (3.11)–(3.14) and a sequence of optimal solutions $\{\bar{x}_K(\cdot), \bar{T}_K\}$ to (P_K) which converges to $\{\bar{x}(\cdot), \bar{T}\}$ in the sense of Theorem 3.2. Now using the necessary optimality conditions for $\{\bar{x}_K(\cdot), \bar{T}_K\}$ in Theorem 5.2, we find sequences of real numbers $\{\lambda_{iK} \mid i = 0, \ldots, q+r\}$ and discrete n-vector functions $\{p_K(t_j) \mid j = 0, \ldots, K\}$ satisfying (5.22)–(5.29). Due to Corollary 5.3, one can always impose the normalization condition (5.47).

We consider the piecewise-linear extensions $\bar{x}_K(t)$ and $p_K(t)$ of the corresponding discrete functions on the continuous-time intervals $[0, \bar{T}_K]$ according to (3.2) and (3.3). We also denote

$$\rho_K(t) := \rho_{jK}/h_K \ \ \text{for} \ \ t \in [t_j, t_{j+1}), \ \ j = 0, \ldots, K-1\ ,$$

where the numbers ρ_{jK} are defined in (5.28). One can see that

$$(6.7)\quad \int_0^{\bar{T}_K} |\rho_K(t)|dt = \sum_{j=0}^{K-1} |\rho_{jK}| \leq 2\sum_{j=0}^{K-1}\int_{t_j}^{t_{j+1}} |\dot{\bar{x}}(t) - \frac{\bar{x}_K(t_{j+1}) - \bar{x}_K(t_j)}{h_K}|dt = 2\int_0^{\bar{T}_K} |\dot{\bar{x}}(t) - \dot{\bar{x}}_K(t)|dt \to 0 \text{ as } K \to \infty$$

by virtue of Theorem 3.2. Considering (if necessary) the piecewise-constant extensions of $\dot{\bar{x}}_K(t)$ and $\rho_K(t)$ on the interval $[0, \bar{T}]$ and using the classical results, we can suppose without any loss of generality that

$$(6.8)\quad \dot{\bar{x}}_K(t) \to \dot{\bar{x}}(t) \text{ and } \rho_K(t) \to 0 \text{ a.e. } t \in [0, \bar{T}] \text{ as } K \to \infty.$$

Now let us obtain an estimate of the adjoint arcs $p_K(\cdot)$ for large K which turns out to be a crucial point of the proof. According to the adjoint discrete inclusion (5.24) and Definition 4.2 of the coderivative D^*F, one can find vectors $(v_{jK}, w_{jK}) \in \partial f(\bar{x}_K(t_j), (\bar{x}_K(t_{j+1}) - \bar{x}_K(t_j)/h_K)$ such that

$$(6.9)\quad \frac{p_K(t_{j+1})-p_K(t_j)}{h_K} - \lambda_{0K}v_{jK} \in D^*F(\bar{x}_K(t_j), \frac{\bar{x}_K(t_{j+1})-\bar{x}_K(t_j)}{h_K})(\lambda_{0K}w_{jK} - \lambda_{0K}\rho_{jK}/h_K - p_K(t_{j+1})) \text{ for } j = 0, \ldots, K-1.$$

By virtue of the Lipschitz continuity of F around $\bar{x}(t)$ with modulus l_F in (3.5) and due to the uniform convergence $\bar{x}_K(\cdot) \to \bar{x}(\cdot)$, we employ Corollary 4.8 in (6.9) and provide the estimate

$$(6.10)\quad |\frac{p_K(t_{j+1})-p_K(t_j)}{h_K} - \lambda_{0K}v_{jK}| \leq l_F|\lambda_{0K}w_{jK} - \lambda_{0K}\rho_{jK}/h_K - p_k(t_{j+1})|$$

for $j = 0, \ldots, K-1$ when K is large enough. Taking into account that $\lambda_{0K} \in [0, 1]$, one gets from (6.10) the following recurrent sequence for $p_K(t_j)$:

$$(6.11)\quad |p_K(t_j)| \leq (1 + l_F h_K)|p_K(t_{j+1})| + h_K(|v_{jK}| + l_F|w_{jK}|) + l_F|\rho_{jK}|$$

for $j = 0, \ldots, K-1$. Now observe that

$$(\bar{x}_K(t_j), \frac{\bar{x}_K(t_{j+1}) - \bar{x}_K(t_j)}{h_K}) \in A \text{ for } j = 0, \ldots, K-1$$

and all large K, where the set A is defined in (6.1). So we can employ Proposition 4.13(i) and obtain the estimates

$$(6.12)\quad |v_{jK}| \leq l_f \;\&\; |w_{jK}| \leq l_f \;\; \forall j = 0, \ldots, K-1 \text{ and large } K$$

in terms of the Lipschitz modulus l_f for the function f in our assumptions. From (6.11), (6.12), and $|p_K(\bar{T}_K)| \le 1$ due to (5.47), one gets

$$(6.13)\quad |p_K(t_j)| \le (1 + h_K l_F)|p_K(t_{j+1})| + h_K l_f(1 + l_F) + l_F|\rho_{jK}| \le \cdots \le$$

$$(1 + l_F\bar{T}_K/K)^K + \bar{T}_K l_f(1 + l_F) + l_F \sum_{j=0}^{K-1} |\rho_{jK}| \;\; \forall j = 0, \ldots, K-1$$

when K is large enough. Taking into account (6.7) and

$$\bar{T}_K \to \bar{T} \;\; \& \;\; (1 + l_F\bar{T}_K/K)^K \to \exp[l_F\bar{T}] \;\; \text{as} \;\; K \to \infty,$$

we establish the uniform boundedness of $\{p_K(t)\}$ on $[0, \bar{T}_K]$ as $K \to \infty$.

Then employing (6.10) and (6.12), one gets the estimate of the extended adjoint velocities $\dot{p}_K(t)$ as follows:

$$(6.14)\quad \begin{aligned} |\dot{p}_K(t)| = |\tfrac{p_K(t_{j+1}) - p_K(t_j)}{h_K}| \le l_f + l_F(l_f + |\rho_K(t)| + \\ |p_K(t_{j+1})|) \;\; \text{for } t_j \le t < t_{j+1}. \end{aligned}$$

Without loss of generality, we can consider $\dot{p}_K(t)$ in the interval $[0, \bar{T}]$ using, if necessary, the piecewise-constant extension on $[\bar{T}_K, \bar{T}]$. By virtue of (6.7), (6.13), and the classical compactness criterion, estimate (6.14) implies that the sequence $\{\dot{p}_K(t)\}$ is weakly compact in $L^1[0, \bar{T}]$. Therefore, one can find an absolutely continuous function $p : [0, \bar{T}] \to \mathbf{R}^n$ such that $p_K(\cdot) \to p(\cdot)$ uniformly in $[0, \bar{T}]$ and $\dot{p}_K(\cdot) \to \dot{p}(\cdot)$ weakly in $L^1[0, \bar{T}]$ (as usual we take all $K = 1,\ 2 \ldots$).

Now we carry out the limiting process in the necessary optimality conditions of Theorem 5.2. Passing to the limit in (5.22) and (5.23) and taking into account that $\alpha_{iK} \to 0$ as $K \to \infty$, one easily gets the sign and complementary slackness conditions (6.2) and (6.3), respectively. Taking the limit in (5.47), we provide the normalization condition

$$(6.15)\quad |p(\bar{T})| + \sum_{i=0}^{q} \lambda_i + \sum_{i=q+1}^{q+r} |\lambda_i| = 1$$

which ensures that $\lambda_0, \ldots, \lambda_{q+r}$ and $p(\cdot)$ are not equal to zero simultaneously. Employing the limiting procedure in (6.13), we can actually conclude that if $p(t_0) = 0$ at some point $t_0 \in [0, \bar{T}]$, then $p(t) \equiv 0$ in the whole interval $[0, \bar{T}]$.

Let us pass to the limit in the discrete Euler-Lagrange inclusion (5.24). Using our continuous-time notation, we represent (5.24) in the following equivalent form:

$$(6.16)\quad \dot{p}_K(t) \in \{u \in \mathbf{R}^n |\, (u,\ p_K(t_{j+1}) + \lambda_{0K}\rho_K(t)) \in \lambda_{0K}\partial f(\bar{x}_K(t_j), \dot{\bar{x}}(t)) +$$

$$N((x_K(t),\dot{\bar{x}}(t));\text{gph } F)\} \text{ for } t\in[t_j,t_{j+1}),\ j=0,\dots,K-1.$$

Employing the classical Mazur theorem, we find a sequence of convex combinations of functions $\dot{p}_k(t)$ on $[0,\bar{T}]$ which converges to $\dot{p}(t)$ for a.e. $t\in[0,\bar{T}]$. Now passing to the limit in (6.15) as $K\to\infty$ and using (6.8) as well as the robustness of the subdifferential $\partial f(\cdot)$ and the normal cone $N(\cdot;\text{gph } F)$, we establish the differential Euler-Lagrange inclusion (6.4).

It remains to prove the endpoint inclusion (6.5) which combines transversality conditions on $(p(0),p(\bar{T}))$ with an additional condition on the optimal time interval $[0,\bar{T}]$. First let us consider the left-hand side of (5.25) as $K\to\infty$. Obviously

$$(6.17)\quad \begin{aligned}&(p_K(0)+2\lambda_{0K}(\bar{x}(0)-\bar{x}_K(0)),\ -p_K(\bar{T}_K),\ 2\lambda_{0K}(\bar{T}-\bar{T}_K))\\ &\qquad\to(p(0),\ -p(\bar{T}),\ 0).\end{aligned}$$

Considering ϱ_K in (5.29) and using (3.2), (5.26) as well as Riemann integrability of $\dot{\bar{x}}(\cdot)$ and $\tau(\dot{\bar{x}},f)\to 0$ when $h\to 0$, one has

$$(6.18)\quad \begin{aligned}\varrho_K=\tfrac{1}{\bar{T}_K}\textstyle\sum_{j=0}^{K-1}[j\int_{t_j}^{t_{j+1}}|\dot{\bar{x}}_K(t)-\dot{\bar{x}}(t_j)|^2dt-(j+1)\int_{t_j}^{t_{j+1}}\\ |\dot{\bar{x}}_K(t)-\dot{\bar{x}}(t_{j+1})|^2dt\sim\end{aligned}$$

$$\frac{1}{\bar{T}_K}\sum_{j=0}^{K-1}\int_{t_j}^{t_{j+1}}|\dot{\bar{x}}_K(t)-\dot{\bar{x}}(t)|^2dt=\frac{1}{\bar{T}_K}\int_0^{\bar{T}_K}|\dot{\bar{x}}_K(t)-\dot{\bar{x}}(t)|^2dt\to 0 \text{ as } K\to\infty$$

by virtue of the convergence (3.18) in Theorem 3.2.

Now let us evaluate $\bar{H}_K$ in (5.27). Using (3.2), (5.26), and the convergences (3.16)–(3.18), (6.8) as well as $p_K(\cdot)\to p(\cdot)$ uniformly in $[0,\bar{T}]$, we get

$$(6.19)\quad \bar{H}_K=\frac{1}{\bar{T}_K}\sum_{j=0}^{K-1}\int_{t_j}^{t_{j+1}}[\langle p_K(t_{j+1}),\dot{\bar{x}}_K(t)\rangle-\lambda_{0K}f(\bar{x}_K(t_j),\dot{\bar{x}}(t))dt]\sim$$

$$\begin{aligned}\tfrac{1}{\bar{T}_K}\textstyle\sum_{j=0}^{K-1}\int_{t_j}^{t_{j+1}}[\langle p_K(t),\dot{\bar{x}}_K(t)\rangle-\lambda_{0K}f(\bar{x}_K(t),\dot{\bar{x}}_K(t))dt]=\\ \tfrac{1}{\bar{T}_K}\textstyle\int_0^{\bar{T}_K}[\langle p_K(t),\dot{\bar{x}}_K(t)\rangle-\end{aligned}$$

$$\begin{aligned}\lambda_{0K}f(\bar{x}_K(t),\dot{\bar{x}}_K(t))dt]\to\tfrac{1}{\bar{T}}\textstyle\int_0^{\bar{T}}[\langle p(t),\dot{\bar{x}}(t)\rangle-\lambda_0 f(\bar{x}(t),\dot{\bar{x}}(t))dt]:=\bar{H}\\ \text{as } K\to\infty\end{aligned}$$

due to the classical Lebesgue limiting theorem.

Finally let us consider the limit of the right-side expression in (5.25). First we note that

$$(6.20)\quad \begin{aligned}&\operatorname{Limsup}_{K\to\infty}\partial(\textstyle\sum_{i=0}^{q+r}\lambda_{iK}\varphi_i)(\bar{x}_K(0),\bar{x}_K(\bar{T}_K),\bar{T}_K)\\ &\qquad=\partial(\textstyle\sum_{i=0}^{q+r}\lambda_i\varphi_i)(\bar{x}(0),\bar{x}(\bar{T}),\bar{T})\end{aligned}$$

due to the robustness of the subdifferential (4.9). Then observe that the set Ω_K in (3.14) is represented in the form

$$(6.21)\ \Omega_K = \{(x_0, x_K, T) \in \mathbf{R}^{2n+1} |\ \mathrm{dist}((x_0, x_K, T), \Omega) \le \gamma_K\}.$$

Thus the geometric constraint (3.14) can be reduced to an inequality constraint with a Lipschitz continuous function. Now using Proposition 4.14 if $(\bar{x}_K(0), \bar{x}_K(\bar{T}_K), \bar{T}_K) \in \Omega$ and just the definition of the normal cone (4.1) if $(\bar{x}_K(0), \bar{x}_K(\bar{T}_K), \bar{T}_K) \notin \Omega$ in (6.21), we get

$$(6.22) \mathrm{Limsup}_{K\to\infty} N((\bar{x}_K(0), \bar{x}_K(\bar{T}_K), \bar{T}_K); \Omega_K) = N((\bar{x}(0), \bar{x}(\bar{T}), \bar{T}); \Omega).$$

Putting together relationships (6.17)–(6.20) and (6.22) we obtain the endpoint inclusion (6.5) by passing to the limit in its discrete counterpart (5.25). This ends the proof of the theorem. □

6.2. COROLLARY. *In the assumptions of Theorem* 6.1, *let the functions* φ_i *and the set* Ω *do not depend on the* T *variable. Then one has all the conclusions of the theorem with changing* (6.5) *for*

$$(6.23)\quad (p(0), -p(\bar{T})) \in \partial(\sum_{i=0}^{q+r} \lambda_i \varphi_i)(\bar{x}(0), \bar{x}(\bar{T})) + N(\bar{x}(0), \bar{x}(\bar{T})); \Omega) \quad \textit{and}$$

$$(6.24)\qquad \int_0^{\bar{T}} [\langle p(t), \dot{\bar{x}}(t)\rangle - \lambda_0 f(\bar{x}(t), \dot{\bar{x}}(t))] dt = 0.$$

Proof. This immediately follows from (6.5) and (6.6) under the additional assumption made. □

6.3. *Remark.* The integrand

$$(6.25)\qquad H(t) := \langle p(t), \dot{\bar{x}}(t)\rangle - \lambda_0 f(\bar{x}(t), \dot{\bar{x}}(t)),\quad 0 \le t \le \bar{T},$$

in (6.6) and (6.24) is the *pseudo-Hamiltonian* of the Bolza problem (P) calculated on the optimal solution $\{\bar{x}(\cdot), \bar{T}\}$ and the corresponding adjoint pair $\{p(\cdot), \lambda_0\}$. If we have the *Weierstrass-Pontryagin maximum condition* in the problem under consideration, then

$$(6.26)\qquad H(t) = \max\{\langle p(t), v\rangle - \lambda_0 f(\bar{x}(t), v)\} \ \text{a.e}\ t \in [0, \bar{T}],$$

i.e., the values of the pseudo-Hamiltonian (6.25) coincides with the corresponding values of the (maximized) *Hamiltonian*

$$\mathcal{H}(x, p, \lambda_0) := \max\{\langle p, v\rangle - \lambda_0 f(x, v) |\ v \in F(x)\}$$

along optimal processes. The latter always-holds (actually it follows from the Euler-Lagrange condition (6.4)) for *convex* problems (P) when the sets $F(x)$ are convex and the function f is convex in the v variable.

Moreover, it is well known that for classical autonomous problems in the calculus of variations and optimal optimal control the pseudo-Hamiltonian (6.25) is *constant:*

$$(6.27) \qquad H(t) \equiv c \ \text{ on } \ [0, \bar{T}]$$

where $c = 0$ if endpoint constraints do not depend on T. An analogue of the constancy relation (6.27) for the maximized Hamiltonian $\mathcal{H}$ is proved by Clarke [11] in the framework of his Hamiltonian conditions for Mayer problems involving convex differential inclusions.

One can easily see that in the case of (6.26) and (6.27), the number $\bar{H}$ in (6.5) and (6.6) is equal to the value of the Hamiltonian at any point $t \in [0, \bar{T}]$. In the general nonconvex setting of Theorem 6.1, the number $\bar{H}$ turns out to be the *averaged value* of the pseudo-Hamiltonian on the optimal interval $[0, \bar{T}]$.

6.4. *Remark.* It is essential in Theorem 6.1 that $\{\bar{x}(\cdot), \bar{T}\}$ is a *feasible* solution to the *original nonconvex* problem (P) but not to its relaxation. Indeed, the Euler-Lagrange inclusion (6.4) for (P) may be different from its counterpart for the relaxed problem (R). It happens because the normal cone to the graph of F and to the graph of its convex hull are not the same (as well as the subdifferentials for f and $\hat{f}_F$).

Theorem 6.1 provides necessary optimality conditions in the Bolza problem (P) for autonomous differential inclusions with the property of relaxation stability. Now we consider a corollary of the theorem where the latter property is automatically fulfilled.

6.5. COROLLARY. *Let $\{\bar{x}(\cdot), \bar{T}\}$ be an optimal solution to the Bolza problem* (2.1), (2.3)–(2.5) *with no differential inclusion. Suppose that for some numbers $\mu > 0$ and open set $U \subset \mathbf{R}^n$ one has:*

$$(6.28) \qquad \bar{x}(t) \in U \ \ \forall t \in [0, \bar{T}] \ \ \& \ \ |\dot{\bar{x}}(t)| < \mu \ \ \text{a.e. } t \in [0, \bar{T}],$$

the set Ω is closed, the functions φ_i are locally Lipschitz around $(\bar{x}(0), \bar{x}(\bar{T}), \bar{T})$, and the function $f = f(x, v)$ is Lipschitz continuous around any point $(x, v) \in U \times (\mu B)$. Then there exist real numbers $\lambda_0, \ldots, \lambda_{q+r}$ and an absolutely continuous function $p : [0, \bar{T}] \to \mathbf{R}^n$, not all zero, such that conditions (6.2), (6.3), *and* (6.5) *are fulfilled and*

$$(6.29) \quad \dot{p}(t) \in \operatorname{co}\{u \mid (u, p(t)) \in \lambda_0 \partial f(\bar{x}(t), \dot{\bar{x}}(t))\} \ \ \text{a.e. } t \in [0, \bar{T}].$$

Proof. It follows from (6.28) that $\bar{x}(\cdot) \in W^{1,\infty}[0, \bar{T}]$. Using Proposition 2.2, we conclude that the problem under consideration possesses the property of relaxation stability. In fact, this problem is equivalent to the problem (P) with the (trivial) differential inclusion

$$(6.30) \qquad \dot{x} \in F(x) := \mu B \ \ t \in [0, T]$$

where $(\bar{x}(t), \dot{\bar{x}}(t)) \in (\text{int gph } F)$ for a.e. $t \in [0, \bar{T}]$. Obviously, all the assumptions in Theorem 6.1 are fulfilled for problem (2.1), (2.3)–(2.5), (6.30) and the Euler-Lagrange inclusion (6.4) is equivalent to (6.29). This proves the corollary. □

In conclusion of this section, we consider problem (P) with a non-autonomous differential inclusion where $F(x,t)$ is Lipschitz continuous with respect to both variables. Such a problem can be reduced to the autonomous case treated above.

6.6. THEOREM. *Let $\{\bar{x}(\cdot), \bar{T}\}$ be an optimal solution to problem (P) in* (2.1)–(2.5) *which possesses the property of relaxation stability. In addition to* (H1) *and* (H7), *we suppose that F is Lipschitz continuous in (t,x) jointly, φ_i are Lipschitz continuous around $(\bar{x}(0), \bar{x}(\bar{T}), \bar{T})$, and f is Lipschitz continuous around any point of the set*

$$\tilde{A} := \{(t,x,v) \in \mathbf{R}^{2n+1} |\ (t,x) \in [0,\bar{T}] \times U,\ v \in F(t,x)\}.$$

Then there exist real numbers $\lambda_0, \ldots, \lambda_{q+r}$ as well as absolutely continuous functions $p^0 : [0,\bar{T}] \to \mathbf{R}$ and $p : [0,\bar{T}] \to \mathbf{R}^n$ such that one has (6.2), (6.3),

$$\{\lambda_0, \ldots, \lambda_{q+r}, p(\cdot)\} \neq 0; \tag{6.31}$$

$$(\dot{p}^0(t), \dot{p}(t)) \in \text{co}\{(u^0,u) \in \mathbf{R}^{n+1} |(u^0,u,p(t)) \in \lambda_0 \partial f(t,\bar{x}(t),\dot{\bar{x}}(t)) + N((t,\bar{x}(t),\dot{\bar{x}}(t)); \text{gph } F)\ \textit{a.e.}\ t \in [0,\bar{T}]; \tag{6.32}$$

$$(p(0), -p(\bar{T}), -p^0(\bar{T})) \in \partial(\sum_{i=0}^{q+r} \lambda_i \varphi_i)(\bar{x}(0), \bar{x}(\bar{T}), \bar{T}) + N((\bar{x}(0), \bar{x}(\bar{T}), \bar{T}); \Omega); \tag{6.33}$$

$$\int_0^{\bar{T}} p^0(t)dt = -\int_0^{\bar{T}} H(t)dt \tag{6.34}$$

where the pseudo-Hamiltonian $H(t)$ is defined in (6.25) *with $f = f(t,\bar{x}(t), \dot{\bar{x}}(t))$.*

Proof. Let us denote $t := x^0$ and introduce a new state variable $z = (x^0, x)$, corresponding endpoint variables $z^0 = (x_0^0, x_0)$ and $z_T = (x_T^0, x_T)$, and a velocity variable $w = (\vartheta, v)$ in $\mathbf{R}^{n+1}$. We define

$$\tilde{F}(z) := (1, F(x^0,x)),\quad \tilde{f}(z,w) := f(x^0,x,v), \tag{6.35}$$

$$\tilde{\varphi}_i(z_0, z_T) := \varphi_i(x_0, x_T, x_T^0), \;\; \tilde{\Omega} := \{(z_0, z_T)|\; x_0^0 = 0, \; (x_0, x_T, x_T^0) \in \Omega\}. \tag{6.36}$$

Then one can see that the arc $\bar{z}(t) = \{(t, \bar{x}(t)), \; 0 \leq t \leq \bar{T}\}$ is an optimal solution to the $n+1$-dimensional autonomous Bolza problem:

$$\text{minimize } \tilde{J}[z, T] := \tilde{\varphi}_0(z(0), z(T)) + \int_0^T \tilde{f}(z(t), \dot{z}(t))dt$$

over all trajectories of the differential inclusion

$$\dot{z}(t) \in \tilde{F}(z(t)) \;\text{ a.e. }\; t \in [0, T]$$

subject to the endpoint constraints

$$\tilde{\varphi}_i(z(0), z(T)) \leq 0 \;\text{ for }\; i = 1, \ldots, q;$$

$$\tilde{\varphi}_i(z(0), z(T)) = 0 \;\text{ for }\; i = q+1, \ldots, q+r;$$

$$(z(0), z(T)) \in \tilde{\Omega} \in \mathbf{R}^{2n+2}.$$

Under the hypotheses of the theorem, the autonomous problem formulated satisfies all the assumptions of Corollary 6.2. Using this corollary and taking into account the special structure of the data (6.35) and (6.36), we get conditions (6.32) and (6.33) directly from (6.4) and (6.23). In this way, (6.34) follows from (6.24) and (6.25), and one can ensure

$$\{\lambda_0, \ldots, \lambda_{q+r}, p^0(\cdot), p(\cdot)\} \neq 0. \tag{6.37}$$

Suppose that (6.31) does not hold, i.e., $\lambda_0, \ldots, \lambda_{q+r}$ are equal to zero simultaneously with $p(\cdot)$. Then due to (4.4) and Corollary 4.8, (6.32) implies that $\dot{p}^0(t) = 0$ a.e. $t \in [0, \bar{T}]$, i.e., $p^0(t)$ is constant on $[0, \bar{T}]$. Now taking (6.34) into account, we can conclude that $p^0(t) \equiv 0$ on $[0, \bar{T}]$. This contradicts (6.37) and completes the proof of the theorem. □

6.7. *Remark.* The additional assertion (6.27) in the framework of Corollary 6.2 would imply that

$$p^0(t) \equiv -H(t) \;\text{ on }\; [0, \bar{T}]$$

and conditions (6.32) and (6.33) are reduced to the forms

$$(-\dot{H}(t), \dot{p}(t)) \in \text{co}\{(u^0, u) \in \mathbf{R}^{n+1}|\; (u^0, u, p(t)) \in \lambda_0 \partial f(t, \bar{x}(t), \dot{\bar{x}}(t)) +$$

$$N((t, \bar{x}(t), \dot{\bar{x}}(t)); \text{gph } F) \;\text{ a.e. }\; t \in [0, \bar{T}];$$

$$(p(0), -p(\bar{T}), H(\bar{T})) \in \partial(\sum_{i=0}^{q+r} \lambda_i \varphi_i)(\bar{x}(0), \bar{x}(\bar{T}), \bar{T}) + N((\bar{x}(0), \bar{x}(\bar{T}), \bar{T}); \Omega).$$

7. Concluding remarks. Now we discuss some possible improvements and generalizations of the principal results obtained in this paper.

7.1. *Remark.* In the main body of the paper, we have used a discrete approximation procedure to obtain necessary optimality conditions in the constrained problem of Bolza for differential inclusions with free time. This method allows to prove the refined Euler-Lagrange conditions for the original problem possessing the property of relaxation stability. An important question consists of discovering settings where the ***property of relaxation stability can be removed.***

In the previous paper [43], we removed the latter property for the case of general (nonconvex) Mayer problems with fixed time. To accomplish it, we developed another procedure for approximating the original problem by a parametric family of unconstrained (but nonsmooth) problems of Bolza with no differential inclusion. This procedure is based on the metric approximation method in Mordukhovich [31, 33, 35] and the usage of Ekeland's variational principle as in Clarke [11] and Kaskosz and Lojasiewicz [27]. In this way, besides removing the property of relaxation stability, we obtained [43] more general transversality conditions with and without Lipschitzian assumptions on φ_i.

To develop the latter approximation procedure for non-fixed time problems, one needs to employ necessary optimality conditions as in Corollary 6.5 for unconstrained problems of Bolza with optimal solutions belonging to $W^{1,\infty}$. However, the results available in Corollary 6.5 are proved so far for the case of optimal solutions whose derivatives are Riemann integrable. This difference seems to be minor from the practical viewpoint but it plays an important role for the realization of the mentioned approximation procedure.

Thus any progress in justifying the results in Corollary 6.5 for the case of $W^{1,\infty}$-optimal solutions allows to remove the property of relaxation stability in the necessary conditions of Theorem 6.1 for general problems of Mayer. The same is true for *local controllability* and related results; cf. [43].

7.2. *Remark.* The discrete approximation procedure and necessary optimality conditions are proved above for global optimal solutions of the Bolza problem (P). Obviously, they hold for *strong local minima* in the space of trajectories $x(\cdot) = \{x(t),\ 0 \le t \le T\}$ with the C-norm

$$||x(\cdot)||_C := T + \max_{t\in[0,T]} |x(t)|.$$

Moreover, the method developed enables us to study directly the so-called *intermediate local minimum* with respect to the $W^{1,p}$-norm

$$||x(\cdot)||_{W^{1,p}} := T + \max_{t\in[0,T]} |x(t)| + (\int_0^T |\dot{x}(t)|^p dt)^{1/p}$$

for any $p \in [1, \infty)$. This notion has been considered in [43] for fixed-time problems involving differential inclusions. Clearly, it takes an intermediate place between the classical concepts of strong and weak ($p = \infty$) local minima.

The reader can check that the constructions of the paper allow to keep the results obtained for feasible solutions to the original problem (P) which provide an intermediate local minimum for the relaxed problem (R) with the same value of the cost functional; cf. [43] in the setting of fixed-time problems.

7.3. *Remark.* It appears that the results obtained work for the *weak local minimum* (in the $W^{1,\infty}$-norm) under some additional assumptions on the time dependence of the reference trajectory. (I am indebted to A. D. Ioffe for drawing my attention to this observation.)

Indeed, if $\bar{x}(\cdot) = \{\bar{x}(t),\ 0 \le t \le \bar{T}\}$ is a weak local minimum to (P), then there exists $\varepsilon > 0$ such that $\bar{x}(\cdot)$ provides the strong minimum to an auxiliary problem (P_ε) of the same structure with the differential inclusion

$$\dot{x} \in \tilde{F}_\varepsilon(t, x) := F(t, x) \cap G_\varepsilon(t, x) \tag{7.1}$$

where $G_\varepsilon(t, x) := \{v \in \mathbf{R}^n |\ |v - \dot{\bar{x}}(t)| \le \varepsilon\}$.

Now one can employ the Euler-Lagrange inclusion and the set of adjacent conditions obtained for problem (P_ε). Taking into account the structure of G_ε in (7.1) and using Proposition 4.16 for representing the normal cone to the graph of $\tilde{F}_\varepsilon$ in the Euler-Lagrange inclusion for (7.1), we arrive at the required Euler-Lagrange inclusion for $\bar{x}(\cdot)$ in (P). This allows to claim the fulfillment of the Euler-Lagrange necessary conditions in [43] and the present paper for the weak local minimum to nonconvex differential inclusion problems.

To justify such a conclusion on the basis of results in [43] and Section 6, one should ensure that (7.1) satisfies the assumptions made therein. Namely, for the case of fixed-time problems in [43], we need to assume that $\dot{\bar{x}}(\cdot)$ is a.e. continuous (Riemann integrable) on the time interval to provide the a.e. Hausdorff continuity of $\tilde{F}_\varepsilon$. In the case of free-time problems, we can employ Theorem 6.6 for (7.1) if $\dot{\bar{x}}(\cdot)$ is assumed to be Lipschitz continuous on $[0, \bar{T}]$.

It seems that these additional assumptions are related to the technique used in discrete approximations and can be avoided in further developments. We also believe that the *pseudo-Hamiltonian constancy relation* (6.27) adjoing to the Euler-Lagrange inclusion turns out to be a necessary condition for the *weak minimum* to nonconvex problems while its Hamiltonian counterpart is related to the Weierstass-Pontryagin maximum condition for the strong minimum.

7.4. *Remark.* Developing the approach of the paper, we can consider a generalization of the Bolza problem (P) where constraints (2.3) and (2.4) are replaced by the following:

$$\text{(7.2)} \quad \varphi_i(x(0), x(T), T) + \int_0^T f_i(t, x(t), \dot{x}(t))dt \le 0 \quad \text{for } i = 1, \ldots, q;$$

$$\text{(7.3)} \quad \varphi_i(x(0), x(T), T) + \int_0^T f_i(t, x(t), \dot{x}(t))dt = 0 \quad \text{for } i = q+1, \ldots, q+r.$$

Note that new constraints (7.2) and (7.3) unify the previous endpoint constraints (2.3) and (2.4) with the *isoperimetric* type constraints in variational problems.

The constructions of Section 3 allow to build a sequence of discrete approximations for problem (2.1), (2.2), (2.5), (7.2), and (7.3) whose solutions strongly $W^{1,2}$-converge to the reference optimal solution to the original problem under the corresponding property of relaxation stability. Following the procedure in Sections 5 and 6, we obtain an analogue of the main Theorem 6.1 where conditions (6.3) (6.4), and (6.6) are changed for, respectively,

$$\lambda_i[\varphi_i(\bar{x}(0), \bar{x}(\bar{T}), \bar{T}) - \int_0^{\bar{T}} f_i(\bar{x}(t), \dot{\bar{x}}(t))dt] = 0 \quad \text{for } i = 1, \ldots, q;$$

$$\dot{p}(t) \in \text{co}\{u |\ (u, p(t)) \in \partial(\sum_{i=0}^{q+r} \lambda_i f_i)(\bar{x}(t), \dot{\bar{x}}(t)) + N((\bar{x}(t), \dot{\bar{x}}(t)); \text{gph } F)$$

for a.e. $t \in [0, \bar{T}]$; and

$$\bar{H} = \frac{1}{\bar{T}} \int_0^{\bar{T}} [\langle p(t), \dot{\bar{x}}(t)\rangle - \sum_{i=0}^{q+r} \lambda_i f_i(\bar{x}(t), \dot{\bar{x}}(t))]dt$$

with $f_0 := f$.

7.5. *Remark.* The discrete approximation approach allows to extend the results of this paper to problems of Bolza where the functions φ_i depend not only on endpoint state positions and a varying time interval but also on *intermediate times* $\tau_s \in (0, T)$ and *intermediate states* $x(\tau_s)$; cf. Clarke and Vinter [13, 14] in the framework of optimal multiprocesses. Our results for such (nonconvex) problems combine the refined Euler-Lagrange inclusion (6.4) with a new condition in the line of (6.5) involving jumps of the adjoint arc.

REFERENCES

[1] J.-P. Aubin and A. Cellina, *Differential Inclusions*, Springer-Verlag, Berlin, 1984.

[2] J.-P. Aubin and H. Frankowska, *Set-Valued Analysis*, Birhäuser, Boston, 1990.
[3] L. D. Berkovitz, *Optimal Control Theory*, Springer-Verlag, New York, 1974.
[4] N. N. Bogoljubov, *Sur quelques methods nouvelles dans le calculus des variations*, Ann. Math. Pura Appl. **7** (1930), 249–271.
[5] J. M. Borwein and D. M. Zhuang, *Verified necessary and sufficient conditions for regularity of set-valued and single-valued maps*, J. Math. Anal. Appl. **134** (1988), 441–459.
[6] J. V. Burke, *Calmness and exact penalization*, SIAM J. Control Optim. **29** (1991), 493–497.
[7] F. H. Clarke, *Admissible relaxation in variational and control problems*, J. Math. Anal. Appl. **51** (1975), 557–576.
[8] F. H. Clarke, *A new approach to Lagrange multipliers*, Math. Oper. Res. **1** (1976), 165–174.
[9] F. H. Clarke, *The generalized problem of Bolza*, SIAM J. Control Optim. **14** (1976), 682–699.
[10] F. H. Clarke, *Optimal solutions to differential inclusions*, J. Optim. Theory Appl. **19** (1976), 469–478.
[11] F. H. Clarke, *Optimization and Nonsmooth Analysis*, Wiley-Interscience, New York, 1983.
[12] F. H. Clarke, *Methods of Dynamic and Nonsmooth Optimizations*, SIAM Publications, Philadelphia, PA, 1989.
[13] F. H. Clarke and R. B. Vinter, *Optimal multiprocesses*, SIAM J. Control Optim. **27** (1989), 1072–1091.
[14] F. H. Clarke and R. B. Vinter, *Applications of optimal multiprocesses*, SIAM J. Control Optim. **27** (1989), 1048–1071.
[15] F. H. Clarke, P. D. Loewen, and R. B. Vinter, *Differential inclusions with free time*, Ann. Inst. H. Poincaré Anal. Non Linéaire **5** (1989), 573–593.
[16] M. G. Crandall, H. Ishii, and P.-L. Lions, *User's guide to viscosity solutions of second order differential equations*, Bull. Amer. Math. Soc. **27** (1992), 1–67.
[17] A. L. Dontchev, *Discrete approximations in optimal control*, this volume.
[18] A. L. Dontchev and E. M. Farkhi, *Error estimates for discretized differential inclusions*, Computing **41** (1989), 349–358.
[19] A. L. Dontchev and F. Lempio, *Difference methods for differential inclusions: a survey*, SIAM Review **34** (1992), 263–294.
[20] A. L. Dontchev and T. Zolezzi, *Well-Posed Optimization Problems*, Springer-Verlag, Berlin, 1993.
[21] I. Ekeland and R. Teman, *Convex Analysis and Variational Problems*, North-Holland, Amsterdam, 1976.
[22] A. D. Ioffe, *Approximate subdifferentials and applications. I: The finite dimensional theory*, Trans. Amer. Math. Soc. **281** (1984), 389–416.
[23] A. D. Ioffe, *Approximate subdifferentials and applications. III: The metric theory*, Mathematika **36** (1989), 1–38.
[24] A. D. Ioffe, *Nonsmooth subdifferentials: their calculus and applications*, Proceed. 1st World Congress of Nonlinear Analysts, Tampa, FL, August 1992; to appear in W. de Gruyter, Berlin, 1994.
[25] A. D. Ioffe and V. M. Tikhomirov, *Extensions of variational problems*, Trans. Moscow Math. Soc. **18** (1968), 207–273.
[26] A. D. Ioffe and V. M. Tikhomirov, *Theory of Extremal Problems*, North-Holland, Amsterdam, 1979.
[27] B. Kaskosz and S. Lojasiewicz Jr., *Lagrange-type extremal trajectories in differential inclusions*, Systems Control Lett. **19** (1992), 241–247.
[28] P. D. Loewen, *Optimal Control via Nonsmooth Analysis*, CRM Proceedings and Lecture Notes **2**, American Math. Soc. Publications, Providence, RI, 1993.
[29] P. D. Loewen and R. T. Rockafellar, *Optimal control of unbounded differential inclusions*, SIAM J. Control Optim. **32** (1994), 442–470.
[30] A. A. Lyapounov, *On complete additive vector functions*, Izvest. AN SSSR, Ser.

Math. **4** (1940), 465–478. (Russian)

[31] B. S. Mordukhovich, *Maximum principle in the problem of time optimal control with nonsmooth constraints*, J. Appl. Math. Mech. **40** (1976), 960–969.

[32] B. S. Mordukhovich, *On difference approximations of optimal control systems*, J. Appl. Math. Mech. **42** (1978), 452-461.

[33] B. S. Mordukhovich, *Metric approximations and necessary optimality conditions for general classes of nonsmooth extremal problems*, Soviet Math. Dokl. **22** (1980), 526–530.

[34] B. S. Mordukhovich, *On necessary conditions for an extremum in nonsmooth optimization*, Soviet Math. Dokl. **32** (1985), 215–220.

[35] B. S. Mordukhovich, *Approximation Methods in Problems of Optimization and Control*, Nauka, Moscow, 1988.

[36] B. S. Mordukhovich, *Optimization and approximation of differential inclusions*, Cybernetics **24** (1988), 781–788.

[37] B. S. Mordukhovich, *Necessary optimality conditions for nonsmooth control problems with non-fixed time*, Different. Equations **25** (1989), 290–299.

[38] B. S. Mordukhovich, *Sensitivity analysis in nonsmooth optimization*, in Theoretical Aspects of Industrial Design (D. A. Field and V.Komkov, eds.), SIAM Publications, Philadelphia, PA, 1992, pp. 32–47.

[39] B. S. Mordukhovich, *On variational analysis of differential inclusions*, in Optimization and Nonlinear Analysis (A. Ioffe, L. Marcus, and S. Reich, eds.), Pitman Res. Notes Math. Ser. **244** (1992), 199-213.

[40] B. S. Mordukhovich, *Complete characterizations of openness, metric regularity, and Lipschitzian properties of multifunctions*, Trans. Amer. Math. Soc. **340** (1993), 1–36.

[41] B. S. Mordukhovich, *Generalized differential calculus for nonsmooth and set-valued mappings*, J. Math. Anal. Appl. **183** (1994), 250–288.

[42] B. S. Mordukhovich, *Stability theory for parametric generalized equations and variational inequalities*, Trans. Amer. Math. Soc. **343** (1994), 609–658.

[43] B. S. Mordukhovich, *Discrete approximations and refined Euler-Lagrange conditions for nonconvex differential inclusions*, IMA Preprint Series # 1115, March 1993; SIAM J. Control Optim. **33** (1995), 882–915.

[44] J.-P. Penot, *Metric regularity, openness and Lipschitzian behavior for multifunctions*, Nonlinear Anal. Theory Methods Appl. **13** (1989), 629–643.

[45] E. Polak, *On the use of consistent approximations in the solution of semi-infinite optimization and optimal control problems*, Math. Programming **62** (1993), 385–414.

[46] L. S. Pontryagin, V. G. Boltyanskii, R. V. Gamkrelidze, and E. F. Mischenko, *The Mathematical Theory of Optimal Processes*, Wiley-Interscience, New York, 1962.

[47] B. N. Pshenichnyi, *Convex Analysis and Extremal Problems*, Nauka, Moscow, 1980. (Russian)

[48] R. T. Rockafellar, *Extensions of subgradient calculus with applications to optimization*, Nonlinear Anal. Theory Methods Appl. **9** (1985), 665–698.

[49] R. T. Rockafellar, *Lipschitzian properties of multifunctions*, Nonlinear Anal. Theory Methods Appl. **9** (1985), 867–885.

[50] R. T. Rockafellar, *Maximal monotone relations and the second derivatives of nonsmooth functions*, Ann. Inst. H. Poincaré Anal. Non Linéaire **2** (1985), 167–184.

[51] R. T. Rockafellar, *Dualization of subgradient conditions for optimality*, Nonlinear Anal. Theory Methods Appl. **20** (1993), 627–646.

[52] R. T. Rockafellar, *Lagrange multipliers and optimality*, SIAM Review **35** (1993), 183–238.

[53] R. T. Rockafellar and R. J-B Wets, *Variational Analysis*, "book in preparation." To be published by Springer-Verlag, New York.

[54] J. D. L. Rowland and R. B. Vinter, *Dynamic optimization problems with free time*

and active state constraints, SIAM J. Control Optim. **31** (1993), 677–697.

[55] B. Sendov and V. A. Popov, *The Averaged Moduli of Smoothness*, Wiley-Interscience, New York, 1988.

[56] G. V. Smirnov, *Discrete approximations and optimal solutions to differential inclusions*, Kibernetika, 1991, 76–79. (Russian). Cybernetics, **27** (1991), pp. 101–107 (English translation).

[57] J. Warga, *Optimal Control of Differential and Functional Equations*, Academic Press, New York, 1972.

[58] J. Warga, *Controllability, extremality, and abnormality in nonsmooth optimal control*, J. Optim. Theory Appl. **41** (1983), 239–260.

[59] L. C. Young, *Lectures on the Calculus of Variations and Optimal Control Theory*, Saunders, Philadelphia, PA, 1969.

SMALL-TIME REACHABLE SETS AND TIME-OPTIMAL FEEDBACK CONTROL

HEINZ SCHÄTTLER*

1. Introduction. This paper describes selected aspects of a direct geometric approach to time-optimal control. The underlying idea is to obtain a regular synthesis of time-optimal controls from a precise knowledge of the structure of the small-time reachable set for an extended system to which time has been adjoined as extra coordinate. We consider a system of the form Σ

$$(1.1) \qquad \dot{x} = f(x) + ug(x), \quad |u| \leq 1, \quad x \in M \subseteq \mathbb{R}^n,$$

where M is a sufficiently small open neighborhood of some reference point p and admissible controls are Lebesgue measurable functions u with values in the closed interval $[-1, 1]$ almost everywhere. Given an admissible control defined on some open interval J which contains 0, there exists a unique absolutely continuous curve $x(\cdot)$, defined on a maximal open subinterval I of J, such that $x(0) = p$, $\dot{x}(t) = f(x(t), u(t))$ holds almost everywhere on I, and $x(t)$ lies in M for $t \in I$. This curve is called the corresponding trajectory. A point q is said to be reachable from p in time t if there exists an admissible control u such that the corresponding trajectory x satisfies $q = x(t)$. If q is not reachable in smaller time, then the trajectory is called time-optimal. The set of all points which are reachable in time t is denoted by $Reach_{\Sigma,t}(p)$. The reachable set of the system Σ from p, $Reach_{\Sigma}(p)$, consists of all points which are reachable from p in nonnegative times and $Reach_{\Sigma,\leq T}(p)$ denotes those points which are reachable in times t, $0 \leq t \leq T$. If T is small, we call this set the small-time reachable set.

The Pontryagin Maximum Principle gives necessary conditions for optimality which are obtained by approximating the reachable set with convex cones. For problem (1.1) concatenations of singular arcs with bang-bang trajectories are singled out as candidates for optimality (see Section 2) and further restrictions on the structure of optimal trajectories are imposed. In general, however, the precise structure of optimal controls is left wide open. If it is possible to analyze the boundary of the reachable set directly, better characterizations of time-optimal trajectories can be given. Though this appears a difficult task in general, it is feasible in low dimensions. Combining the analysis of high order conditions with explicit geometric considerations, we attempt to construct the reachable set. The procedure is to analyze extremal trajectories corresponding to simple structures suggested

* Department of Systems Science and Mathematics, Campus Box 1040, Washington University, One Brookings Drive, St. Louis, Missouri 63130-4899. Supported in part by the National Science Foundation under Grant No. DMS-9100043.

by the necessary conditions, such as bang-bang trajectories with a small number of switchings or simple concatenations of bang and singular arcs and then choosing increasingly more complex sequences while analysing the geometric structure of the resulting surfaces. If this can be done, then a regular synthesis of time-optimal trajectories can be constructed by projection into the state space along the 'time' coordinate.

In general, even for systems of the form (1.1), the structure of the (small-time) reachable set can be arbitrarily complicated making such an analysis prohibitively difficult. Therefore it is reasonable, and this is our guiding principle, to proceed from the "general" to the "special". Like the local behavior of an analytic function near a point p is determined by its Taylor coefficients, the local behavior of an analytic system is determined by the values of the control vector fields and their Lie-brackets at p [15]. Let us loosely call this the *Lie bracket configuration* of the system at p. In this sense, the most general situation arises then if all the vector fields f, g and their relevant (typically low-order) Lie brackets are in general position, that is, if no linear dependencies (equality relations) exist. We refer to this as codimension 0 conditions and depending on how many nontrivial independent equality relations exist, we distinguish cases of positive codimension. The concept of *codimension* is used to organize the Lie bracket conditions into groups of increasing degrees of degeneracy. In this sense, the results in [7] give the precise structure of the small-time reachable set from a reference point p in dimension 4 under codimension 0 conditions and in [14] the codimension 1 case is analyzed. The corresponding time-optimal synthesis for the problem of stabilizing an equilibrium point p in dimension 3 under codimension 0 conditions is given in [12] and is stated for the codimension 1 case in [14]. In recent years a computational framework based on a Lie-algebraic formalism has been developed which allows to perform the necessary calculations efficiently [16,18]. In this paper the general ideas and some aspects of the technical problems encountered in these constructions will be highlighted. However, we need to refer the reader to the cited literature for complete proofs.

2. The maximum principle. Time-optimal trajectories lie in the boundary of the reachable set for the extended system where 'time' has been added as extra coordinate, $\dot{x}_0 \equiv 1$. Rather than to consider this particular situation, we consider the extended system directly and our aim is to analyze boundary trajectories. The *Pontryagin Maximum Principle* [9] gives first order necessary conditions for a trajectory to lie on the boundary of the small-time reachable set. Let $\gamma = (x(\cdot), u(\cdot))$ be an admissible pair defined over the interval $[0, T]$ and suppose $x(t)$ lies in the boundary of the reachable set from $x(0)$ for all $t \in [0, T]$. Then there exists an absolutely continuous curve $\lambda(\cdot) : [0, T] \to \mathbb{R}^n$, called the adjoint variable, which is not identically zero, such that in local coordinates we have almost everywhere on $[0, T]$

$$\dot{\lambda}(t) = -\lambda(t)\left(Df(x(t)) + Dg(x(t))u(t)\right) \tag{2.1}$$

$$< \lambda(t), g(x(t)) > u(t) \equiv \min_{|v|\leq 1} < \lambda(t), g(x(t)) > v \tag{2.2}$$

$$H =< \lambda(t), f(x(t)) + u(t)g(x(t)) > \equiv 0. \tag{2.3}$$

(We write λ as a row vector, $< \cdot, \cdot >$ denotes the Euclidean inner product, and Df and Dg are the Jacobian matrices of f and g, respectively.) We call trajectories which satisfy these conditions *extremal*. Note that the absolutely continuous function $\Phi_\gamma(t) :=< \lambda(t), g(x(t)) >$ uniquely determines the control as

$$u(t) = -sgn\ \Phi_\gamma(t) \tag{2.4}$$

away from the set $Z(\gamma) = \{t \in [0, T] : \Phi_\gamma(t) = 0\}$. Therefore Φ_γ is called the *switching function*. But all that can be said about $Z(\gamma)$ is that it is a closed subset of $[0, T]$. In principle it can be *any* closed subset. This leaves questions of regularity properties of u wide open. For instance, the Fuller phenomenon [8] shows that optimal controls can have an infinite number of switchings in arbitrarily small intervals. In order to investigate the structure of the set $Z(\gamma)$ further, it is a natural next step to consider the derivatives of the switching function. For instance, if $\Phi_\gamma(\bar{t}) = 0$, but $\dot{\Phi}_\gamma(\bar{t}) \neq 0$, then $\bar{t}$ is an isolated point of $Z(\gamma)$ and the control has a bang-bang switch at time $\bar{t}$.

LEMMA 2.1. *Let $\gamma = (x(\cdot), u(\cdot)) : [0, T] \to M \times U$ be an extremal pair with adjoint variable λ. Let Z be a smooth vector field and define*

$$\Psi_\gamma(t) =< \lambda(t), Z(x(t)) > . \tag{2.5}$$

Then

$$\dot{\Psi}_\gamma(t) =< \lambda(t), [f, Z](x(t)) + u(t)[g, Z](x(t)) > . \tag{2.6}$$

where $[A, B]$ denotes the Lie-bracket of two vector fields A and B.

Proof. The proof is a direct computation. Omitting the variable t and writing the inner product as dot product, we have

$$\begin{aligned}
\dot{\Psi}_\gamma &= \dot{\lambda}Z(x) + \lambda DZ(x)\dot{x} \\
&= -\lambda(Df(x) + uDg(x))Z(x) + \lambda DZ(x)(f(x) + ug(x)) \\
&= \lambda(DZ(x)f(x) - Df(x)Z(x)) + u\lambda(DZ(x)g(x) - Dg(x)Z(x)) \\
&= \lambda[f, Z](x) + u\lambda[g, Z](x)
\end{aligned}$$

where the last line gives a definition of the Lie-bracket. □

For the switching function Φ_γ we get therefore

$$\begin{aligned}
(2.7)\qquad \dot{\Phi}_\gamma(t) &= < \lambda(t), [f,g](x(t)) + u(t)[g,g](x(t)) > \\
&= < \lambda(t), [f,g](x(t)) > \\
(2.8)\qquad \ddot{\Phi}_\gamma(t) &= < \lambda(t), [f,[f,g](x(t)) + u(t)[g,[f,g]](x(t)) > .
\end{aligned}$$

Even though the control u is in principle undetermined on $Z(\gamma)$, if $Z(\gamma)$ contains an open interval I, then also all the derivatives of the switching function vanish on I and this may determine the control. For instance, if in equation (2.8) the quantity $< \lambda(t), [g,[f,g](x(t)) >$ does not vanish on I, then this determines the control as

$$u(t) = -\frac{< \lambda(t), [f,[f,g](x(t)) >}{< \lambda(t), [g,[f,g](x(t)) >}. \tag{2.9}$$

Controls of this type, which are calculated by equating derivatives of the switching function with zero, are called *singular*. However, this expression only defines an admissible control if $u(t)$ takes values in the control set. Once a singular control violates the control constraint, it is no longer admissible and must be terminated. We call this *saturation*.

The precise structure of optimal controls, like what kind of concatenations between bang and singular controls are optimal, can typically not be settled using only these results. Several high order necessary conditions for optimality have been developed to restrict the number of candidates further. They include the Legendre-Clebsch condition for singular arcs [6] or the only recently developed theory of envelopes [17,18], to mention just a couple. They are almost exclusively variational in nature. Here we argue by means of several explicit examples that a direct geometric construction of the small-time reachable set is a natural way to complement variational arguments in the solution to time-optimal control problems and that depending on the circumstances it can actually make some of the more technical and difficult variational arguments unnecessary.

Some remarks on our notation follow. We use exponential notation for the flow of a vector field, i.e. we write $q_t = q_0 e^{tZ}$ for the point at time t on the integral curve of the vector field Z which starts at q_0 at time 0. Note that we let the diffeomorphisms act on the right. This agrees with standard Lie-algebraic conventions [5] and simplifies formal calculations. Accordingly, we use the same notation for the flow of Z, that is $q_0 X e^{tZ}$ denotes the vector X transported from q_0 to q_t along the flow of Z. But we freely use $X(q_0)$ for $q_0 X$ and we switch between $q_0 X e^{tZ} = q_t e^{-tZ} X e^{tZ}$ and the more standard notation $e^{-t\,adZ} X(q_0)$. Recall that $adZ(X) = [Z,X]$ and that $e^{-t\,adZ}$ has the asymptotic Taylor-series representation [3]

$$e^{-t\,adZ} \asymp \sum_{n=0}^{\infty} \frac{(-t)^n}{n!} ad^n Z(X). \tag{2.10}$$

We also recall the following commutator formula [3]:

COROLLARY 2.2.

$$e^{rA}e^{sB} = \cdots e^{\frac{1}{2}rs^2[B,[A,B]]}e^{\frac{1}{2}r^2s[A,[A,B]]}e^{rs[A,B]}e^{sB}e^{rA} \tag{2.11}$$

Our notation for concatenations of different types of trajectories is as follows: $\mathcal{C}_0 = \mathcal{S}_0 = \{p\}$,

$$\mathcal{C}_X = \{pe^{sX} : s \geq 0\}, \ \ \mathcal{S}_X = \{pe^{sX} : s > 0\},$$

$$\mathcal{C}_{XY} = \{pe^{sX}e^{tY} : s,t \geq 0\}, \ \ \mathcal{S}_{XY} = \{pe^{sX}e^{tY} : s,t > 0\},$$

and analogously for other concatenations of pieces. Also we denote intersections of sets with coordinate cubes of length ϵ by a superscript ϵ.

3. Small-time reachable sets as CW-complexes. We want to motivate the constructions with a simple derivation of a classical result of Lobry [4] about the structure of the small-time reachable set in dimension 3 under codimension 0 conditions. Let $X = f - g$, $Y = f + g$ and assume that X, Y and their Lie bracket $[X,Y]$ are linearly independent on a neighborhood M of a reference point $p \in \mathbb{R}^3$. Choose the following canonical coordinates of the second type on M:

$$(\xi_0, \xi_1, \xi_2) \mapsto pe^{\xi_2[X,Y]}e^{\xi_1 Y}e^{\xi_0 X}. \tag{3.1}$$

In these coordinates we have trivially

$$\begin{aligned} \mathcal{C}_X^\epsilon &= \{(\xi_0, 0) : 0 \leq \xi_0 \leq \epsilon\} \\ \mathcal{C}_Y^\epsilon &= \{(0, \xi_1) : 0 \leq \xi_1 \leq \epsilon\} \\ \mathcal{C}_{YX}^\epsilon &= \{(\xi_0, \xi_1) : 0 \leq \xi_i \leq \epsilon, i = 1, 2\} \end{aligned}$$

For a XY-trajectory it follows from Corollary 2.2 that

$$pe^{sX}e^{tY} = pe^{st(1+\mathcal{O}(T))[X,Y]}e^{t(1+\mathcal{O}(T^2))Y}e^{s(1+\mathcal{O}(T^2))X} \tag{3.2}$$

where $\mathcal{O}(T^k)$ denotes terms which can be bounded by CT^k for some constant $C < \infty$. Thus

$$\xi_0 = s(1 + \mathcal{O}(T^2)), \ \ \xi_1 = t(1 + \mathcal{O}(T^2)), \ \ \xi_2 = st(1 + \mathcal{O}(T)). \tag{3.3}$$

The equations for ξ_0 and ξ_1 can be solved for s and t near $(0,0)$ and hence for ϵ sufficiently small we can write ξ_2 as a function of ξ_0 and ξ_1 on $D_2 = \{(\xi_0, \xi_1) : 0 \leq \xi_i \leq \epsilon, i = 1, 2\}$. In fact,

$$\xi_2 = \xi_0\xi_1(1 + \cdots) \tag{3.4}$$

and this expression is positive for $0 < \xi_i \leq \epsilon$, i=1,2. Hence the surface $\mathcal{S}^{\epsilon}_{XY}$ can be described as the graph of a smooth function

$$\xi_2^{XY}(\xi_0, \xi_1) = \phi_2^{2,+}(\xi_0, \xi_1), \quad (\xi_0, \xi_1) \in D_2,$$

which lies entirely above $\mathcal{S}^{\epsilon}_{YX}$. Note that the latter is given as the graph over D_2 of the trivial function

$$\xi_2^{YX}(\xi_0, \xi_1) = \phi_2^{2,-}(\xi_0, \xi_1) \equiv 0.$$

We claim that the small-time reachable set $Reach_{\Sigma, \leq T}(p)$ restricted to $C(\epsilon)$ (see Figure 3.1), is given by

$$(3.5) \quad Reach^{\epsilon}_{\Sigma, \leq T}(p) = \{(\xi_0, \xi_1, \xi_2) : (\xi_0, \xi_1) \in D_2,\ 0 \leq \xi_2 \leq \xi_2^{XY}(\xi_0, \xi_1)\}.$$

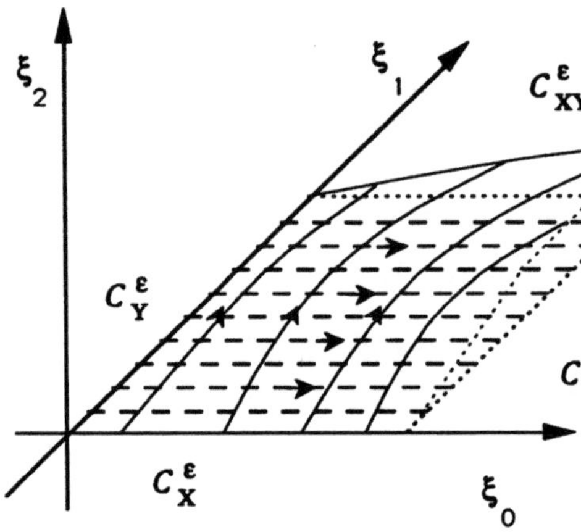

FIG. 3.1. $f \wedge g \wedge [f, g] \neq 0$ *in dimension 3*

To see this, temporarily call this set D_3. Take an arbitrary point on $\mathcal{S}^{\epsilon}_{XY}$ and integrate the vector field X forward in time, i.e. consider $\mathcal{S}^{\epsilon}_{XYX}$. In our coordinates $X = (1, 0, 0)^T$ and thus, also taking into account equation (3.4), it follows that the flow of X starting from $\mathcal{S}^{\epsilon}_{XY}$ covers the relative interior of D_3, i.e.

$$\mathcal{S}^{\epsilon}_{XYX} = \{(\xi_0, \xi_1, \xi_2) : (\xi_0, \xi_1) \in D_2, 0 < \xi_2 < \xi_2^{XY}(\xi_0, \xi_1)\}.$$

Hence all these points are reachable and, in fact, $D_3 = \mathcal{C}^{\epsilon}_{XYX}$. (Using $Y = (*, 1, *)^T$, where the asterisk denotes terms of order T, it can similarly be shown that also $D_3 = \mathcal{C}^{\epsilon}_{YXY}$.) It remains to show that no other points in M^ϵ are reachable. One way to see this is geometrically. It is easy to verify (using the explicit formulas) that X points inside D_3 at points in $\mathcal{S}^\epsilon_{XY} \setminus \{\xi_0 = \epsilon\} \cup \{\xi_1 = \epsilon\}$ and that Y points inside D_3 at points in $\mathcal{S}^\epsilon_{YX} \setminus \{\xi_0 = \epsilon\} \cup \{\xi_1 = \epsilon\}$. Since admissible controls are convex combinations of ± 1, this implies that any admissible trajectory which starts at a boundary point of $D_3 \setminus \{\xi_0 = \epsilon\} \cup \{\xi_1 = \epsilon\}$ either is tangent to or enters into the interior of D_3. Hence the small-time reachable set restricted to $\{\xi_0 \leq \epsilon, \xi_1 \leq \epsilon\}$ must lie in D_3. This proves the claim.

An alternative approach, which will be useful in higher dimensions, is to prove that the construction exhausts all boundary trajectories. To this end suppose γ is a boundary trajectory and that $\Phi_\gamma(t_1) = 0$. Then it follows from equation (2.3) that also $< \lambda(t_1), f(x(t_1)) >= 0$ and thus

$$< \lambda(t_1), X(x(t_1)) >= 0 \quad \text{and} \quad < \lambda(t_1), Y(x(t_1)) >= 0. \tag{3.6}$$

But f, g, and $[f,g]$ are linearly independent on M and so $\lambda(t_1)$ cannot vanish against $[f,g]$ at $x(t_1)$. Hence $\dot{\Phi}_\gamma(t_1) \neq 0$ and t_1 is a bang-bang switch. Thus boundary trajectories are bang-bang with isolated switchings. Suppose there is a second switching at time $t_2 > t_1$ and without loss of generality assume that $u \equiv -1$ on (t_1, t_2). Then, as in equation (3.6), $\lambda(t_2)$ vanishes against both X and Y at $x(t_2)$. Let $q_1 = x(t_1)$, $q_2 = x(t_2)$ and move the vector $Y(q_2)$ back to q_1 along the flow of X. This is done by integrating the variational equation. In exponential notation this simply reads

$$\begin{aligned} q_2 Y e^{(t_1-t_2)X} &= q_1 e^{(t_2-t_1)X} Y e^{(t_1-t_2)Y} = e^{(t_2-t_1)adX} Y(q_2) \\ &= Y(q_1) + (t_2 - t_1)[X,Y](q_2) + \mathcal{O}((t_1-t_2)^2) \end{aligned} \tag{3.7}$$

where we made use of the asymptotic expansion (2.10). The covector λ is moved backward along the flow of X by integrating the covariational equation. But this is the adjoint equation (2.1) and so we simply get $\lambda(t_1)$. Furthermore, by the definition of the adjoint operator,

$$\begin{aligned} < \lambda(t_1), q_2 Y e^{(t_2-t_1)X} > &= < \lambda(t_1)(e^{(t_2-t_1)X})^*, \quad q_2 Y > \\ &= < \lambda(t_2), Y(q_2) > = 0. \end{aligned} \tag{3.8}$$

Hence $\lambda(t_1)$ vanishes against X, Y and $e^{(t_2-t_1)adX}Y$, all evaluated at q_1. By the nontriviality of λ these vectors must therefore be linearly dependent. Hence

$$\begin{aligned} 0 &= X \wedge Y \wedge e^{(t_2-t_1)adX} Y \\ &= X \wedge Y \wedge Y + (t_2-t_1)[X,Y] + \mathcal{O}((t_2-t_1)^2) \\ &= (t_2-t_1)(1+\mathcal{O}(t_2-t_1))\,(X \wedge Y \wedge [X,Y]). \end{aligned} \tag{3.9}$$

But this contradicts the linear independence of f, g and $[f,g]$. Hence boundary trajectories are bang-bang with at most one switching. Thus

D_3 exhausts all possible boundary trajectories and no additional reachable points can exist in small time. (It follows from general results that the small-time reachable set $Reach_{\Sigma,\leq T}(p)$ is compact).

This 3-dimensional example was first analyzed by Lobry in a landmark paper on nonlinear control [4]. The proof given here shows the power of using a good set of coordinates. This is a pervasive feature in all constructions. Another one is the inductive nature of the construction. Lobry's example contains the structure of the trivial 2-dimensional case in the form of $\mathcal{C}_{XY}$ and the 3-dimensional reachable set is constructed on top of this structure. Similar reductions occur in every step to a higher dimension [7,14]. This is particularly useful since it allows to utilise earlier results as the dimension is increased.

This example motivates to construct the small-time reachable set as a stratified CW-complex by successively attaching cells of increasing dimension to the 0-dimensional cell consisting of the initial condition. In Lobry's example it was possible to carry out this construction in a very special and simple way realizing time-slices as stratified spheres. More generally, this will be the structure of the small-time reachable set if, inductively, given two k-dimensional cells which have common relative boundary and whose projections into a $(k+1)$-dimensional submanifold of M do not intersect, two $(k+1)$-dimensional cells can be attached which are parametrized over the set "enclosed" by the k-dimensional cells. If the projections again "enclose" a $(k+1)$-dimensional set in a $(k+2)$-dimensional submanifold of M, then the construction can be iterated, otherwise the intersection is analyzed. We formalize this as a general principle as follows: We postulate that on M there exists a coordinate chart $(\xi_0, \ldots, \xi_n)$ centered at p such that for sufficiently small $\epsilon > 0$ the small-time reachable set restricted to the coordinate cube $C(\epsilon) = \{(\xi_0, \ldots, \xi_n) : |\xi_i| \leq \epsilon \text{ for } i = 0, \ldots n\}$,

$$Reach_{\Sigma,\leq T}(p) \cap C(\epsilon),$$

can be constructed as a CW-complex in the following way by successively attaching higher-dimensional cells which consist of admissible trajectories:

- $\mathcal{C}_0 = \{p\}$
- Attach to $\mathcal{C}_0$ two 1-dimensional cells $\mathcal{C}_{-,1}$ and $\mathcal{C}_{+,1}$ which are parametrized over $D_1 = [0, \epsilon]$. Specifically, construct continuous maps

  $$\Phi^{1,\pm} : D_1 \to M, \quad \alpha \mapsto (\phi_0^{1,\pm}(\alpha), \phi_1^{1,\pm}(\alpha), \ldots, \phi_n^{1,\pm}(\alpha))$$

  which are smooth on the open interval $(0, \epsilon)$, satisfy $\Phi_1^{\pm}(0) = p$, and have the property that the projections of these curves into (ξ_0, ξ_1)-space do not intersect otherwise.
- Connect the 1-dimensional cells $\mathcal{C}_{-,1}$ and $\mathcal{C}_{+,1}$ by two 2-dimensional cells $\mathcal{C}_{-,2}$ and $\mathcal{C}_{+,2}$ which are parametrized over the set D_2 between

the projections of $\mathcal{C}_{-,1}$ and $\mathcal{C}_{+,1}$ into (ξ_0, ξ_1)-space. More precisely, construct continuous maps

$$\Phi^{2,\pm} : D_2 \to M, \quad (\xi_0, \xi_1) \mapsto (\xi_0, \xi_1, \phi_2^{2,\pm}(\xi_0, \xi_1), \ldots, \phi_n^{2,\pm}(\xi_0, \xi_1))$$

which are smooth on the interior of D_2, satisfy

$$\Phi^{2,\pm}(\phi_0^{1,\pm}(\alpha), \phi_1^{1,\pm}(\alpha)) = \Phi^{1,\pm}(\alpha) \quad \text{for } \alpha \in [0, \epsilon],$$

and for which $\phi_2^{2,-}(\xi_0, \xi_1) < \phi_2^{2,+}(\xi_0, \xi_1)$ holds for (ξ_0, ξ_1) in the interior of D_2.

- Inductively continue this procedure until two $(n-1)$-dimensional cells $\mathcal{C}_{+,n-1}$ and $\mathcal{C}_{-,n-1}$ have been constructed.
- Connect the cells $\mathcal{C}_{-,n-1}$ and $\mathcal{C}_{+,n-1}$ by two n-dimensional cells $\mathcal{C}_N$ and $\mathcal{C}_S$ which are parametrized over

$$\begin{aligned} D_n &= \{(\xi_0, \ldots, \xi_{n-1}) : (\xi_0, \ldots, \xi_{n-2}) \in D_{n-1}, \\ &\qquad \phi_{n-1}^{n-1,-}(\xi_0, \ldots, \xi_{n-2}) \le \xi_{n-1} \le \phi_{n-1}^{n-1,+}(\xi_0, \ldots, \xi_{n-2})\}. \end{aligned}$$

Specifically, construct continuous functions

$$\phi_n^{n,\pm} : D_n \to M, \quad (\xi_0, \ldots, \xi_{n-1}) \mapsto \phi_n^{n,\pm}(\xi_0, \ldots, \xi_{n-1})),$$

which are piecewise smooth on the interior of D_n, with the properties that their graphs

$$\begin{aligned} &\Phi^{n,\pm} : D_n \to M, \\ &\qquad (\xi_0, \ldots, \xi_{n-1}) \mapsto (\xi_0, \ldots, \xi_{n-1}, \phi_n^{n,\pm}(\xi_0, \ldots \xi_{n-1})) \end{aligned}$$

satisfy

$$\Phi_n^{n,\pm}(\xi_0, \ldots \xi_{n-2}, \phi_{n-1}^{n-1,\pm}(\xi_0, \ldots \xi_{n-2})) = \Phi^{n-1,\pm}(\xi_0, \ldots, \xi_{n-2})$$

and that $\phi_n^{n,-}(\xi_0, \ldots, \xi_{n-1}) < \phi_n^{n,+}(\xi_0, \ldots, \xi_{n-1})$ holds for $(\xi_0, \ldots, \xi_{n-1})$ in the interior of D_n.

- The small-time reachable set restricted to $C(\epsilon)$ is given as the set of points between the graphs of these functions:

$$\begin{aligned} \mathit{Reach}_{\Sigma,\le T}(p) \cap C(\epsilon) = \{&(\xi_0, \ldots, \xi_n) : (\xi_0, \ldots, \xi_{n-1}) \in D_n, \\ &\phi_n^{n,-}(\xi_0, \ldots, \xi_{n-1}) \le \xi_n \le \phi_n^{n,+}(\xi_0, \ldots, \xi_{n-1})\} \end{aligned}$$

We do *not* suggest that this is a generally valid structure, but we have persistently encountered it in low dimensional situations. If the small-time reachable set can be constructed in this way, we say that it is a *regular* $(n+1)$-*dimensional conical cell complex* in the coordinates $(\xi_0, \ldots, \xi_n)$. Under these conditions there exists a direction $v = (v_0, v_1)$ in (ξ_0, ξ_1)-space such that all slices $\{v_0\xi_0 + v_1\xi_1 = c\}$ for sufficiently small c are stratified n-dimensional disks with S^{n-1} as boundary. These spheres can be described as union of a lower hemisphere S_c, an equator E_c, and an upper hemisphere N_c. The equator consists of points on the lower dimensional cells $\mathcal{S}_{\pm,i}$ for $i = 1, \ldots, n-1$. Our motivation for considering a qualitative structure of this kind, which can be considered the most regular structure possible, simply is the guiding principle to analyse nondegenerate cases first. For nondegenerate Lie-bracket configurations there seems to be reasonable expectation of a simple structure and our outline above simply provides a precise framework to get started with. If necessary, it is possible to deviate from this stringent set-up at any moment in the construction by analyzing the geometric properties of the strata which are being attached.

We now present two examples of small-time reachable sets in dimension 4 (hence corresponding to time-optimal control in dimension 3) with low codimension which exhibit the structure described above.

4. Time-optimal control in $\mathbb{R}^3$: the codimension 0 case. We give the structure of the small-time reachable set for the codimension 0 case in dimension 4 [7]. Assume that the vector fields X, Y, $[X,Y]$ and $[Y,[X,Y]]$ are linearly independent on M and choose as coordinates

(4.1) $$(\xi_0, \xi_1, \xi_2, \xi_3) \mapsto pe^{\xi_3[Y,[X,Y]]}e^{\xi_2[X,Y]}e^{\xi_1 Y}e^{\xi_0 X}.$$

The projection onto $\{\xi_3 = 0\}$ gives Lobry's example analyzed above. It follows therefore that

(4.2) $$E_2 = \mathcal{C}_{XY} \cup \mathcal{C}_{YX} = \mathcal{S}_0 \cup \mathcal{S}_X \cup \mathcal{S}_Y \cup \mathcal{S}_{XY} \cup \mathcal{S}_{YX}$$

is a regular 2-dimensional conical cell-complex in these coordinates. Set

(4.3) $$D_3 = \{(\xi_0, \xi_1, \xi_2) : (\xi_0, \xi_1) \in D_2, 0 \le \xi_2 \le \xi_2^{XY}(\xi_0, \xi_1)\}.$$

We have seen above that the small-time reachable set for Lobry's example can be described as either $\mathcal{C}_{XYX}$ or as $\mathcal{C}_{YXY}$. For the 4-dimensional problem these surfaces are now separated in the direction of ξ_3. Continuing the inductive construction it can be shown that both $\mathcal{S}_{XYX}$ and $\mathcal{S}_{YXY}$ are graphs over D_3.

LEMMA 4.1. *The cells $\mathcal{C}^{\epsilon}_{XYX}$ and $\mathcal{C}^{\epsilon}_{YXY}$ are graphs of continuous functions $\phi_3^{3,\mp} : D_3 \to M$. These functions are smooth on the relative interior of D_3, have smooth extensions to the boundary strata $\mathcal{S}_{XY}$ and $\mathcal{S}_{YX}$, and to $\mathcal{S}_Y$ for $\mathcal{S}_{XYX}$, respectively to $\mathcal{S}_X$ for $\mathcal{S}_{YXY}$. Furthermore, these maps attach the 3-dimensional cells to E_2^{ϵ}.* □

Write the vector field $[X,[X,Y]]$ as a linear combination of the basis vector fields as

(4.4) $$[X,[X,Y]] = \alpha X + \beta Y + \gamma[X,Y] + \delta[Y,[X,Y]]$$

where α, β, γ and δ are smooth functions on M. We henceforth assume that $\delta \ne 0$ on M. A direct application of the Maximum Principle yields:

LEMMA 4.2. [7] *If δ is positive on M, then bang-bang trajectories with more than two switchings are not boundary trajectories. All bang-bang trajectories with at most two switchings are extremals.* $\square$

If $\delta > 0$, then it can indeed be shown that if $q \in \mathcal{S}_{XYX}$ and $r \in \mathcal{S}_{YXY}$ have the same (ξ_0, ξ_1, ξ_2) coordinates, then the ξ_3 coordinate of r is always smaller than the ξ_3 coordinate of q. (This can also be seen from the cut-locus calculation given below.) Therefore $\mathcal{S}_{YXY}$ entirely lies below $\mathcal{S}_{XYX}$. It is not difficult to verify that the small-time reachable set is a regular 4-dimensional conical cell complex in the coordinates $(\xi_0, \xi_1, \xi_2, \xi_3)$. Indeed, its structure and this proof are a direct extension of Lobry's 3-dimensional example to dimension 4.

Now assume $\delta < 0$. Then necessary conditions for optimality determine the lengths of successive arcs. Even in this case, it follows from results of Agrachev and Gamkrelidze [1] and of Sussmann [18] that bang-bang trajectories with more than two switchings are not optimal and so $\mathcal{C}_{XYX}$ and $\mathcal{C}_{YXY}$ exhaust all possible bang-bang boundary trajectories. We will not need to use these results in our construction, however. In this case it becomes necessary to investigate the geometric structure of these surfaces further and we will see that there is a nontrivial intersection of these surfaces which we call the *cut-locus*. It is determined by the (nontrivial) nonnegative solutions to the equation

$$pe^{s_1X}e^{s_2Y}e^{s_3X} = pe^{t_1Y}e^{t_2X}e^{t_3Y}. \tag{4.5}$$

(Trivial solutions are obtained if some of the times are zero and they correspond to trajectories in $E = \mathcal{C}_{XY} \cup \mathcal{C}_{YX}$).

As an illustration for the Lie-algebraic formalism, we calculate this cut-locus using Corollary 2.2. We have

$$\begin{aligned} & pe^{s_1X}e^{s_2Y}e^{s_3X} \\ (4.6)\quad &= p\cdots e^{\frac{1}{2}s_1s_2^2[Y,[X,Y]]}e^{\frac{1}{2}s_1^2s_2[X,[X,Y]]}e^{\frac{1}{2}s_1s_2[X,Y]]}e^{s_2Y}e^{(s_1+s_3)X} \\ &= pe^{\frac{1}{2}s_1s_2(s_1\delta+s_2+\cdots)[Y,[X,Y]]}e^{\frac{1}{2}s_1s_2(1+\cdots)[X,Y]]}e^{(s_2+\cdots)Y}e^{(s_1+s_3+\cdots)X} \end{aligned}$$

and

$$\begin{aligned} & pe^{t_1Y}e^{t_2X}e^{t_3Y} \\ &= pe^{t_1Y}\ldots e^{\frac{1}{2}t_2t_3^2[Y,[X,Y]]}e^{\frac{1}{2}t_2^2t_3[X,[X,Y]]}e^{t_2t_3[X,Y]}e^{t_3Y}e^{t_2X} \\ &= p\cdots e^{\frac{1}{2}t_2t_3^2[Y,[X,Y]]}e^{\frac{1}{2}t_2^2t_3[X,[X,Y]]}\ldots e^{t_1t_2t_3[Y,[X,Y]]}e^{t_2t_3[X,Y]}e^{(t_1+t_3)Y}e^{t_2X} \\ &= pe^{\frac{1}{2}t_2t_3(2t_1+t_2\delta+t_3+\cdots)[Y,[X,Y]]}e^{t_2t_3(1+\cdots)[X,Y]}e^{(t_1+t_3+\cdots)Y}e^{(t_2+\cdots)X}. \end{aligned}$$
(4.7)

Equating coordinates gives the following equations:

$$(4.8)\quad (\xi_0) \qquad s_1 + s_3 + O(S^3) = t_2 + O(T^3)$$

$$(4.9)\quad (\xi_1)\qquad s_2 + O(S^3) = t_1 + t_3 + O(T^3)$$

$$(4.10)\quad (\xi_2)\qquad s_1 s_2(1 + O(S))) = t_2 t_3(1 + O(T))$$

$$(4.11)\quad (\xi_3)\qquad \frac{1}{2} s_1 s_2(s_1\delta + s_2 + O(S)) = \frac{1}{2} t_2 t_3(2t_1 + t_2\delta + t_3 + O(T))$$

where $S = s_1 + s_2 + s_3$ and $T = t_1 + t_2 + t_3$. A closer examination of the calculations shows that $s_1 s_2$ actually divides the left-hand side of the equation for ξ_3 and that $t_2 t_3$ divides the right-hand side. Dividing ξ_3 by ξ_2 we get therefore

$$(4.12)\qquad s_1\delta + s_2 + O(S) = 2t_1 + t_2\delta + t_3 + O(T)$$

Equations (4.8), (4.9) and (4.12) can be solved uniquely in terms of s or t. We have, for instance,

$$(4.13)\qquad \begin{aligned} s_1 &= \tfrac{1}{\delta} t_1 + t_2 + O(T^2), \quad s_2 = t_1 + t_3 + O(T^3), \\ s_3 &= -\tfrac{1}{\delta} t_1 + O(T^2) \end{aligned}$$

All these times (and also t_3 calculated below) are nonnegative for extremals [7]. Now substitute these functions for s into equation (4.10) to obtain

$$(4.14)\qquad t_1\ (\frac{1}{\delta} t_1 + t_2 + \frac{1}{\delta} t_3)\ +\ O(T^3)\ = 0$$

In general, the quadratic terms need not dominate the cubic remainders. However, if the times t_i satisfy a relation of the type

$$(4.15)\qquad t_1 \geq \varepsilon T = \varepsilon(t_1 + t_2 + t_3) \quad \Leftrightarrow \quad t_1 \geq \frac{\varepsilon}{1-\varepsilon}\ (t_2 + t_3)\ ,$$

then this equation can be solved for t_3 as

$$(4.16)\qquad t_3 = -t_1 - \delta t_2 + O(T^2).$$

Note that this solution is well-defined near $\{t_3 = 0\}$ and thus a nontrivial cut-locus Γ,

$$(4.17)\qquad \Gamma = \mathcal{S}_{YXY} \cap \mathcal{S}_{XYX},$$

extends beyond $\mathcal{S}_{YX}$ near $t_3 = 0$. Similarly, by solving equations (4.8), (4.9) and (4.12) for t as a function of s, we can show that Γ also extends beyond $\mathcal{S}_{XY}$ near $t_1 = 0$. The curves Γ_{XY} and Γ_{YX} of intersection of Γ with $\mathcal{S}_{YX}$ and $\mathcal{S}_{XY}$ can be characterized precisely [7] and are called the curves of conjugate points. They correspond to XY respectively YX trajectories for which the times along X and Y are such that extremal trajectories must have switchings at both the initial and terminal points. Except for the stated transversality, which is easily verifiable, we have therefore proved the following result:

PROPOSITION 4.3. [7] *If $\delta < 0$, then $\mathcal{S}_{YXY}$ and $\mathcal{S}_{XYX}$ intersect transversally along a 2-dimensional surface Γ. This surface extends smoothly across $\mathcal{S}_{YX}$ and $\mathcal{S}_{XY}$ and the intersections with these surfaces are the curves Γ_{YX} and Γ_{XY} of conjugate points.* □

The cut-locus Γ is the decisive structure for local optimality of bang-bang trajectories. Note that, as subset of $\mathcal{S}_{XYX}$ (or $\mathcal{S}_{YXY}$), Γ can be described as the graph of the function $\xi_3 = \xi_3^{XYX}(\xi_0, \xi_1, \xi_2)$ which describes $\mathcal{S}_{XYX}$ over a 2-dimensional submanifold D_Γ of D_3. D_Γ divides D_3 into two connected components $D_{3,+}$ and $D_{3,-}$ which have the property that $\mathcal{S}_{XYX}$ lies above $\mathcal{S}_{YXY}$ in direction of ξ_3 over $D_{3,-}$ and below $\mathcal{S}_{YXY}$ over $D_{3,+}$. If we denote the corresponding substrata by a superscript $\pm$, then only the trajectories in

$$N = \mathcal{S}^-_{XYX} \cap \Gamma \cap \mathcal{S}^+_{YXY} \tag{4.18}$$

maximize the ξ_3-coordinate over D_3 (see Figure 4.1). The surfaces $\mathcal{S}^+_{XYX}$

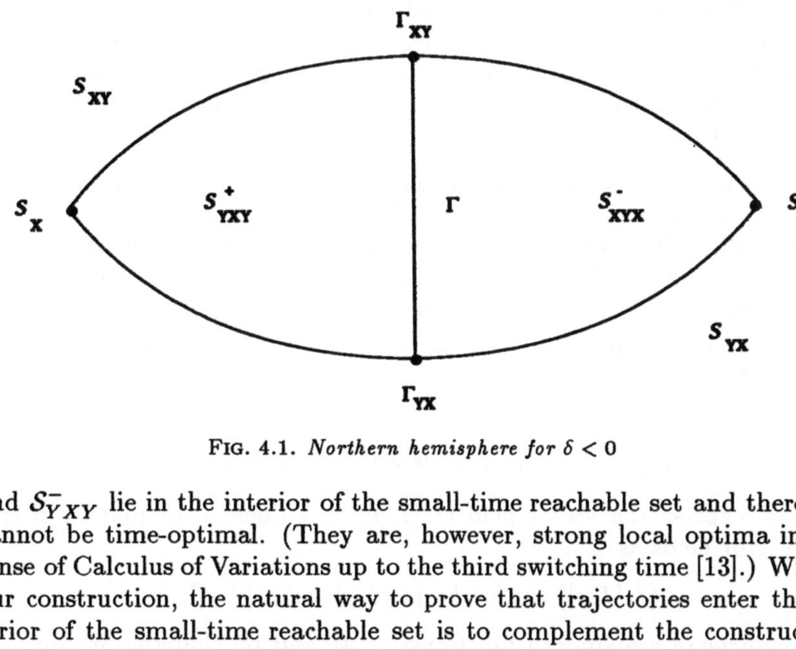

FIG. 4.1. *Northern hemisphere for $\delta < 0$*

and $\mathcal{S}^-_{YXY}$ lie in the interior of the small-time reachable set and therefore cannot be time-optimal. (They are, however, strong local optima in the sense of Calculus of Variations up to the third switching time [13].) Within our construction, the natural way to prove that trajectories enter the interior of the small-time reachable set is to complement the construction with a stratified hypersurface S such that the small-time reachable set is realized as a regular 4-dimensional conical cell complex with upper hemisphere N and lower hemisphere S in the coordinates $(\xi_0, \xi_1, \xi_2, \xi_3)$. This is indeed possible. The surface S consists of all trajectories which are concatenations of a bang arc (X or Y), followed by a singular arc and followed by one more bang arc with no restrictions on the lengths of the individual pieces. A qualitative sketch of this structure is given in Figure 4.2.

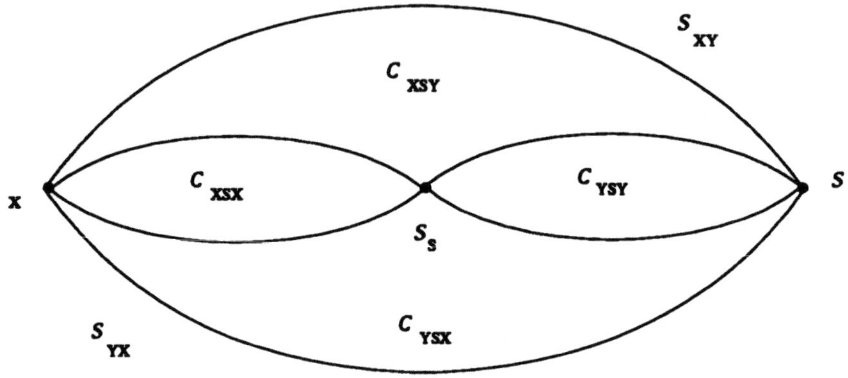

FIG. 4.2. *Southern hemisphere for* $\delta < 0$

Together, Figures 4.1 and 4.2 give the precise structure of the small-time reachable set for the case $\delta < 0$. Using these results a local regular synthesis was derived for the problem of stabilizing an equilibrium point in $\mathbb{R}^3$ time-optimally under corresponding conditions [12] and these trajectories are indeed the optimal ones.

5. Time-optimal control in $\mathbb{R}^3$: the codimension 1 case. We now analyze the structure of extremals which contain singular arcs for the more difficult codimension 1 case. Again we describe the small-time reachable set for the 4-dimensional system where time has been added as extra coordinate. Here we assume that

(A1) the vector fields f, g, $[f,g]$ and $[g,[f,g]]$ are linearly independent on M

and choose as coordinates

$$(\xi_0, \xi_1, \xi_2, \xi_3) \mapsto p e^{\xi_3 [g,[f,g]]} e^{\xi_2 [f,g]} e^{\xi_1 g} e^{\xi_0 X}. \tag{5.1}$$

Then we express the vector field $[f,[f,g]]$ as a linear combination of the basis vector fields as

$$[f,[f,g]] = af + bg + c[f,g] + d[g,[f,g]]. \tag{5.2}$$

The coefficients are smooth functions on M and by changing g into $-g$, if necessary, we may assume that $d(p) \geq 0$. Note that with δ defined in equation (4.4) we have

$$d = \frac{1+\delta}{1-\delta}$$

and the exceptional case $\delta = 0$ here corresponds to $d = 1$. The next lemma follows from equation (2.9).

LEMMA 5.1. **[14]** *If $\gamma = (x(\cdot), u(\cdot))$ is an extremal pair defined over the interval $[0, T]$ with the property that the control is singular on an open subinterval I, then necessarily $u(t) = -d(x(t))$ for $t \in I$.* □

Thus the singular control (2.9) is a smooth feedback control, $u_{sin} = -d$, defined on M. However, it is admissible only on $\{q \in M : d(q) \leq 1\}$, not on $\Delta_+ = \{q \in M : d(q) > 1\}$. Hence we have a smooth vector field $S = f - dg$ defined on all of M, whose integral curves are singular arcs as long as they lie in $\Delta = \{q \in M : d(q) = 1\}$ or $\Delta_- = \{q \in M : d(q) < 1\}$, but have to be terminated as they cross over into Δ_+. We call Δ a *saturation boundary*. If $d(p) \neq 1$, then, by choosing M small enough, it can always be assumed that d is bounded away from 1 on M. The structure of the small-time reachable set from p in these cases has been given above. Now we assume that

(A2) $d(p) = 1$ and the Lie-derivatives of d along X and Y, $L_X(d)$ and $L_Y(d)$, do not vanish on M.

It follows that Δ is a smooth 3-dimensional, embedded submanifold of M. Since admissible control values are convex combinations of $+1$ and -1, if X and Y point to the same side of Δ, then, except for the initial point p, the small-time reachable set will lie to one side of Δ. It is geometrically clear that the structure of the small-time reachable set is then exactly as for initial points p in Δ_- or Δ_+, depending on whether X and Y point into Δ_- or Δ_+. This leaves the case where X and Y point to opposite sides of Δ. Without loss of generality (more precisely, by applying an input symmetry which reverses time along the trajectories), we assume that

(A3) $L_X(d)(p) > 0$ and $L_Y(d)(p) < 0$

Under these conditions we shall give the precise structure of the small-time reachable set from p.

It is shown in [14] that $\mathcal{S}_{XYX}$ entirely lies below $\mathcal{S}_{YXY}$ in the direction of the coordinate vector field $\frac{\partial}{\partial \xi_3}$ and that $\mathcal{S}_{XYX}$ defines the lower hemisphere S of the small-time reachable set, i. e. consists of those trajectories, which for a given point (ξ_0, ξ_1, ξ_2) minimize the ξ_3-coordinate. The upper hemisphere N consists of some (but not all) YXY-trajectories and trajectories which contain singular arcs.

Let us analyze the structure of extremal trajectories which contain singular arcs. This is an application of the concept of *conjugate points* (see also [10]). Consider an arbitrary SX-trajectory of the form $q_2 = q_0 e^{rS} e^{sX}$ and let $q_1 = q_0 e^{rS}$ denote the junction. If the trajectory is extremal, then there exists an adjoint multiplier which vanishes against X, Y, and $[X, Y]$ at q_1. We call any point q_1 where this holds a *singular junction*, regardless of whether a singular arc is actually present or not. If q_2 is now another junction, then λ vanishes also against X and Y at q_2. By moving Y backward along the trajectory to q_1, again four conditions are imposed on λ. By the nontriviality of the multiplier these four vectors must be linearly dependent (see also Section 3) and this will typically restrict the time s along X. We call such a relation a *singular conjugate point relation* and we call q_1 and q_2 *singular conjugate points*. In our example, we get

$$(5.3) \qquad \begin{aligned} 0 &= X \wedge Y \wedge [X,Y] \wedge e^{sadX} Y \\ &= \tfrac{1}{2}s^2 \left(X \wedge Y \wedge [X,Y] \wedge \tfrac{e^{sadX} - (1+sadX)}{\frac{1}{2}s^2} Y \right). \end{aligned}$$

Define a function $\Psi_{*X}(\cdot;s)$ by writing

$$(5.4) \qquad \begin{aligned} & X \wedge Y \wedge [X,Y] \wedge \tfrac{e^{sadX} - (1+sadX)}{\frac{1}{2}s^2} Y \\ & = \Psi_{*X}(\cdot;s)\,(X \wedge Y \wedge [X,Y] \wedge [g,[f,g]]). \end{aligned}$$

Then it is a necessary condition for the trajectory to be extremal that

$$(5.5) \qquad \Psi_{*X}(q_1;s) = 0.$$

Using equation (2.10), an expansion for Ψ_{*X} can readily be calculated. Define a smooth function μ on M as the coefficient at $[g,[f,g]]$ of

$$(5.6) \qquad [X,[X,[X,Y]] = \cdots + \mu[g,[f,g]].$$

A simple calculation shows that the Lie-derivative of d along X, $L_X(d)$, is given by

$$(5.7) \qquad L_X(d) = \frac{1}{2}\mu + (d-1)\nu$$

where ν is a smooth function on M whose precise form is irrelevant for our problem. Note that by assumptions (A2) and (A3) μ is positive. Then we have

$$\begin{aligned} X &\wedge Y \wedge [X,Y] \wedge \frac{e^{sadX} - (1+sadX)}{\frac{1}{2}s^2} Y \\ &= X \wedge Y \wedge [X,Y] \wedge [X,[X,Y]] + \frac{1}{3}s[X,[X,[X,Y]]] + \cdots \\ &= X \wedge Y \wedge [X,Y] \wedge \left(2(d-1) + \frac{1}{3}\mu s(1+\cdots)\right)[g,[f,g]] \end{aligned}$$

and thus

$$(5.8) \qquad \Psi_{*X}(q_1;s) \doteq 2(d-1) + \frac{1}{3}\mu s.$$

Here the notation $\doteq$ indicates that (different) factors $(1+\mathcal{O}(T))$ have been dropped at (possibly all) terms in an equation. If the singular junction lies in Δ_- (which is necessarily the case if q_1 is the endpoint of a singular arc which has not been saturated), then the equation $\Psi_{*X}(q_1;s)=0$ has a unique positive solution

$$\sigma=\sigma(q_1)\doteq\frac{6(1-d)}{\mu}. \tag{5.9}$$

The trajectory is extremal only if $s\le\sigma(q_1)$, and at time $s=\sigma(q_1)$ a switch to Y must occur. After the trajectory switches to Y, the time until the next switching can be similarly calculated in terms of another conjugate point relation for the $\cdot XY\cdot$ portion of the trajectory. This gives [14]

$$-(d+1)t-(d-1)s-\frac{1}{3}\mu s^2\doteq 0. \tag{5.10}$$

If we substitute equation (5.9) into equation (5.10) for the conjugate point along XY-trajectories, we see that the time τ until the next switching is given by

$$\tau(q_1,s)=\frac{1}{2}\left((1-d)(q_1)s-\frac{1}{3}\mu(q_1)s^2+\cdots\right)\doteq-\frac{1}{12}\mu(q_1)s^2<0. \tag{5.11}$$

Therefore the time along Y is in fact unrestricted.

This singular conjugate point relation σ plays a crucial role in the synthesis of extremal trajectories which lie in the boundary of the small-time reachable set. For a trajectory of the form $pe^{tY}e^{rS}e^{sX}$ define a function $\bar{s}$ as

$$\bar{s}(t,r)=\sigma(pe^{tY}e^{rS}). \tag{5.12}$$

By Taylor's theorem we get that

$$\bar{s}(t,r)\doteq-3\left(\frac{L_Y(d)(p)}{L_X(d)(p)}t+r\right). \tag{5.13}$$

Note in particular, that for $r=0$, we have $pe^{tY}e^{sX}\in\mathcal{S}_{YX}$. We denote the curve of points $pe^{tY}e^{\bar{s}(t,0)X}\in\mathcal{S}_{YX}$ by $\mathcal{S}_{Y\bar{X}}$ and call these points singular conjugate points as well. Also define

$$\mathcal{S}_{Y\bar{X}Y}=\{pe^{tY}e^{\bar{s}(t,0)X}e^{uY}:u,t>0\}.$$

Since no conjugate point restrictions exist on the time u in the definition of $\mathcal{S}_{Y\bar{X}Y}$, all these trajectories are extremal. It is precisely this surface $\mathcal{S}_{Y\bar{X}Y}$, where a part of $\mathcal{S}_{YXY}$ is glued together with strata consisting of concatenations of bang and singular arcs to form the upper hemisphere N.

For the structure of singular extremals we also need to take into account that saturation occurs as the trajectory hits Δ. Since

$$S=f-dg=X+(1-d)g, \tag{5.14}$$

for any trajectory of the form $pe^{tY}e^{rS}$, there exists a unique smooth function $\bar{r} = \bar{r}(t)$ such that

$$pe^{tY}e^{\bar{r}(t)S} \in \Delta. \tag{5.15}$$

Since

$$d(pe^{tY}e^{\bar{r}(t)S}) \doteq 1 + L_Y(d)(p)t + L_X(d)(p)\bar{r}(t)$$

we get

$$\bar{r}(t) \doteq -\frac{L_Y(d)(p)}{L_X(d)(p)}t. \tag{5.16}$$

The time r along the singular arc must satisfy $r \leq \bar{r}(t)$. We call $\bar{r}$ the *saturation constraint*. It is easy to see that

LEMMA 5.2. **[14]** *If a singular arc is saturated at time $\bar{r}$, then for any $\epsilon > 0$, the control $u \equiv -1$ is not optimal on $(\bar{r}, \bar{r}+\epsilon)$. Hence, if an extremal trajectory is a concatenation of bang and singular arcs, then at saturation a switch to Y occurs.* □

The next lemma summarizes the structure of extremals which contain singular arcs.

LEMMA 5.3. **[14]** *Let γ be an extremal trajectory which starts from p, contains a singular arc and is a finite concatenation of bang and singular arcs. Then γ is of the form YS, YSY, YSX or $YSXY$. The times along Y-trajectories are unrestricted. If t is the time along the first Y-arc, then the time along the singular arc is restricted by the saturation constraint $r \leq \bar{r}(t)$. The time s along the X-arc is restricted by the singular conjugate point conditions $s \leq \bar{s}(t, r)$ for YSX, respectively $s = \bar{s}(t, r)$ for $YSXY$.* □

Let

$$\begin{aligned}
\mathcal{S}_{YS} &= \{pe^{t_1Y}e^{t_2S} : t_1 > 0, 0 < t_2 < \bar{r}(t_1)\} \\
\mathcal{S}_{Y\bar{S}} &= \left\{pe^{t_1Y}e^{\bar{r}(t_1)S} : t_1 > 0\right\} \\
\mathcal{S}_{Y\bar{S}Y} &= \left\{pe^{t_1Y}e^{\bar{r}(t_1)S}e^{t_3Y} : t_1, t_3 > 0\right\} \\
\mathcal{S}_{YSY} &= \{pe^{t_1Y}e^{t_2S}e^{t_3Y} : t_1, t_3 > 0, 0 < t_2 < \bar{r}(t_1)\} \\
\mathcal{S}_{YSX} &= \{pe^{t_1Y}e^{t_2S}e^{t_3X} : t_1 > 0, 0 < t_2 < \bar{r}(t_1), 0 < t_3 < \bar{s}(t_1, t_2)\} \\
\mathcal{S}_{YS\bar{X}} &= \left\{pe^{t_1Y}e^{t_2S}e^{\bar{s}(t_1,t_2)X} : t_1 > 0, 0 < t_2 < \bar{r}(t_1)\right\} \\
\mathcal{S}_{YS\bar{X}Y} &= \left\{pe^{t_1Y}e^{t_2S}e^{\bar{s}(t_1,t_2)X}e^{t_3Y} : t_1, t_3 > 0, 0 < t_2 < \bar{r}(t_1)\right\}
\end{aligned}$$

It is easy to verify that all these sets are smooth embedded submanifolds and that they have the obvious dimensions. Recall from Section 2 that we denote the closures of these submanifolds by $\mathcal{C}$, i.e. $\mathcal{C}_{YS} = \{pe^{t_1Y}e^{t_2S} : t_1 \geq 0, 0 \leq t_2 \leq \bar{r}(t_1)\}$ etc. Also let

$$\mathcal{S}^{+}_{YX} = \{pe^{t_1Y}e^{t_2X} : t_1 > 0, 0 < t_2 < \bar{s}(t_1,0)\}$$

be the substratum of $\mathcal{S}_{YX}$ which lies in the boundary of $\mathcal{S}_{YSX}$ and recall that

$$\begin{aligned} \mathcal{S}_{Y\bar{X}} &= \left\{pe^{t_1Y}e^{\bar{s}(t_1,0)X} : t_1 > 0\right\} \subset \mathcal{S}_{YX}, \\ \mathcal{S}_{Y\bar{X}Y} &= \left\{pe^{t_1Y}e^{\bar{s}(t_1,0)X}e^{t_3Y} : t_1, t_3 > 0\right\} \subset \mathcal{S}_{YXY}. \end{aligned}$$

Combine these strata with their frontier strata into

$$\begin{aligned} \mathcal{C}_{Sing} &= \mathcal{S}_0 \cup \mathcal{S}_Y \cup \mathcal{S}^{+}_{YX} \cup \mathcal{S}_{Y\bar{X}} \cup \mathcal{S}_{Y\bar{X}Y} \\ &\quad \mathcal{S}_{YS} \cup \mathcal{S}_{Y\bar{S}} \cup \mathcal{S}_{YSY} \cup \mathcal{S}_{Y\bar{S}Y} \cup \mathcal{S}_{YSX} \cup \mathcal{S}_{YS\bar{X}} \cup \mathcal{S}_{YS\bar{X}Y}. \end{aligned} \tag{5.17}$$

LEMMA 5.4. [14] *$\mathcal{C}^{\epsilon}_{Sing}$ is the graph of a continuous function $\phi_3^{sing} : D_3^{sing} \to M$. The restrictions of ϕ_3^{sing} to the domains of the 3-dimensional submanifolds $\mathcal{S}_{YSY}$, $\mathcal{S}_{YSX}$, and $\mathcal{S}_{YS\bar{X}Y}$ are smooth and have smooth extensions to their 2-dimensional boundary strata.* □

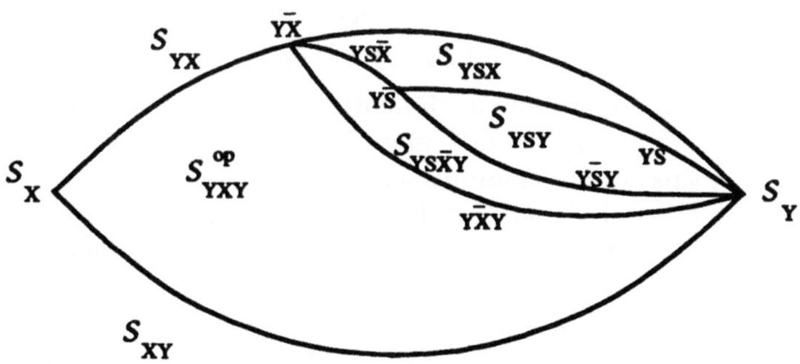

FIG. 5.1. *Stratification of the northern hemisphere.*

A qualitative sketch of $\mathcal{C}_{Sing}$ projected into a 2-dimensional slice $\{\xi_0 = \epsilon,\ \xi_3 = 0\}$ is given in Figure 5.1. Now bisect $\mathcal{S}_{YXY}$ along $\mathcal{S}_{Y\bar{X}Y}$ into the connected components

$$\mathcal{S}^{op}_{YXY} = \{pe^{t_1Y}e^{t_2X}e^{t_3Y} : t_1, t_3 > 0, t_2 > \bar{s}(t_1,0)\}$$

and

$$\mathcal{S}^{nop}_{YXY} = \{pe^{t_1Y}e^{t_2X}e^{t_3Y} : t_1, t_3 > 0, 0 < t_2 < \bar{s}(t_1,0)\}.$$

Observe that the closure of $\mathcal{S}^{nop}_{Y\bar{X}Y}$ and $\mathcal{C}_{Sing}$ have the same relative boundary and are, as functions of (ξ_0, ξ_1, ξ_2), defined over the same domain D_3^{sing}. The next lemma concludes the construction. Its proof is another explicit, though quite a bit more technical calculation as the calculation of the cut-locus in codimension 0.

LEMMA 5.5. **[14]** *$\mathcal{C}_{Sing}$ entirely lies above $\mathcal{S}^{nop}_{Y\bar{X}Y}$ over points in the interior of D_3^{sing}. The cells coincide on the boundary.* □

Summarizing, boundary trajectories are

- all bang-bang trajectories with at most one switching which form the equator: $E = \mathcal{S}_0 \cup \mathcal{S}_X \cup \mathcal{S}_{XY} \cup \mathcal{S}_Y \cup \mathcal{S}_{YX}$
- all bang-bang trajectories with two switchings which start with $u = -1$ which form the lower hemisphere: $S = \mathcal{S}_{XYX}$
- all extremal trajectories which contain a singular arc as described in Lemma 5.3 and all bang-bang trajectories with two switchings which start with $u = +1$ and lie in $\mathcal{S}^{op}_{YXY} \cup \mathcal{S}_{Y\bar{X}Y}$ (i.e. if t_1 and t_2 denote the times along the first and second arc respectively, then $t_2 \geq \bar{s}(t_1, 0)$, where $\bar{s}$ is the *singular* conjugate point relation). Together these trajectories form the upper hemisphere N.

As a Corollary a time-optimal synthesis for stabilizing an equilibrium point p for a 3-dimensional system Σ of the form (1.1) on a small neighborhood M of p can be constructed. Here the assumptions are that

(B1) g, $[f, g]$ and $[g, [f, g]]$ are linearly independent on M

and, writing $[f, [f, g]] = bg + c[f, g] + d[g, [f, g]]$, also

(B2) $d(p) = 1$, $L_X(d) < 0$ and $L_Y(d) > 0$ on M.

Define $\Sigma^{e,rev}$ as the system in dimension 4 where time has been added as fourth coordinate and where in addition the time direction has been reversed. For this system assumptions (A1)–(A3) made above hold at $(0, p)$. (Time adds a fourth independent direction. The signs of the Lie-derivatives of d change due to time-reversal, but d itself is left unchanged.) Hence from the structure of the small-time reachable set from $(0, p)$ a regular synthesis can be constructed simply by projecting the boundaries of the coordinate slices $Reach_{\Sigma^{e,rev}, \leq T}(0, p) \cap \{\xi_0 = c\}$ for $0 \leq c \leq \epsilon$ into the original state-space. Note that Σ satisfies the Hermes-conditions and thus Σ is small-time locally controllable near p. Hence this results in a time-optimal feedback synthesis of stabilizing controls in a neighborhood of p. We briefly describe this synthesis. Let

$$\mathcal{T}_0 = \{p\}, \quad \mathcal{T}_X = \{pe^{-sX} : s > 0\}, \quad \mathcal{T}_Y = \{pe^{-tY} : t > 0\},$$

$$\mathcal{T}_{XY} = \{pe^{-sX}e^{-tY} : s, t > 0\}, \quad \mathcal{T}_{YX} = \{pe^{-tY}e^{-sX} : s, t > 0\}.$$

Then

$$E = \mathcal{T}_0 \cup \mathcal{T}_X \cup \mathcal{T}_Y \cup \mathcal{T}_{XY} \cup \mathcal{T}_{YX}$$

is a stratified surface which bisects a sufficiently small ball centered at p into two connected components N and S. The surface E consists of extremal trajectories and on E a feedback u^* is defined in the obvious way. On S the optimal control is constant and given by $u^* = -1$. These are precisely the XYX-trajectories on the lower hemisphere,

$$S = \mathcal{T}_{XYX} = \{pe^{-s_1X}e^{-s_2Y}e^{-s_3X} : s_i > 0\}.$$

The trajectories in the other region N correspond to the trajectories in the northern hemisphere of the small-time reachable set. Let

$$\mathcal{T}_{YS} = \{pe^{-tY}e^{-rS} : t > 0,\ \ 0 < r < \bar{r}(t)\}$$

and analogously define all the other strata $\mathcal{T}_{YSX}$, $\mathcal{T}_{YSY}$, etc. Integrating X backward from $\mathcal{T}_{YS}$ until the singular conjugate point is reached defines another open region $\mathcal{T}_{YSX}$ where $u^* = -1$ is optimal. In the complement of the closure of these sets the optimal control is $u^* = 1$. More precisely, if we define

$$\mathcal{T}_{YX}^{-} = \{pe^{-t_1Y}e^{-t_2X} : t_1 > 0,\ 0 < t_2 < \bar{s}(t_1, 0)\}$$

and

$$\mathcal{T}_{YX}^{+} = \{pe^{-t_1Y}e^{-t_2X} : t_1 > 0,\ t_2 > \bar{s}(t_1, 0)\},$$

then we have the following result:

THEOREM 5.6. *A sufficiently small connected open neighborhood M of p is bisected by the stratified surface*

(5.18) $$SW = \mathcal{T}_0 \cup \mathcal{T}_X \cup \mathcal{T}_Y \cup \mathcal{T}_{XY} \cup \mathcal{T}_{YX}^{+} \cup \mathcal{T}_{Y\bar{X}} \cup \mathcal{T}_{YS\bar{X}} \cup \mathcal{T}_{Y\bar{S}} \cup \mathcal{T}_{YS}$$

into two open and connected components

(5.19) $$M^+ = \mathcal{T}_{YXY}^{op} \cup \mathcal{T}_{Y\bar{X}Y} \cup \mathcal{T}_{YS\bar{X}Y} \cup \mathcal{T}_{Y\bar{S}Y} \cup \mathcal{T}_{YSY}$$

and

(5.20) $$M^- = \mathcal{T}_{XYX} \cup \mathcal{T}_{YX}^{-} \cup \mathcal{T}_{YSX}.$$

The optimal control u^* is given by

$$u^*(x) = \begin{cases} +1 & \text{for } x \in M^+ \\ -1 & \text{for } x \in M^- \end{cases}$$

Except for $\mathcal{T}_{YS\bar{X}}$, all strata in SW consist of extremal trajectories and there u^* is defined in the obvious way. In particular, $u^*(x) = -d(x)$ is singular on the surface $\mathcal{T}_{YS}$. Trajectories cross $\mathcal{T}_{YS\bar{X}}$ transversally and switch from -1 to $+1$.

In the terminology of Boltyansky $\mathcal{T}_{YS\bar{X}}$ is a cell of the second kind [2]. All other strata are of the first kind. This construction even yields a regular synthesis as defined in [2], i.e. all the strong technical postulates made in Boltyansky's original definition of a regular synthesis hold. In particular, this proves optimality of the synthesis.

6. Conclusion. The examples given here try to make the point that the construction of the small-time reachable set is an effective method to solve local time-optimal control problems in low dimensions for control systems which are affine in the controls and have bounded control values. Necessary conditions for optimality single out the constant controls ± 1 and singular controls as prime candidates, but typically do not give the precise structure. In this paper we outlined selected aspects of a construction of the small-time reachable set as a cell complex by attaching inductively cells of increasing dimensions. These cells consisted of extremal trajectories formed by concatenations of increasing lengths of possible candidates for optimal trajectories like bang-bang or singular trajectories. The construction is entirely geometric and relies on a Lie-algebraic framework to perform explicit calculations to establish the geometric properties of the strata. Structural features which were found in this way, like cut-loci, played decisive roles in the analysis of the low-dimensional examples considered so far. From the precise structure of the small-time reachable set a local regular synthesis of time-optimal controls was derived. For more details on the technical aspects and the proofs we need to refer the reader to [7,12,14].

REFERENCES

[1] A.A. Agrachev and R.V. Gamkrelidze, *Symplectic geometry for optimal control*, in *Nonlinear Controllability and Optimal Control* (H. Sussmann, ed.), Marcel Dekker, 1990, pp. 263–277.

[2] V.G. Boltyansky, *Sufficient conditions for optimality and the justification of the dynamic programming method*, SIAM Journal of Control, Vol.4, No.2 (1966), pp. 326–361.

[3] N. Bourbaki, *Elements of Mathematics, Lie Groups and Lie Algebras*, Chapters 1–3, Springer–Verlag, Berlin, 1989.

[4] C. Lobry, *Contrôlabilité des Systèmes nonlinéaires*, SIAM J. Control, Vol.8, (1970), pp. 573–605.

[5] N. Jacobson, *Lie Algebras*, Dover, New York, 1979.

[6] A.J. Krener, *The high order maximum principle and its application to singular extremals*, SIAM J. Control and Optimization, Vol.15, (1977), pp. 256–293.

[7] A.J. Krener and H. Schättler, *The structure of small time reachable sets in low dimension*, SIAM J. Control and Optimization, Vol.27, No.1, (1989), pp. 120–147.

[8] I.A.K. Kupka, *The ubiquity of Fuller's phenomenon*, in *Nonlinear Controllability and Optimal Control*, (H. Sussmann, ed.), Marcel Dekker, New York, 1990, pp. 313–350.

[9] L. Pontryagin, V. Boltyansky, R. Gamkrelidze, and E. Mishchenko, *The mathematical theory of optimal processes*, Wiley-Interscience, New York, 1962.

[10] H. Schättler, *The local structure of time-optimal trajectories in dimension* 3 *under generic conditions*, SIAM J. Control and Optimization, Vol.26, No.4, (1988), pp. 899–918.

[11] H. Schättler, *Conjugate points and intersections of bang-bang trajectories*, Proceedings of the 28th IEEE Conference on Decision and Control, Tampa, Florida, (1989), pp. 1121–1126.

[12] H. Schättler, *A local feedback synthesis of time-optimal stabilizing controls in dimension three*, Mathematics of Control, Signals and Systems, Vol.4, (1991),

pp. 293–313.

[13] H. SCHÄTTLER, *Extremal trajectories, small-time reachable sets and local feedback synthesis: a synopsis of the three-dimensional case*, in *Nonlinear Synthesis*, Proceedings of the IIASA Conference on Nonlinear Synthesis, Sopron, Hungary, June 1989, (C. I. BYRNES, A. KURZHANSKY, eds.), Birkhäuser, Boston, 1991, pp. 258–269.

[14] H. SCHÄTTLER AND M. JANKOVIC, *A synthesis of time-optimal controls in the presence of saturated singular arcs*, Forum Mathematicum.

[15] H. SUSSMANN, *Lie brackets and real analyticity in control theory*, in *Mathematical Control Theory*, Banach Center Publications, Vol.14, Polish Scientific Publishers, Warsaw, Poland, 1985, pp. 515–542.

[16] H. SUSSMANN, *A product expansion for the Chen series*, in *Theory and Applications of Nonlinear Control Systems* (C. BYRNES, A. LINDQUIST, eds.) North-Holland, Amsterdam, 1986, pp. 323–335.

[17] H. SUSSMANN, *Envelopes, conjugate points, and optimal bang-bang extremals*, in *Proceedings of the 1985 Paris Conference on Nonlinear Systems* (M. FLIESS, M. HAZEWINKEL, eds.) Reidel Publishing, Dordrecht, 1987.

[18] H. SUSSMANN, *Envelopes, high order optimality conditions and Lie brackets*, Proceedings of the 28th IEEE Conference on Decision and Control, Tampa, Florida, (1989), pp. 1107–1112.

HIGHER ORDER VARIATIONS: HOW CAN THEY BE DEFINED IN ORDER TO HAVE GOOD PROPERTIES?

GIANNA STEFANI*

Abstract. Good properties of variations in connection with their use are discussed. A definition for variations at a point of a trajectory is proposed with properties sufficiently good to state a higher order maximum principle for the minimum time problem and for an optimal control problem with constraints on the end-point. The definition allows to define variations on the base of the relations at a point in the Lie Algebra associated to the system.

Key words. Tangent vectors, higher order variations, maximum principles.

1. Introduction. Consider a control system on an n-dimensional manifold M

$$\dot{x}(t) = f(x(t), u(t)) \ , \ x(0) = \xi_0, \tag{Σ}$$

where $f : M \times R^m \rightarrow TM$ is a smooth function and the set $\mathcal{U}$ of admissible controls is the class of those integrable maps u from the compact interval $J = [0, T]$ into a subset Ω of R^m for which the time dependent vector field $(t, \xi) \mapsto f(\xi, u(t))$ and its first derivative with respect to the state is locally L^1-bounded. Let $t \rightarrow x(t, \xi_0, u)$ denote the solution of (Σ) relative to the control u and let $R(\xi_0, t)$ be the reachable set at time t.

Consider a reference control $\hat{u}$ and suppose that the corresponding reference trajectory $t \rightarrow \hat{x}(t) \equiv x(t, \xi_0, \hat{u})$ is defined on J. There are classical problems in optimal control theory linked to the study of the reachable sets "near" a point $\hat{x}(t)$ of the reference trajectory.

A first problem, related to minimum time, is to decide whether the point $\hat{x}(t)$ belongs to the interior of the reachable set at time t. In fact if it does, then the trajectory cannot be of minimum time after t.

A second problem is to decide whether $\hat{x}$ minimizes a cost of the final point, say $\beta_0(x(T))$, possibly in presence of constraints $\beta_1(x(T)) = \cdots = \beta_r(x(T)) = 0$. This problem can be transformed in the following way. If we define $\xi_1 = \hat{x}(T)$ and $\beta \equiv (\beta_0, \ldots, \beta_r) : M \rightarrow R^{r+1}$, then the reference trajectory $\hat{x}$, solves the original problem if and only if it satisfies the constraints and $\beta(R(\xi_0, T))$ does not intersect the half-line $\{\beta(\xi_1) - \lambda(1, 0, \ldots, 0) \ : \ \lambda > 0\}$. In particular $\beta(\xi_1)$ cannot be interior to $\beta(R(\xi_0, T))$.

If ξ_0 is an equilibrium point and $\hat{x}(t) \equiv \xi_0$, then the first problem reduces to decide whether $\xi_0 \in int\, R(\xi_0, T)$ and, if we allow T to vary, then we deal with the so called small time local controllability (STLC) property, i.e. ξ_0 belongs to the interior of the reachable set at each time $t > 0$.

---

* University of Naples, Dipartimento di Matematica e Applicazioni, Via Mezzocannone 8, 80134 Napoli, Italia. email: stefani@facec.cce.unifi.it

A way to attempt the above problems is to study the *tangent vectors* to a subset A of a manifold N at a point y. The simplest way to define tangent vectors is to define them as tangent directions to curves starting at y and lying in A.

DEFINITION 1.1. We say that $v \in T_yN$ is tangent to A at y and we write $v \in T_yA$, if and only if there exists $\gamma : [0, \bar{\epsilon}] \to A$, such that $\gamma(\epsilon) = y + \epsilon v + o(\epsilon)$.

With this definition T_yA is a cone, possibly nonconvex. In general T_yA has not the following "good property"

$$T_yA = T_yN \Rightarrow y \in int\, A, \tag{1.1}$$

so that we have to select particular subsets of T_yA to obtain it. The choice of such sets may depend on the problem.

In our cases we have to study the sets $R(\xi_0, T)$ and $\beta(R(\xi_0, T))$. It is clear that if $v \in T_{\xi_1}R(\xi_0, T)$ then $T_{\xi_1}\beta(v) \in T_{\beta(\xi_1)}\beta(R(\xi_0, T))$, therefore we are mainly interested in studying subsets of tangent vectors to $R(\xi_0, T)$.

A natural way to obtain tangent vectors in the reachable sets is to consider control variations

$$\epsilon \to u_\epsilon \ , \ u_0 = \hat{u}$$

of the reference control $\hat{u}$ and the curve

$$\epsilon \mapsto \gamma(\epsilon) = x(T, \xi_0, u_\epsilon) \in R(\xi_0, T).$$

If $\gamma(\epsilon) \equiv \xi_1 + \epsilon v + o(\epsilon)$, then $v \in T_{\xi_1}R(\xi_0, T)$. Such vectors v with "good properties" are sometimes called *variations*.

In particular consider the so-called "Pontryagin variations" obtained in the following way. Take the controls $u_{\epsilon t \omega} : [0, T] \to \Omega$ obtained replacing the reference control $\hat{u}$ by the constant value ω in the interval $[t, t+\epsilon]$. The final point of the relative trajectory gives rise to a curve

$$\gamma : \epsilon \mapsto x(T, \xi_0, u_{\epsilon t \omega})$$

in the reachable set $R(\xi_0, T)$. If t is a Lebesgue point of $t \mapsto f(\hat{x}(t), \hat{u}(t))$, see [10], then the tangent vector to this curve can be thought as the transport along the reference trajectory of a "trajectory's variation" v_t produced at time t, given by

$$v_t = f(\hat{x}(t), \omega) - f(\hat{x}(t), \hat{u}(t)).$$

Namely, let $\xi \mapsto \hat{\Phi}_{t,T}$ be the flow from time t to time T of the reference time-dependent vector field $(t, \xi) \mapsto f(\xi, \hat{u}(t))$. If t is a Lebesgue point of $\hat{u}$, then

$$T_{\hat{x}(t)}\hat{\Phi}_{t,T}\Big(f(\hat{x}(t), \omega) - f(\hat{x}(t), \hat{u}(t))\Big)$$

belongs to $T_{\xi_1}R(\xi_0,T)$ for all $\omega\in\Omega$.

The Pontryagin variations are a subclass of the so-called *needle variations*, for which the support (i.e. the set in which the control differs from the reference one) goes to zero with the variational parameter ϵ.

DEFINITION 1.2. We call $v_t\in T_{\hat{x}(t)}$ a needle trajectory variation of order k at time t if there exists a control variation $\epsilon\mapsto u_{\epsilon t}$ such that the relative trajectory gives rise to a continuous curve

$$\gamma:\epsilon\mapsto\hat{\Phi}_{t+\epsilon^{1/k},t}(x(\epsilon^{1/k},\hat{x}(t),u_{\epsilon t}))\equiv\hat{x}(t)+\epsilon v_t+o(\epsilon).$$

We easily obtain as before that $T_{\hat{x}(t)}\hat{\Phi}_{t,T}(v_t)$ is a tangent vector to the reachable set $R(\xi_0,T)$.

Let us review some results which apply in the above considered cases.

1.1. STLC problem. In [6] a sufficient condition of STLC for a multivalued inclusion is given using a suitable definition of variations. If we apply that definition to system (Σ) we obtain the following:

DEFINITION 1.3. We say that $v\in T_{\xi_0}M$ is a STLC variation of order k if and only if for all ϵ sufficiently small

$$\xi_0+\epsilon v+o(\epsilon)\in R(\xi_0,\epsilon^{1/k}).$$

The result in [6], applied to our case is the following:

THEOREM 1.4. *Let Ω be bounded. If the convex cone generated by the STLC variations of any order defined above is the whole tangent space, then*

$$\xi_0\in int\,R(\xi_0,[0,t])\ ,\ \forall t>0.$$

Notice that no continuity far from zero is required in the definition of STLC variations. Moreover the above variations are tangent vectors to the sets $R(\xi_0,[0,t])$, for all $t>0$, but only particular tangent vectors are variations.

A suitable generalization to unbounded control of the above definition and result has been obtained recently by R.M. Bianchini [3].

1.2. Minimum cost without constraints. It is clear that if (-1) is tangent to $\beta_0(R(\xi_0,T))$ at $\beta_0(\xi_1)$, then $\hat{x}$ does not minimize $\beta_0(x(T))$ if there are no constraints on the final point. In this case all the tangent vectors to $\beta_0(R(\xi_0,T))$ have to be considered and we obtain:

THEOREM 1.5. *If $\hat{x}$ minimizes $\beta_0(x(T))$ over all the solutions of the system (Σ), then*

$$-T_{\xi_1}\beta_0(v)\leq 0\ ,\ \forall v\in T_{\xi_1}R(\xi_0,T).$$

In particular if we apply the above result to the Pontryagin variations, then we obtain

$$-T_{\xi_1}\beta_0 \circ T_{\hat{x}(t)}\hat{\Phi}_{t,T}\Big(f(\hat{x}(t),\omega)-f(\hat{x}(t),\hat{u}(t))\Big) \leq 0\ ,\ \forall\omega\in\Omega\ ,\ a.e.\ t\in[0,T].$$
(1.2)

The covector $\hat{p}(t) \equiv -T_{\xi_1}\beta_0 \circ T_{\hat{x}(t)}\hat{\Phi}_{t,T}$ is the solution of the adjoint equation, given in a chart by

$$\dot{p}(t) = -p(t)\,\frac{\partial f}{\partial x}(\hat{x}(t),\hat{u}(t))\ , \tag{1.3}$$

with boundary condition

$$p(T) = -T_{\xi_1}\beta_0. \tag{1.4}$$

Therefore the inequality (1.2) leads to the Pontryagin Maximum Principle, i. e.

If $\hat{x}$ minimizes $\beta_0(x(T))$ over all the solutions of the system (Σ), then the solution $\hat{p}\ :\ [0,T] \to (R^n)^*$ of the adjoint equation (1.3) with boundary condition (1.4) is such that , for all $\omega\in\Omega$ and almost all $t\in[0,T]$,

$$\hat{p}(t)f(\hat{x}(t),\hat{u}(t)) \geq \hat{p}(t)f(\hat{x}(t),\omega).$$

Every further tangent vector to $R(\xi_0,T)$ at ξ_1 gives a strengthening of the Pontryagin Maximum Principle. In particular needle variations give conditions along the trajectory, see also [5].

In this case where no constraints are imposed on the end point one can obtain tangent vectors to $\beta_0(R(\xi_0,T))$ besides the ones coming from $R(\xi_0,T)$ (see for example [7], [9]).

1.3. Minimum time and minimum cost with constraints. The case of minimum time and the one of minimum cost with constraints are very similar because in both cases we want to select a subset $\mathcal{K}$ of tangent vectors to A at y with the property (1.1). We shall deal with a property little stronger than (1.1), but which is basically equivalent to Maximum Principle when applied to needle variations. Consider the following property which says that $y \in int\,A$ if the convex cone generate by $\mathcal{K}$ is the whole tangent space T_yN

$$y\in\partial A\ \Rightarrow\ \exists p^*\in T^*_yN\ s.t.\ p^*\neq 0\ and\ p^*v\leq 0\ ,\ \forall v\in\mathcal{K}. \tag{1.5}$$

For sake of simplicity, in what follows we consider the case when $N = M\ ,\ y=\xi_1\ ,\ A=R(\xi_0,T)$.

A nice example in [4] shows that tangent vectors generated by general control variations do not have the property (1.5). Following [4], we introduce the summability property which is easily seen to imply (1.5).

DEFINITION 1.6. A subset $\mathcal{K}$ of $T_{\xi_1}R(\xi_0,T)$ has the summability property if for all finite subset $\{v_1,\cdots,v_p\}$ of $\mathcal{K}$ there exists a constant $\bar{\epsilon}>0$ and a continuous map

$$\varphi:[0,\bar{\epsilon}]\times\left\{(c_1,\ldots,c_p):c_i\geq 0\ ,\ \sum c_i=1\right\}\rightarrow R(\xi_0,T)$$

such that

$$\varphi(\epsilon,c_1,\ldots,c_p)=\xi_1+\epsilon(c_1v_1+\cdots+c_pv_p)+o(\epsilon).$$

In [4] systems linear with respect to the control are considered and an order for tangent vectors is defined in connection with the L^1-norm of the control variation. It is shown that the subsets consisting of tangent vectors of order one and at most one tangent vector of higher order have the summability property. Notice that the control variations considered in [4] have no restriction on the support.

In general we can obtain a Maximum Principle each time that we can select a set $\mathcal{V}_t$, $t\in[0,T)$, of needle variations in such a way that the set

$$\mathcal{K}=\left\{T_{\hat{x}(t)}\hat{\Phi}_{t,T}(\mathcal{V}_t):t\in[0,T)\right\}$$

has the summability property. Namely, if this is the case we have that if $\xi_1\in\partial R(\xi_0,T)$ then there exists a nonzero solution $\hat{p}:[0,T]\rightarrow(R^n)^*$ of the adjoint equation (1.3) such that

$$\hat{p}(t)v_t\leq 0\ ,\ \forall v_t\in\mathcal{V}_t\ ,\ t\in[0,T).$$

It is possible to prove that the needle variations of order one (like the Pontryagin variations) have the sommability property, by concatenating the controls. They give rise to the Pontryagin Maximum Principle for this case.

For nonlinear systems it is possible that Pontryagin Maximum Principle is not sufficient to single out a unique candidate and additional conditions may be required. To obtain such additional conditions one can add some new variations.

The main difficulties one meets in order to prove the "summability property" arise when two variations are obtained at the same time. In fact if v_t and v_{t_1} are variations at different times, then it is possible to sum them simply concatenating the two control variations. If v and v_1 are higher order variations at the same time, concatenating their control variations, a new principal part may appear.

To our knowledge there is no example which proves that there are needle variations which do not have the summability property (the example in [4] concerns variations with finite support). Nevertheless this property has not been proved for general higher order needle variations.

The higher order maximum principles developed in [8] and [7] are based on needle variations that are required to be obtained continuously on a small but finite arc of trajectory, so that each variation can be thought as produced at different times. Basically if $v \in T_{\hat{x}(t)}M$ is a variation at t as defined either in [8] or in [7], then there is a positive constant h and a continuous map $w : [t-h, t+h] \to TM$ such that $w(t) = v$ and $w(\tau)$ is a variation at τ, for all $\tau \in [t-h, t+h]$. In particular if the convex cone generated by the variations at a time is the whole tangent space, then the same holds for neighboring times.

In the next section a definition of variations developed in [2] is presented (see also [12]). Such variations verify the summability property and may occur at an isolated time. This aim is obtained by requiring that a variation at time t can be obtained with control variations of time length ϵ and starting times $t + \gamma\epsilon$, depending on a new parameter γ.

The original motivation to this definition was to extend the known results of controllability at a point to the case of nonstationary trajectories.

In particular our aim was to obtain higher order variations at time t using the relations at $\hat{x}(t)$ in the Lie Algebra of the vector fields associated to the system, see [13] and [1]. This kind of "thin conditions" which depend only on a finite number of derivatives at $\hat{x}(t)$, cannot give variations on an arc of trajectory (see Example 3.3 below). Of course if the relations in the Lie Algebra are verified on an arc of trajectory, then the variations are produced in the same arc and some additional properties can be developed.

The variations defined in the next section are more general than the ones defined in [8]. The comparison with the ones defined in [7] is more difficult, nevertheless the variation which appear in Example 3.3 is not a variation as defined in [7].

2. The variational cone. In this section we give a definition of variation which is a slight modification of the one given in [2], where more general time-dependent control systems and admissible control functions are considered.

DEFINITION 2.1. A vector $v \in T_{\hat{x}(t)}M$ is a *right variation* of order k of $(\hat{u}, \hat{x})$ at time $t \in [0, T)$ if there are positive numbers $\bar{\gamma}, \bar{c}, \bar{\epsilon}$ and a three-parameters control variation

$$\eta : [0, \bar{\gamma}] \times [0, \bar{c}] \times [0, \bar{\epsilon}] \to \mathcal{U}$$

such that the maps

$$(\xi, \gamma, c, \epsilon) \mapsto \varphi_{\gamma,c,\epsilon}(\xi) \equiv x(\epsilon^{1/k}, \xi, \eta(\gamma, c, \epsilon))$$

$$(\xi, \gamma, c, \epsilon) \mapsto T_\xi \varphi_{\gamma,c,\epsilon}$$

are continuous for ξ belonging to a neighborhood of $\hat{x}(t)$ and

$$\hat{\Phi}_{t+\epsilon^{1/k}(1+\gamma),t}(\varphi_{\gamma,c,\epsilon}(\hat{x}(t + \gamma\epsilon^{1/k}))) = \hat{x}(t) + \epsilon\, c\, v + o(\epsilon)$$

where

$$\lim_{\epsilon \to 0} \frac{o(\epsilon)}{\epsilon} = 0$$

uniformly with respect to (γ, c).

Remark 2.2. Here only right variations are considered, i.e. relative to control variations which take place after t. In an analogous way left variations could be considered, see [2].

Remark 2.3. Notice that with this definition a topology on $\mathcal{U}$ is not needed. In [2] the continuity of the map φ is obtained through the continuity of η with respect to a suitable topology introduced in the set of admissible control $\mathcal{U}$. This topology may depend on the map f. For example, if f is not linear with respect to the control, then in general the L^1-topology cannot be chosen.

If we denote by $\mathcal{V}_t$ the set of variations at time t, it is easy to see that $\mathcal{V}_t$ is a positive cone which may be not convex. For the properties of $\mathcal{V}_t$ we refer to [2] where the proofs of the results described in this section are also given. It is easily seen that the slight different definition of variation does not change the proofs.

To explain the meaning of the parameters γ and c let us remark that the role of γ is to allow "place" to concatenate the control variations at the same time. Basically we require the same trajectory variation in an interval of time which goes to zero as the time-length of the control variation. The role of c is more technical and it is the same as the one of the c_i's in the summability property.

The following one is the main result on the variations proved in [2] in a little different form.

THEOREM 2.4. *The subset $\mathcal{K}$ of $T_{\xi_1} R(\xi_0, T)$ defined by*

$$\mathcal{K} = \textit{convex hull} \left\{ T_{\hat{x}(t)} \hat{\Phi}_{t,T} \left(\mathcal{V}_t \right) : t \in [0, T) \right\}, \tag{2.1}$$

has the summability property. Therefore if

$$\mathcal{K} = T_{\xi_1} M$$

then

$$\xi_1 \in int R(\xi_0, T).$$

Remark 2.5. Pontryagin variations are of this type, namely

$$f(\hat{x}(t), \omega) - f(\hat{x}(t), \hat{u}(t))$$

is a right (left) variation of order one at each time t that is a right (left) Lebesgue point for $\hat{u}$. Therefore this kind of variations gives a strengthening of the Pontryagin Maximum Principle.

3. Higher order variations induced by the relations in the Lie Algebra. In what follows we give a result proved in [2] which states that suitable relations in the Lie Algebra associated to (Σ) at a point $\hat{x}(t)$ generate variations at t. The result concerns affine systems and a trajectory relative to a control taking values in the interior of Ω.

Let (Σ) be of the form

$$\dot{x} = f_0(x) + \sum_{i=1}^{m} u_i f_i(x),$$

and let Ω be the hypercube

$$H_\rho = \Big\{(\omega_1, .., \omega_m) \in R^m : |\omega_i| \leq \rho, i = 1, .., m\Big\},$$

possibly $\rho = +\infty$, i.e. $H_\infty = R^m$. We choose $\hat{u} \equiv 0$ so that $\hat{x}(t) = exp\, t f_0(\xi_0)$.

The same arguments can be used to analyze trajectories relative to C^∞ controls, by modifying the system in a standard way.

In order to state the result we need the notations introduced in [1], [2]. Let $LieX$ be the free Lie Algebra on R generated by the noncommutative indeterminate $X = \Big\{X_0, \ldots, X_m\Big\}$. $\mathcal{S}$ will denote the ideal of $LieX$ generated by $\{X_1, \ldots, X_m\}$. Replacing X_i by f_i in an element $\chi \in LieX$, we obtain a vector field which will be denoted by χ_f. For any subset A of $LieX$, $A_f = \Big\{\chi_f : \chi \in A\Big\}$ will denote a subset of the Lie Algebra $Lief$ generated by $\Big\{f_0, \ldots, f_m\Big\}$.

By means of a set $l = \{l_0, \ldots, l_m\}$ of nonnegative integers we define a weight on $LieX$ which will induce a weight on $Lief$. Let Λ be a bracket in $\mathcal{S}$. We denote the "length" of Λ with respect to X_i (i.e. the number of times that X_i appears in Λ) by $|\Lambda|_i$. The weight of Λ is defined by

$$||\Lambda||_l = \sum_{i=0}^{m} l_i |\Lambda|_i,$$

and the subspace of $\mathcal{S}_f$ consisting of the elements of weight not greater than i is given by

$$V_i^l = span\Big\{\Lambda_f : \Lambda \in \mathcal{S} \ , \ ||\Lambda||_l \leq i\Big\}.$$

An element $\chi \in \mathcal{S}$ is called $l-$ *homogeneous* if it is a linear combination of brackets with the same weight, which will be called the weight of χ.

Following [13] we say that an $l-$ *homogeneous* element $\chi \in \mathcal{S}$ is $l-$ *neutralized* for (Σ) at ξ if χ_f is a linear combination at ξ of brackets with less weight. In other words

DEFINITION 3.1. χ is $l-$ *neutralized* at ξ if there is an $i < \|\chi\|_l$ such that $\chi_f(\xi) \in V_i^l(\xi)$.

For more details and examples see [1], [2].

To characterize the elements of $Lief$ which give rise to variations, we introduce the set of obstructions relative to a weight l, see [1], [2]. Let

$$\mathcal{B} = span\Big\{\Lambda \in \mathcal{S} : |\Lambda|_0 \text{ is odd },\ |\Lambda|_i \text{ is even },\ i = 1, \dots, m\Big\}$$

$$\mathcal{B}_S^l = \Big\{\chi \in \mathcal{B} : \chi \text{ is symmetric w.r.t. the } X_i^l s \text{ with the same weight}\Big\}.$$

The set of obstructions relative to the weight l is the set:

$$\mathcal{B}_l^* = Lie\{X_0, \mathcal{B}_S^l)\} \cap \mathcal{S}.$$

The following result states that an element of $\mathcal{S}_f$ determines a variation at time t if every obstruction with less or equal weight is neutralized at $\hat{x}(t)$.

THEOREM 3.2. *[2]. Let* $l = \Big\{l_0, \dots, l_m\Big\}$ *be a set of nonnegative weights. If* $\rho < +\infty$, *then we also suppose* $l_0 \le l_i$, $i = 1, \dots, m$. *If* Φ *is a bracket in* $\mathcal{S}$ *such that each* $\Lambda \in \mathcal{B}_l^*$ *with* $\|\Lambda\|_l \le \|\Phi\|_l$ *is* $l-$*neutralized at* $\hat{x}(t)$, *then* $\Phi_f(\hat{x}(t))$ *is a variation at* t.

Applying the above theorem it is possible to prove that particular brackets induce variations at any time, see [1], [2]. As an example we have that for all $t \in [0, T)$

$$\mathcal{C}_t \equiv \big\{ad_{f_0}^k f_i(\hat{x}(t)) : k \ge 0\ ,\ i = 1 \dots m\big\} \subset \mathcal{V}_t$$

and

$$[f_i, f_j]\,(\hat{x}(t)) \subset \mathcal{V}_t\ ,\ i, j = 1 \dots m.$$

Such vectors are variations also in the sense of [8] and [7].

We are going now to provide an example of a vector field which defines a variation v at $t = 0$, but the set of variations at positive times is separated from v. In particular, otherwise than what happens in [8] and [7], the convex cone generated by the variations is the whole tangent space at an isolated point.

Example 3.3. Let us consider the following system on R^3

$$\begin{aligned} \dot{x} &= 1 + u \\ \dot{y} &= ux \\ \dot{z} &= u(x^3 y + y^2) \end{aligned}$$

with $\xi_0 = (0, 0, 0)$ and $|u| \le 1$. Let the reference control be $\hat{u} \equiv 0$ so that the reference trajectory is given by $\hat{x}(t) = (t, 0, 0)$. We have that the convex cone generated by $\mathcal{C}_0$ is given by

$$span\Big\{\frac{\partial}{\partial x},\frac{\partial}{\partial y}\Big\}.$$

Consider the bracket

$$\chi_f=[ad_{f_0}f_1,ad^2_{f_1}f_0]=(2-6x)\frac{\partial}{\partial z}.$$

$\pm\chi_f(\xi_0)$ define variations at 0. In fact they satisfy the hypotheses of Theorem 3.2 with $l=\{2,3\}$, for details see [1],[2]. Therefore

$$convex\ cone\ \mathcal{V}_0=R^3.$$

To show that the convex cone generated by $\mathcal{V}_t$, t "near 0", cannot be R^3, consider $\bar{\xi}=\hat{x}(\bar{t})$, $0<\bar{t}<1/3$ and define

$$\hat{y}:t\mapsto exp\,t f_0(\bar{\xi}).$$

The convex cone generated by the variations at $\bar{t}$ for $(\hat{u},\hat{x})$ is the same as the one generated by the variations at 0 for $(\hat{u},\hat{y})$. If this cone would be R^3, then $\hat{y}(t)$ would be interior to $R(\bar{\xi},t)$ for all $t>0$. But

$$ad^2_{f_1}f_0=-(2x^3+6xy-2y)\frac{\partial}{\partial z},$$

moreover also for the couple $(\hat{u},\hat{y})$ the convex cone generated by $\mathcal{C}_0$ is equal to $span\left\{\frac{\partial}{\partial x},\frac{\partial}{\partial y}\right\}$ for all $t>0$. Therefore

$$ad^2_{f_1}f_0(\bar{\xi})=(6-2\bar{t}^3)\frac{\partial}{\partial z}\notin span\,\mathcal{C}_0.$$

The results in [11] imply that for each $t>0$ sufficiently small $\hat{y}(t)=exp\,t\,f_0(\bar{\xi})$ belongs to the boundary of $R(\bar{\xi},t)$. A contradiction.

REFERENCES

[1] R.M. Bianchini, G. Stefani, *Graded approximations and controllability along a trajectory*, SIAM J. Control and Optim. 28 (1990), pp. 903–924.
[2] R.M. Bianchini, G. Stefani, *Controllability along a trajectory: a variational approach*, SIAM J. Control and Optim. 31 (1993), pp. 900–927.
[3] R.M. Bianchini, *Variation of a control process at the initial point*, J. of Opt. Theory and Appl. 81 (1994), pp. 249–258.
[4] A. Bressan, *A high order test for optimality of bang-bang controls*, SIAM J. Control and Optim. 23 (1985), pp. 38–48.
[5] H. Frankowska, *Contingent cones to reachable sets of control systems*, SIAM J. Control and Optim. 27 (1989), pp. 170–198.
[6] H. Frankowska, *Local controllability of control systems with feedback*, J. of Opt. Theory and Appl. 60 (1989), pp. 277–296.

[7] H.W. KNOBLOCH, *Higher order necessary conditions in optimal control theory*, in *Lecture Notes in Control and Inf. Sci.*, 34, Springer–Verlag, Berlin, 1981.

[8] A. KRENER, *The higher order maximal principle and its applications to singular extremals*, SIAM J. Control and Optim. 15 (1977), pp. 256–292.

[9] F. LAMNABHI LAGARRIGUE, G. STEFANI, *Singular optimal problems: on the necessary conditions of optimality*, SIAM J. Control and Optim. 28 (1990), pp. 823–840.

[10] E.B. LEE, L. MARKUS, *Foundation of optimal control theory*, John Wiley, New York, 1967.

[11] G. STEFANI, *On the local controllability of a scalar-input system*, in *Theory and Applications of Nonlinear Control Systems*, North Holland, Amsterdam, 1986, pp. 167–179.

[12] G. STEFANI, *On maximum principles* in *Analysis of Controlled Dynamical Systems*, (B. BONNARD, K. GAUTHIER, eds.) Progress in Systems and Control Theory, vol. 8, Birkhäuser, Boston, 1991.

[13] H.J. SUSSMANN, *A general theorem on local controllability*, SIAM J. Control and Optim. 25 (1987), pp. 158–194.

WELL POSED OPTIMAL CONTROL PROBLEMS: A PERTURBATION APPROACH

TULLIO ZOLEZZI*

1. Tikhonov and Hadamard well posedness. Let X be a convergence space and

$$J : X \to (-\infty, +\infty]$$

a proper extended-real valued function on X. The (global) minimization problem (X, J) is called **Tikhonov well posed** iff there exists exactly one global minimizer x^* and every minimizing sequence for (X, J) converges to x^*; **Hadamard well posed** iff there exists exactly one global minimizer x^* and, roughly speaking, x^* depends continuously upon problem's data.

In most applications, X is a subset of a real Banach space equipped with the strong convergence.

The concept of Tikhonov well posedness was firstly isolated in [1]. In [2] Tikhonov pointed out that many optimal control problems (involving ordinary differential equations) are ill posed with respect to the uniform convergence of the minimizing sequences.

Example. The minimization of

$$\int_0^1 (x^2 + u^2)dt$$

subject to

$$(*) \qquad \dot{x} = u \ , \ x(0) = 0 \ , \ |u(t)| \le 1 \text{ a.e.}$$

has the unique optimal control $u^*(t) = 0$ a.e., and it is a Tikhonov well posed optimal control problem with respect to the strong convergence in $L^2(0,1)$. However, if one strengthens the convergence to the strong one in $L^\infty(0,1)$, the problem becomes ill posed. Indeed, given any $c \in (0,1)$, consider the minimizing sequence

$$u_n(t) = 0 \text{ if } t > 1/n, u_n(t) = c \text{ if } 0 \le t \le 1/n.$$

Then in the L^∞ -norm one has

$$||u_n - u^*|| = c \text{ for every } n.$$

* Dipartimento di Matematica, Universita' di Genova, Via L.B. Alberti 4, 16132 Genova, Italy. Supported in part by MURST, funded 40%, and the Institute for Mathematics and its Applications.

Uniqueness of the global solution to (X, J) does not imply Tikhonov well posedness, as we see from the following example: minimize

$$\int_0^1 x^2 dt$$

subject to (*), and consider the strong convergence of $L^2(0,1)$. Then the unique optimal control is $u^*(t) = 0$ a.e., however the minimizing sequence $u_n(t) = \sin(nt)$ does not converge.

The (naive) notion of Hadamard well posedness reminds us of the analogous concept in the framework of boundary value problems of the mathematical physics, which goes back to [3]. More important than the mere similarity, there are significant results, showing that many linear operator equations, or variational inequalities, are well posed in the classical sense of Hadamard if and only if an associated minimization problem has a unique optimal solution, which depends continuously on problem's data: see [4,5].

Both notions of well posed optimization problems are significant as far as the numerical solution is involved. Ill posed problems in the sense of Tikhonov or Hadamard should be handled with special care, since numerical methods of solution will fail in general, and regularization techniques will be required. There are many links between Tikhonov and Hadamard well posedness in optimization. For a survey, see [6].

2. A perturbation approach to well posedness, and the role of the value function. Given the minimization problem (X, J), we embed it into a smoothly parameterized family $(X, I(\cdot, p))$ of minimization problems, in such a way that $p = p^*$, say, is the parameter of interest, to which the given (unperturbed) problem corresponds. Then we analyze the behavior of the small perturbations close to p^*, in order to find necessary and sufficient conditions of well posedness of (X, J).

The setting is the following: X is a fixed convergence space, P is a real Banach space, L is a ball in P around a given point $p^* \in P$, and

$$I : X \times L \to (-\infty, +\infty]$$

in such a way that $I(\cdot, p^*) = J(\cdot)$. For every $p \in L$ we consider the optimal **value function**

$$V(p) = \inf \{I(x,p) : x \in X\}.$$

We denote by *problem* (p) the minimization problem $(X, I(\cdot, p))$, $p \in L$. Then problem (p^*) will be called here **well posed** iff there exist an unique global minimizer

$$x^* = \arg\min(p^*)$$

and for every sequence $p_n \to p^*$ in P, every sequence $x_n \in X$ such that

$$I(x_n, p_n) - V(p_n) \to 0$$

necessarily converges to x^* (in X).

Thus, well posedness of problem (p^*) requires Tikhonov well posedness, and in addition convergence to arg min (p^*) of every asymptotically minimizing sequence, corresponding to any convergent perturbation of p^*. Hence we impose an intrinsic form of well posedness of problem (p^*), independent of the embedding, and in addition a form of Hadamard well posedness, depending on the particular embedding we have chosen. This amounts to upper semicontinuity at $(0, p^*)$ of the multifunction

$$(\varepsilon, p) \to \varepsilon - \arg\min(p)$$

from $[0, +\infty) \times L$ to the subsets of X, if X is a topological space.

The above approach may be used to get necessary and sufficient conditions of well posedness of problem (p^*), as follows.

We posit the following assumption:

- I is lower semicontinuous on $X \times L$, and for every x in X, $I(x, \cdot)$ is Gâteaux differentiable on L with a continuous gradient at arg min $(p^*) \times \{p^*\}$.

THEOREM 1 (necessary condition). *Let V be finite on L. If problem (p^*) is well posed, then V is Fréchet differentiable at p^*.*

THEOREM 2 (sufficient condition). *Problem (p^*) is well posed provided that the following conditions hold:*

(1) V is upper semicontinuous and Gâteaux differentiable on L, with a continuous gradient at p^*;

(2) $\nabla I(\cdot, p^*)$ is one-to-one on arg min(p^*)*;*

(3) for any sequence $p_n \to p^*$, every sequence $x_n \in X$ such that $I(x_n, p_n) - V(p_n) \to 0$ and $\nabla I(x_n, p_n)$ converges strongly in P^*, has a convergent subsequence.

The above theorems give explicit conditions for well posedness of (X, J). They deal with *free* problems, i.e. the effective domain of $I(\cdot, p)$ is independent of the parameter p. Similar results hold for *constrained* problems, i.e. the effective domain of $I(\cdot, p)$ depends on p. (The proof of Theorem 2 makes essential use of the differentiability properties of $|(x, \cdot)$, so that the usual trick of absorbing the constraints by adding infinite penalties to I is of no use here).

The Fréchet differentiability of the value function at p^* is a key property for well posedness of $(X, I(\cdot, p^*))$, as emphasized by Theorems 1 and 2 (the latter requires slightly more than that, according to (1)). There are known interesting cases in optimization, where the link between differentiability of the value function and well posedness is significant, as follows.

Example 1. Let X be a real Banach space, and let

$$J : X \to (-\infty, +\infty]$$

be a lower semicontinuous, proper, convex function. Consider the conjugate function

$$J^*(p) = \sup\{< p, x > -J(x) \ : \ x \in X\}.$$

THEOREM 3 (Asplund-Rockafellar [7]). *(X, J) is Tikhonov well posed with respect to the strong convergence iff J^* is Fréchet differentiable at* 0.

Example 2. Let X be a real Banach space, and K a proper closed subset of X. Given $p^* \in X \backslash K$, we consider the best approximation problem of minimizing the distance of p^* from K. Write

$$\text{dist }(p, K) = \inf\{\|p - u\| : u \in K\}, \ p \in X.$$

THEOREM 4 (Fitzpatrick [8]). *Let X and X^* be strongly smooth. Then the following are equivalent properties:*

- *the best approximation problem for p^* is Tikhonov well posed;*
- *the function dist $(\cdot, K)$ is Fréchet differentiable at p^*;*
- *there exists a unique best approximation x^* to p^* from K and x^* depends continuously on p^* (in the strong topology).*

EXAMPLE 3. Given the integrand $f = f(x, u) \in C^\infty(R^{2n})$, a fixed point $x^* \in R^n$, a number $T > 0$ and points $(t, p) \in [0, T] \times R^n$, consider the problem (t, p) in the calculus of variations, to minimize

$$\int_t^T f[x(s), \dot{x}(s)]ds$$

subject to

$$(**) \qquad x \in W^{1,2}(t, T), x(t) = p, x(T) = x^*.$$

Write

$$V(t, p) = \text{infimum of } \int_t^T f[x(s), \dot{x}(s))]\, ds \,\text{subject to } (**)$$

Given $p^* \in R^n$, consider the problem $(0, p^*)$. Assume that, everywhere, $f_{uu} > 0$ and $f(x, u) \geq h(|u|)$ for a suitable continuous function h such that $h(z)/z \to +\infty$ as $z \to +\infty$.

THEOREM 5 (Kutznetzov-Siskin, Fleming [9]). *The following are equivalent conditions:*

- *problem $(0, p^*)$ has a unique solution and every minimizing sequence with pointwise equibounded derivatives converges strongly in $W^{1,2}(0, T)$;*
- *V is differentiable at $(0, p^*)$.*

Theorems 1 and 2 are a common extension of Theorems 3,4,5. They show that the Fréchet differentiability of the value function at the relevant parameter value is necessary for well posedness, and becomes sufficient under the assumptions (2), (3), if strengthened as in (1). Nonsmooth behavior (at p^*) of the value function is thereby unavoidable, under suitable conditions, iff the corresponding optimization problem is ill posed. Hence nonsmooth analysis plays there a significant role, while for well posed problems classical differentiability of the value function is relevant.

3. Well posed optimal control problems. We consider the problem of minimizing the integral performance

$$\int_0^T f[y(s), u(s)]ds \tag{3.1}$$

subject to the state equation

$$\dot{y}(s) = g[y(s), u(s)] \text{ a.e. in } (0, T), \tag{3.2}$$

$$y(0) = x^*, \tag{3.3}$$

and to the control constraints

$$u(s) \in U \text{ a.e. in } (0, T). \tag{3.4}$$

Here the state vector $y(s) \in R^n$, the control $u(s) \in R^m$; $T > 0$ and x^* are fixed.

We posit the following assumptions:

- U is compact and $f(x, u)$, $g(x, u)$, f_x, g_x, are continuous on $R^n \times U$;
- $|g(x, u)| \leq a + b|x|$ everywhere, for suitable constants $a, b > 0$.

Among the feasible embeddings of this problem, which are relevant for a well posedness analysis, we select the following three.

(I) Perturbations of the initial state: problem (p) is defined by minimizing (3.1) subject to (3.2) and (3.4), with (3.3) replaced by

$$y(0) = p \tag{3.5}$$

Then $p^* = x^*$ defines the parameter of interest.

(II) A dynamic programming approach defines the parameter $p = (t, x)$, $0 \leq t \leq T$, $x \in R^n$, and replaces the time interval $[0, T]$ by $[t, T]$ in (3.1), (3.2), (3.4); the initial condition (3.5) is modified to

$$y(t) = x, \tag{3.6}$$

so that $p^* = (0, x^*)$.

(III) Additive perturbations of the dynamics (3.2), replaced by

$$\dot{y}(s) = g[y(s), u(s)] + p(s) \text{ a.e. in } (0, T). \tag{3.7}$$

Here the parameter $p \in L^2(0, T)$, and $p^* = 0$.

We fix X to be the set of all admissible (open loop) control laws equipped with the strong convergence of $L^2(0,T)$. Hence well posedness of the given optimal control problem (3.1),(3.2),(3.3),(3.4) (for any embedding) means existence and uniqueness of the optimal control u^*, together with strong convergence in $L^2(0,T)$ toward u^* of every asymptotically minimizing sequence corresponding to the convergent perturbations of p^*.

As a corollary to Theorem 1, for any embedding of the problem, the corresponding value function must be Fréchet differentiable at p^*, while a set of conditions guaranteeing well posedness comes from Theorem 2. The embedding described in (II) is particularly relevant here, because it allows us to make the connection with well posedness in the calculus of variations (Theorem 5) and to take advantage from the available theory of the Hamilton-Jacobi-Bellman equation in order to check the differentiability of the value function. Here we see that the fulfillment of the Hamilton-Jacobi-Bellman equation in a nonsmooth way (viscosity sense), as opposed to the classical one, is really relevant for ill posed problems.

Consider the embedding (I) along with its variant (II), and the corresponding value functions

$$W(p) = \text{ infimum of (3.1) subject to (3.2),(3.4) and (3.5)}$$

$V(t,x) =$ infimum of $\int_t^T f(y,u)$ ds subject to (3.2) and (3.4) on $[t,T]$, and (3.6).

Write $Q(x) = \{(z, g(x,u)) \in R^{n+1} : z \geq f(x,u), u \in U\}$,

and for any optimal trajectory (u^*, y^*) consider the Hamiltonian

$$H(s,u) = f(y^*(s),u) - q(s)'g(y^*(s),u)$$

with the adjoint state q defined by

$$\dot{q}(s) = f_x[y^*(s), u^*(s)] - g_x[y^*(s), u^*(s)]'q(s), q(T) = 0.$$

THEOREM 6 (necessary condition). *If the given optimal control problem is well posed with respect to the embedding (II), then the value function V is Fréchet differentiable at $(0, x^*)$.*

THEOREM 7 (sufficient condition). *The given optimal control problem is well posed with respect to the embedding (I), provided that the following conditions hold:*

- *$Q(x)$ is convex for every $x \in R^n$;*
- *the value function W is Gâteaux differentiable near x^*, with a continuous gradient at x^*;*
- *for any optimal trajectory, $(U, H(s,\cdot))$ is well posed for a.e. $s \in (0,T)$;*
- *if u_1, u_2 are optimal controls with corresponding adjoint states q_1, q_2, then $q_1(0) = q_2(0)$ implies $u_1 = u_2$ a.e.*

Quite similar results hold when the other embeddings are chosen.

Remarks. 1) The convexity assumption about $Q(x)$ can be relaxed to lower closure, in the following sense: for any asymptotically minimizing sequence (y_n, u_n), with y_n converging weakly (for a subsequence) to y^* in $W^{1,1}(0,T)$ and $y_n(0) \to x^*$, then y^* is an optimal state.
2) The well posedness of $(U, H(s,\cdot))$ is equivalent to the uniqueness of its minimizer, owing to compactness.
3) Write

$$z = (r, v) \in R^{2n}, h(z,u) = f(r,u) - v'g(r,u), u_o(z) = \arg\min(U, h(z,\cdot)).$$

If u_o is locally Lipschitz continuous, then the injectivity assumption (last of Theorem 7) is fulfilled.
4) Theorems 6 and 7 contain, as a particular case, the results of [10,ch.VI, §9] on the regular points of deterministic optimal control problems. See also [12].
5) Extensions to nonautonomous problems, unbounded control regions and performances containing a final state term require only routine modifications.

By applying Theorem 7, we get the following examples of well posed optimal control problems according to the embedding (I).

Examples. 1) Let U be convex, $g(x,u) = A(x) + B(x)u$, with A, B Lipschitz continuous on bounded sets and with linear growth, $f(x,\cdot)$ strictly convex, x^* arbitrary, with a unique optimal control. This is the best known example of a well posed optimal control problem (see[10]).
2) Let $f(x,u) = ax$, $g(x,u) = cx + k(u)$, $m = n = 1$, U a compact interval, $k \in C^2(U)$ and $k''(u) \geq b > 0$ in U, with $c \neq 0$ and $a > 0$, x^* arbitrary.
3) Let $U = [0,1]$, $g(x,u) = ux$, $f(x,u) = ux - x$, $m = n = 1$, $x^* > 0$.
4) Let U be compact convex in R^m, $g(x,u) = Ax + Bu$, $f(x,u) = C'x + D'u$, x^* arbitrary, $\dot{q} = -A'q + C$, $q(T) = 0$. The problem is well posed if for a.e.t, arg max $(U, z(t,\cdot))$ is a singleton, where $z(t,u) = (q'B - D')u$.

Results similar to Theorem 7 can be obtained by selecting the embedding (III). (In this case, a connection between (strict) differentiability of the value function and uniqueness of the optimal solution and multiplier is obtained in [11]. See also [15].) See [13] for results related to Theorem 6 with the embedding which takes a time delay as the parameter, and [10], [12] for the embedding obtained by small stochastic perturbations of the dynamics.

By using the constrained extensions of Theorems 1 and 2, one obtains well posedness criteria for optimal control problems with constrained end point, provided one assumes local controllability of the given problem.

The one-dimensional Lagrange problem is treated in [14], where the proof of Theorems 1 and 2 can be found.

REFERENCES

[1] A. Tikhonov, *On the stability of the functional minimization method*, USSR Comput. Math. and Math. Phys. 6 (1966), 26–33.

[2] A. Tikhonov, *Methods for the regularization of optimal control problems*, Soviet Math. Dokl. 6 (1965), 761–763.

[3] J. Hadamard, *Sur le problèmes aux dérivées partielles et leur signification physique*, Bull. Univ. Princeton 13 (1902), 49–52.

[4] R. Lucchetti, F. Patrone, *A characterization of Tikhonov well posedness for minimum problems with application to variational inequalities*, Numer. Funct. Anal. Optim. 3 (1981), 461–476.

[5] R. Lucchetti, F. Patrone, *Some properties of "well-posed" variational inequalities governed by linear operators.*, Numer. Funct. Anal. Optim. 5 (1982-83), 349–361.

[6] A. Dontchev, T. Zolezzi, *Well posed optimization problems*, Lecture Notes in Math. 1543 (1993), Springer, Berlin.

[7] E. Asplund, R. Rockafellar, *Gradients of convex functions*, Trans. Amer. Math. Soc. 139 (1969), 443–467.

[8] S. Fitzpatrick, *Metric projection and the differentiability of distance functions*, Bull. Austral. Math. Soc. 22 (1980), 291–312.

[9] W. Fleming, *The Cauchy problem for a nonlinear first-order partial differential equation*, J. Differential Equations 5 (1969), 515–530.

[10] W. Fleming, R. Rishel, *Deterministic and stochastic optimal control*, Springer, New York, 1975.

[11] F. Clarke, P. Loewen, *The value function in optimal control: sensitivity, controllability, and time-optimality*, SIAM J. Control Optim. 24 (1986), 243–263.

[12] W. Fleming, M. Soner, *Controlled Markov processes and viscosity solutions*, Springer, New York, 1993.

[13] F. Clarke, P. Wolenski, *The sensitivity of optimal control problems to time delay*, SIAM J. Control Optim. 29 (1991), 1176–1215.

[14] T. Zolezzi, *Well posedness criteria in optimization with application to the calculus of variations*, Submitted.

[15] F. Clarke, *Perturbed optimal control problems*, IEEE Trans. Autom. Control AC-31 (1986), 535–542.

IMA SUMMER PROGRAMS

1987 Robotics
1988 Signal Processing
1989 Robustness, Diagnostics, Computing and Graphics in Statistics
1990 Radar and Sonar (June 18 - June 29)
New Directions in Time Series Analysis (July 2 - July 27)
1991 Semiconductors
1992 Environmental Studies: Mathematical, Computational, and Statistical Analysis
1993 Modeling, Mesh Generation, and Adaptive Numerical Methods for Partial Differential Equations
1994 Molecular Biology
1995 Large Scale Optimizations
1996 Emerging Applications of Number Theory

SPRINGER LECTURE NOTES FROM THE IMA:

The Mathematics and Physics of Disordered Media
Editors: Barry Hughes and Barry Ninham
(Lecture Notes in Math., Volume 1035, 1983)

Orienting Polymers
Editor: J.L. Ericksen
(Lecture Notes in Math., Volume 1063, 1984)

New Perspectives in Thermodynamics
Editor: James Serrin
(Springer-Verlag, 1986)

Models of Economic Dynamics
Editor: Hugo Sonnenschein
(Lecture Notes in Econ., Volume 264, 1986)

The IMA Volumes in Mathematics and its Applications

Current Volumes:

1 **Homogenization and Effective Moduli of Materials and Media** J. Ericksen, D. Kinderlehrer, R. Kohn, and J.-L. Lions (eds.)

2 **Oscillation Theory, Computation, and Methods of Compensated Compactness** C. Dafermos, J. Ericksen, D. Kinderlehrer, and M. Slemrod (eds.)

3 **Metastability and Incompletely Posed Problems** S. Antman, J. Ericksen, D. Kinderlehrer, and I. Muller (eds.)

4 **Dynamical Problems in Continuum Physics** J. Bona, C. Dafermos, J. Ericksen, and D. Kinderlehrer (eds.)

5 **Theory and Applications of Liquid Crystals** J. Ericksen and D. Kinderlehrer (eds.)

6 **Amorphous Polymers and Non-Newtonian Fluids** C. Dafermos, J. Ericksen, and D. Kinderlehrer (eds.)

7 **Random Media** G. Papanicolaou (ed.)

8 **Percolation Theory and Ergodic Theory of Infinite Particle Systems** H. Kesten (ed.)

9 **Hydrodynamic Behavior and Interacting Particle Systems** G. Papanicolaou (ed.)

10 **Stochastic Differential Systems, Stochastic Control Theory, and Applications** W. Fleming and P.-L. Lions (eds.)

11 **Numerical Simulation in Oil Recovery** M.F. Wheeler (ed.)

12 **Computational Fluid Dynamics and Reacting Gas Flows** B. Engquist, M. Luskin, and A. Majda (eds.)

13 **Numerical Algorithms for Parallel Computer Architectures** M.H. Schultz (ed.)

14 **Mathematical Aspects of Scientific Software** J.R. Rice (ed.)

15 **Mathematical Frontiers in Computational Chemical Physics** D. Truhlar (ed.)

16 **Mathematics in Industrial Problems** A. Friedman

17 **Applications of Combinatorics and Graph Theory to the Biological and Social Sciences** F. Roberts (ed.)

18 ***q*-Series and Partitions** D. Stanton (ed.)

19 **Invariant Theory and Tableaux** D. Stanton (ed.)

20 **Coding Theory and Design Theory Part I: Coding Theory** D. Ray-Chaudhuri (ed.)

21 **Coding Theory and Design Theory Part II: Design Theory** D. Ray-Chaudhuri (ed.)

22 **Signal Processing Part I: Signal Processing Theory** L. Auslander, F.A. Grünbaum, J.W. Helton, T. Kailath, P. Khargonekar, and S. Mitter (eds.)

23 **Signal Processing Part II: Control Theory and Applications of Signal Processing** L. Auslander, F.A. Grünbaum, J.W. Helton, T. Kailath, P. Khargonekar, and S. Mitter (eds.)
24 **Mathematics in Industrial Problems, Part 2** A. Friedman
25 **Solitons in Physics, Mathematics, and Nonlinear Optics** P.J. Olver and D.H. Sattinger (eds.)
26 **Two Phase Flows and Waves** D.D. Joseph and D.G. Schaeffer (eds.)
27 **Nonlinear Evolution Equations that Change Type** B.L. Keyfitz and M. Shearer (eds.)
28 **Computer Aided Proofs in Analysis** K. Meyer and D. Schmidt (eds.)
29 **Multidimensional Hyperbolic Problems and Computations** A. Majda and J. Glimm (eds.)
30 **Microlocal Analysis and Nonlinear Waves** M. Beals, R. Melrose, and J. Rauch (eds.)
31 **Mathematics in Industrial Problems, Part 3** A. Friedman
32 **Radar and Sonar, Part I** R. Blahut, W. Miller, Jr., and C. Wilcox
33 **Directions in Robust Statistics and Diagnostics: Part I** W.A. Stahel and S. Weisberg (eds.)
34 **Directions in Robust Statistics and Diagnostics: Part II** W.A. Stahel and S. Weisberg (eds.)
35 **Dynamical Issues in Combustion Theory** P. Fife, A. Liñán, and F.A. Williams (eds.)
36 **Computing and Graphics in Statistics** A. Buja and P. Tukey (eds.)
37 **Patterns and Dynamics in Reactive Media** H. Swinney, G. Aris, and D. Aronson (eds.)
38 **Mathematics in Industrial Problems, Part 4** A. Friedman
39 **Radar and Sonar, Part II** F.A. Grünbaum, M. Bernfeld, and R.E. Blahut (eds.)
40 **Nonlinear Phenomena in Atmospheric and Oceanic Sciences** G.F. Carnevale and R.T. Pierrehumbert (eds.)
41 **Chaotic Processes in the Geological Sciences** D.A. Yuen (ed.)
42 **Partial Differential Equations with Minimal Smoothness and Applications** B. Dahlberg, E. Fabes, R. Fefferman, D. Jerison, C. Kenig, and J. Pipher (eds.)
43 **On the Evolution of Phase Boundaries** M.E. Gurtin and G.B. McFadden
44 **Twist Mappings and Their Applications** R. McGehee and K.R. Meyer (eds.)
45 **New Directions in Time Series Analysis, Part I** D. Brillinger, P. Caines, J. Geweke, E. Parzen, M. Rosenblatt, and M.S. Taqqu (eds.)

46 **New Directions in Time Series Analysis, Part II**
D. Brillinger, P. Caines, J. Geweke, E. Parzen, M. Rosenblatt, and M.S. Taqqu (eds.)
47 **Degenerate Diffusions**
W.-M. Ni, L.A. Peletier, and J.-L. Vazquez (eds.)
48 **Linear Algebra, Markov Chains, and Queueing Models**
C.D. Meyer and R.J. Plemmons (eds.)
49 **Mathematics in Industrial Problems, Part 5** A. Friedman
50 **Combinatorial and Graph-Theoretic Problems in Linear Algebra**
R.A. Brualdi, S. Friedland, and V. Klee (eds.)
51 **Statistical Thermodynamics and Differential Geometry of Microstructured Materials**
H.T. Davis and J.C.C. Nitsche (eds.)
52 **Shock Induced Transitions and Phase Structures in General Media** J.E. Dunn, R. Fosdick, and M. Slemrod (eds.)
53 **Variational and Free Boundary Problems**
A. Friedman and J. Spruck (eds.)
54 **Microstructure and Phase Transitions**
D. Kinderlehrer, R. James, M. Luskin, and J.L. Ericksen (eds.)
55 **Turbulence in Fluid Flows: A Dynamical Systems Approach**
G.R. Sell, C. Foias, and R. Temam (eds.)
56 **Graph Theory and Sparse Matrix Computation**
A. George, J.R. Gilbert, and J.W.H. Liu (eds.)
57 **Mathematics in Industrial Problems, Part 6** A. Friedman
58 **Semiconductors, Part I**
W.M. Coughran, Jr., J. Cole, P. Lloyd, and J. White (eds.)
59 **Semiconductors, Part II**
W.M. Coughran, Jr., J. Cole, P. Lloyd, and J. White (eds.)
60 **Recent Advances in Iterative Methods**
G. Golub, A. Greenbaum, and M. Luskin (eds.)
61 **Free Boundaries in Viscous Flows**
R.A. Brown and S.H. Davis (eds.)
62 **Linear Algebra for Control Theory**
P. Van Dooren and B. Wyman (eds.)
63 **Hamiltonian Dynamical Systems: History, Theory, and Applications**
H.S. Dumas, K.R. Meyer, and D.S. Schmidt (eds.)
64 **Systems and Control Theory for Power Systems**
J.H. Chow, P.V. Kokotovic, R.J. Thomas (eds.)
65 **Mathematical Finance**
M.H.A. Davis, D. Duffie, W.H. Fleming, and S.E. Shreve (eds.)
66 **Robust Control Theory** B.A. Francis and P.P. Khargonekar (eds.)
67 **Mathematics in Industrial Problems, Part 7** A. Friedman
68 **Flow Control** M.D. Gunzburger (ed.)

69 **Linear Algebra for Signal Processing**
A. Bojanczyk and G. Cybenko (eds.)
70 **Control and Optimal Design of Distributed Parameter Systems**
J.E. Lagnese, D.L. Russell, and L.W. White (eds.)
71 **Stochastic Networks** F.P. Kelly and R.J. Williams (eds.)
72 **Discrete Probability and Algorithms**
D. Aldous, P. Diaconis, J. Spencer, and J.M. Steele (eds.)
73 **Discrete Event Systems, Manufacturing Systems, and Communication Networks**
P.R. Kumar and P.P. Varaiya (eds.)
74 **Adaptive Control, Filtering, and Signal Processing**
K.J. Åström, G.C. Goodwin, and P.R. Kumar (eds.)
75 **Modeling, Mesh Generation, and Adaptive Numerical Methods for Partial Differential Equations** I. Babuska, J.E. Flaherty, W.D. Henshaw, J.E. Hopcroft, J.E. Oliger, and T. Tezduyar (eds.)
76 **Random Discrete Structures** D. Aldous and R. Pemantle (eds.)
77 **Nonlinear Stochastic PDEs: Hydrodynamic Limit and Burgers' Turbulence** T. Funaki and W.A. Woyczynski (eds.)
78 **Nonsmooth Analysis and Geometric Methods in Deterministic Optimal Control** B.S. Mordukhovich and H.J. Sussmann (eds.)
79 **Environmental Studies: Mathematical, Computational, and Statistical Analysis** M.F. Wheeler (ed.)

FORTHCOMING VOLUMES

1992-1993: *Control Theory*

Robotics

1993-1994: *Emerging Applications of Probability*

Mathematical Population Genetics

Image Models (and their Speech Model Cousins)

Stochastic Models in Geosystems

Classical and Modern Branching Processes

1994 Summer Program: *Molecular Biology*

Genetic Mapping and DNA Sequencing

Mathematical Approaches to Biomolecular Structure and Dynamics

1994-1995: *Waves and Scattering*

Computational Wave Propagation

Wavelet, Multigrid and Other Fast Algorithms (Multiple, FFT) and Their Use in Wave Propagation

Waves in Random and Other Complex Media

Inverse Problems in Wave Propagation

Singularities and Oscillations

MIX
Papier aus verantwortungsvollen Quellen
Paper from responsible sources
FSC® C105338

If you have any concerns about our products,
you can contact us on
ProductSafety@springernature.com

In case Publisher is established outside the EU,
the EU authorized representative is:
Springer Nature Customer Service Center GmbH
Europaplatz 3, 69115 Heidelberg, Germany

Printed by Libri Plureos GmbH
in Hamburg, Germany